"MODAL RULES"

S-5 INFERENCE RULES

M1	M2	M3	M4	M5	M6		M7	M8
□P	P	P→Q	P→Q	P→Q	P↔Q	P↔Q	P→Q	P→Q
P	◊P	P	~Q	Q→R	P	or Q	Q→P	◊P
		Q	~P	P→R	Q	P	P↔Q	□ Q

(Q must be modally closed)

S-5 REPLACEMENT RULES

E1
P→Q // □(P⊃Q)

E2
P↔Q // □(P≡Q)

DIAMOND EXCHANGE RULE
1. Change diamond to box or box to diamond
2. Take opp of each side.

ADDITIONAL S5 RULES

S5 Nec:
Indent, reiterate only modally closed formulas, derive P, end indentation, assert □P.

S5 M. C. P.
Indent, assume P, reiterate only modally closed formulas, derive Q, end indentation, assert P→Q.

S5 Reduction
Any sequence of iterated modal operators may be reduced to the last operator in the sequence.

Taut Nec
Prove that P is a tautology and then assert □P.

To my logical friend Mike Odell —

Paul Herrick

THE MANY WORLDS OF LOGIC

A Philosophical Introduction

Paul Herrick

Shoreline Community College

Harcourt Brace College Publishers

Fort Worth Philadelphia San Diego New York Orlando Austin
San Antonio Toronto Montreal London Sydney Tokyo

Editor in Chief	Ted Buchholz
Acquisitions Editor	David Tatom
Developmental Editor	Mary K. Bridges
Senior Project Editor	Steve Welch
Production Manager	Cynthia Young
Art Director	Jim Dodson

Address for Orders:
Harcourt Brace & Company,
6277 Sea Harbor Drive, Orlando, FL 32887.
1-800-782-4479, or 1-800-433-0001 (in Florida).

Address for Editorial Correspondence:
Harcourt Brace College Publishers,
301 Commerce Street, Suite 3700,
Fort Worth, TX 76102.

Printed in the United States of America

3 4 5 6 7 8 9 0 1 2 016 9 8 7 6 5 4 3 2 1

Library of Congress Catalog Card Number: 93-79046

ISBN: 0-15-500358-5

To the Instructor

This text includes three units not found in any other symbolic logic text.

- Chapter 5 constitutes an introduction to the philosophy of logic and also lays the conceptual foundation for the unit on modal logic. This chapter may also be used in conjunction with any of the other chapters and was written in such a way that the serious student should understand most of it on his or her own. Also, there is no need to cover the entire chapter. Individual sections may be selected and used to supplement other chapters.
- Following Chapter 6 is a "philosophical interlude" containing a fairly large number of philosophical arguments. The interlude offers students the opportunity to apply logical theory to real philosophical issues. There is no need to cover the entire interlude. Individual segments may be selected and assigned for homework, extra credit, or independent study. The interlude may be used in conjunction with any of the other chapters and is written in such a way that the serious student should be able to understand most of the material on his or her own—with little or no classroom instruction.
- Chapters 6 and 7 constitute a thorough introduction to modal logic. This unit on modal logic parallels the units on truth-functional and quantificational logic. Thus, a conceptual foundation is laid, symbolic language is developed, and natural deduction and truth-tree techniques are set out. In addition, the differences between the various systems of modal logic are presented and discussed.

Although this text features an extensive treatment of modal logic and a fair amount of philosophical material, I have been careful not to slight the units on truth-functional and quantificational logic. The sections on those two branches of our subject contain an ample quantity of examples, problem sets, and explanatory material. Each chapter is written so that the serious student can learn much of the material on his or her own.

The Many Worlds of Logic gives the instructor a lot of flexibility as far as the design of the course goes. I have aimed to organize the text so that a large number of alternative symbolic logic courses can be taught from it. This makes possible some extra variety for the instructor who teaches a large number of symbolic logic courses each year. The following is a chapter outline of 18 different symbolic logic courses that can be taught with this text.

1. A standard symbolic logic course covering truth-functional and quantificational logic:
   With trees: 1, 2, 4, 8, 9
   With natural deduction: 1, 2, 3, 8, 9

2. Variations on the courses in 1:
   (a) Add philosophical applications by selecting portions of the interlude.
   (b) Add a unit on philosophy of logic by selecting portions of Chapter 5.
3. A course covering truth-functional and modal logic:
   With trees: 1, 2, 4, 5, 6, 7
   With natural deduction: 1, 2, 3, 5, 6, 7
4. Variations on the courses in 3:
   (a) Add philosophical applications by selecting portions of the interlude.
   (b) De-emphasize philosophy of logic by covering only part of Chapter 5.
5. A course covering truth-functional, modal, and quantificational logic.
   With trees: 1, 2, 4, 5, 6, 7, 8, 9
   With natural deduction: 1, 2, 3, 5, 6, 7, 8, 9
6. Variations on the courses in 5:
   (a) Add philosophical applications selected from the interlude.
   (b) De-emphasize philosophy of logic by covering only part of Chapter 5.

Several beliefs guided me as I wrote *The Many Worlds of Logic: A Philosophical Introduction*. Ordinarily, the introductory symbolic logic course contains almost no philosophy and the introductory philosophy class contains almost no symbolic logic. As a result of this separation, students in the introductory logic course often wonder why the course is called a "philosophy" course and students in the introductory philosophy course are not exposed to the rigor and insights that modern formal logic makes possible in philosophy. I believe it is both desirable and possible to bring substantial philosophical arguments into the logic course.

It is standard practice within our profession to cover only truth-functional and quantificational logic in the introductory symbolic logic class and to reserve modal logic for an advanced logic course. Thus, modal logic is typically not covered in introductory logic texts, and when it is, it is given a relatively abbreviated treatment within the confines of a single chapter. I believe that modal logic may be made accessible (no pun intended) to the student in the introductory class. I also believe that because of advances made in recent years, modal logic is now as relevant to our field as is quantificational logic. Consequently, I see no reason why an introductory logic course shouldn't contain a unit on modal logic.

Symbolic logic texts typically avoid issues in the philosophy of logic. I believe that a short unit on philosophy of logic illuminates all branches of logical theory and helps the student make better sense of the formal aspects of our subject.

I tried to write the philosophical sections so that no particular point of view is favored. However, there was simply not enough room to cover all the important arguments on all the issues. I'm sure that many instructors will want to supplement the Philisophical Interlude in places with arguments of their own. The philosophical discussions in the text serve primarily to lay the foundation on each issue.

The back of the book contains answers to selected exercises. A solutions manual is available containing answers to the rest of the exercises. If you have any comments or suggestions regarding this text, I would appreciate hearing from you. My address is: Shoreline Community College, 16101 Greenwood Ave. North, Seattle, WA 98133

TO THE STUDENT

This textbook is an introduction to modern symbolic logic. In the following pages, three symbolic logical languages are presented, each designed to represent the logical properties of a general type of logical argument. Languages such as the ones you will study are essential to work in computer science, artificial intelligence, linguistics, mathematics, and philosophy. Furthermore, this book presents some of the philosophical issues surrounding the subject of symbolic logic. In addition, in the middle of the book you will find a philosophical interlude containing a number of important philosophical arguments. After you have learned some logical theory, you should be able to *apply* that theory to the arguments in the interlude.

I have four recommendations for those taking a course in symbolic logic. First, each portion of a symbolic logic course builds on previous portions. Consequently, if you fail to understand one section, you will probably be unable to understand the following sections, and if you miss several days of class, it may be difficult for you to catch up. Do not let yourself fall behind!

Second, symbolic logic courses typically place an emphasis on problem solving. With most people, problem solving is a skill that only develops with practice. Thus, the instructor can show you how to solve logic problems in class, and you can understand the examples in class, and yet if you do not regularly go home and practice problems on your own, you will probably do poorly on the tests. So remember the cliché: Practice makes perfect.

Incidentally, I believe that study groups are a very good way to learn. If two or more students get together regularly to work logic problems, then when one person gets stuck, another member of the group can help that person out. It is good to help another person. But also, your understanding of logical theory will deepen as you explain something to a fellow student.

Third, regular attendance is especially important in a class such as symbolic logic, and I strongly recommend that you attend every class session.

Fourth, as you progress through the course, I suggest you build a glossary of new terms. This will help you keep the terminology straight and it will also help you study for tests.

Why study symbolic logic? Symbolic logic is a field of study that has engaged the minds of many of the world's greatest thinkers since ancient times. It has been considered an important ingredient of a university education since the formation of the world's first universities in the middle ages. A basic understanding of symbolic logic will therefore acquaint you with an important part of humankind's intellectual history. Furthermore, the study of symbolic logic should contribute to the development of

1. your problem solving skills.

2. your ability to think abstractly.

3. your ability to evaluate the logical arguments others present to you.

4. your ability to think systematically and in a disciplined, precise way.

5. your ability to learn a technical field of thought.

6. your ability to work with a symbolic language.

7. your understanding of the logical aspects of language.

8. your understanding of an important branch of modern philosophy.

ACKNOWLEDGMENTS

A number of individuals have helped me at various times as I've worked on this book and I feel indebted to them all. My friend and colleague Dr. Steven Duncan read the interlude and gave me very helpful comments. Steve also used a draft of the manuscript in the classroom and gave valued suggestions based on his classroom experience. Dr. S. Marc Cohen of the University of Washington Philosophy Department read the truth-functional and quantificational chapters and gave me many extremely valuable and helpful comments and suggestions. Professor Cohen's remarks convinced me to switch to the use of metalinguistic variables, a change that I believe greatly improves the presentation of logical theory. Dr. Robert Coburn, also of the University of Washington Philosophy Department, read a draft of the philosophy of logic chapter and gave me very insightful and helpful comments and suggestions. After reading a draft of the interlude, Andy Kildow wrote valuable comments and then discussed them with me. I also thank Wayne McGuire, Professor of English at Shoreline Community College, for the many times I went to him for advice. His suggestions were always helpful.

I am also indebted to a number of persons at Shoreline Community College who helped at one time or another in the typing and preparation of the manuscript. Mrs. Dolores Kleinholz, administrative secretary for the Science Division, very accurately and efficiently typed the bulk of the manuscript. Merlin Talbot and Jean Thompson from our Social Sciences Division also helped type part of the manuscript. Teri Raynor and Alex Berylov from Shoreline's computer lab helped type the exercise sets. Teri Raynor also helped enormously with a pile of "last minute" typing at the end of the project. Fred Watterson and Dorothy Cerelli from our Instructional Computing Division and Dave Holmes from our media lab patiently helped me out many times when I had computer problems. Professor Sally Rollman helped me in the computer lab, and Sandra Masters and Bill Benjamin of Shoreline's Business Division provided much-needed assistance at several key moments. I thank each of these individuals.

Finally, students who have taken my classes since January 1990 have had to use rough drafts of my manuscript as their texts. I thank them for putting up with "lecture notes" packets that were—as I look back on them now—too rough and too incomplete. And I thank Shoreline's tireless printer, Mark Durfee, for cheerfully printing copies of the manuscript for class use every quarter. Finally, I am indebted to Shoreline Community College, in the persons of our President, Dr. Ronald Bell; Dr. Marie Rossenwasser, Vice President; Professor Howard Vogel, Social Sciences Division Chair; and Social Sciences administrative secretary Mrs. Merlin Talbot for encouraging me and backing me in my work on this manuscript.

In writing this book, I am indebted to a number of other textbooks that I have used in the classroom over the years. My greatest debt is to the text *Possible Worlds* by Raymond Bradley and Norman Swartz. Many of my pedagogical ideas developed as I taught with that excellent book, and the authors were kind enough to discuss their book and their approach to logic with me. I am also very indebted to Daniel Bonevac's *Deduction*. I learned much about the teaching of modal logic as I taught with that superb text. Furthermore, my tree systems are all based on the systems Bonevac develops in his book. I am also indebted to Bangs Tapscott's excellent text, *Symbolic Logic*. Much of my approach to quantificational logic is influenced by his presentation. In addition, I am indebted to the fine texts by Irving Copi (*Symbolic Logic*), Robert Hurley (*Logic*), Gustafson and Ulrich (*Symbolic Logic*), and *The Logic Book* by Bergman, Moor, and Nelson.

I also wish to thank Jo-Anne Weaver, Acquisitions Editor at Harcourt Brace College Publishers, for having faith in this project and issuing me the initial contract. I thank Acquisitions Editor David Tatom for all of his efforts on behalf of the book and Mary K. Bridges, the Developmental Editor on this book, for her constant enthusiasm, encouragement, and help. I also want to express my sincere appreciation to Steve Welch, the Senior Project Editor. Throughout our association he has been thoroughly patient and helpful. Production Manager Cindy Young has done an excellent job coordinating the composition and printing of the book, and Art Director Jim Dodson has created an attractive and easy-to-use design of the text and cover.

Thanks are also due to the reviewers: Paul B. L. Anderson, Seattle Central Community College; Bredo Johnsen, University of Houston; Eric R. Stietzel, Foothill College; and Richard J. Van Iten, Iowa State University.

I also want to thank Richard Purtill, Professor of Philosophy at Western Washington University, and Robert Adams, Professor of Philosophy at UCLA, for discussing the overall conception of the text and giving me advice on it.

Finally, my greatest debt of all is to my mother and father. Without them, this book would not have been possible. I thank them for all the ways in which they have helped me.

CONTENTS

FUNDAMENTALS

● — REASON and CIVILIZATION

Try to imagine what it would be like to live without any of the benefits of civilization. With relatively little in the way of language, tools, clothing, shelter, transportation, medical care, or social order, your existence would probably resemble the state of nature described by the British philosopher Thomas Hobbes as "solitary, poor, nasty, brutish, and short." Down through the centuries, each succeeding generation has built upon the accomplishments of previous generations, resulting in the gradual accumulation of a rich and complex deposit of culture. We each benefit every day from the discoveries and creations of past generations, and we in turn are adding our own contributions to a deposit of culture that will be passed on to the next generation, which will in turn add its own contribution to the continuing process, the process we call *civilization*.

Millions of species have inhabited the Earth, yet ours is the only one that has built civilizations. Why are we human beings so different in this respect from the other creatures of our planet? What do we have that the rest of nature doesn't have? The philosophers of ancient Greece, who began the first systematic studies of mathematics, the sciences, and philosophy, suggested that it is the human ability to reason that distinguishes our species from the rest of the living world.

Human reasoning has certainly played an important role in the development of civilization. The great bodies of knowledge, such fields as physics, chemistry, biology, mathematics, astronomy, history, economics, sociology, psychology, and anthropology, are the products of human reasoning. In addition, it was the application of human reason to practical concerns that led to such things as the domestication of plants and animals, metallurgy, architecture, engineering, medicine, electronics, and eventually to the production of cars and trucks, skyscrapers, airplanes, and computers.

Everywhere one turns in the human world, one sees the results of reasoning. Look upon any of the many things created by human beings and you are, in a sense, looking upon an embodiment of a reasoning process. Civilization is the direct result of countless acts of human reasoning.

REASON and DAILY LIFE

Reasoning is a natural activity. Every time you solve a problem, make a decision, or try to get at the truth of some matter, you are reasoning. The proof of this is that in such circumstances, if someone asks you "Why?", you are naturally able to respond with the *reasons* for your solution, decision, or conclusion. Since life is a continuous round of decision making, problem solving, and inquiry, this means that you are reasoning just about every waking minute of every day.

In addition to being a natural human activity, reasoning is a vital activity. Our day-to-day survival depends on the use of large quantities of reasoning. Just think, for example, about the amount of reasoning that goes into the organization of a city's water and power supply, or into the design of its phone system, or into the treatments for the thousands of diseases that attack the human body. Reasoning is therefore an important and fundamental part of our daily lives.

REASON and THE FUTURE OF HUMANKIND

A good case can be made that reason holds our only hope for peace. Our world is full of fundamental disagreements, many of which result in violence and war. When people disagree over important things, there are only two ways to reach an agreement: (1) someone resorts to force in an attempt to impose a solution; or (2) the opposing sides talk things out, *reasoning* their way to some sort of solution. Can you think of any other alternative?

A program aimed at reducing the number of fights on the playground has been instituted in a number of grade schools. As soon as a confrontation begins, student monitors intervene and, on the spot, give each party time to state his or her side of the issue. The students go on to discuss the problem logically, with the monitor moderating the discussion. By trying to understand the other person's point of view, and by weighing the merits of each side's case, the students are reasoning instead of fighting. Imagine what the world would be like if everyone learned this lesson.

An important part of our future will involve establishing peaceful alternatives to violence and war. An important part of our future will therefore involve turning our affairs further in the direction of reason.

THE SUBJECT MATTER OF LOGIC

It is easy to talk, as we have been doing, in general terms about reasoning and the important role it plays in our lives. But what precisely is this activity we call "reasoning"? Let's look a little more deeply into the matter. Suppose you were asked to explain *how* to reason. What would you say? Could you put a set of step-by-step instructions into words? Furthermore, is any one instance of reasoning as good as any other? Or are there standards of good and bad reasoning such that some reasoning counts as *correct* reasoning while other reasoning counts as *incorrect* reasoning? If so, are there principles that can help us to reason better? Logic is the discipline that offers answers to such questions. For the subject matter of logic is, to put it simply, *reasoning*.

●— THE REASONING PROCESS: A CLOSER LOOK

So, just what is this activity we call reasoning? To answer this, let's actually do some reasoning and then we'll reflect upon what it is we have done. Consider the following story.

> An attorney was talking with a group of friends at a luncheon when someone asked her about her very first case. She explained that the first person she ever defended in court was a man who had robbed a bank. However, thanks to a clever defense, the defendant was found not guilty. As the luncheon was ending, a tall man entered the room, greeted the attorney, and said, "I was the first person this fine lawyer ever defended in court!"

Most people, upon listening to this story, would draw a conclusion about the tall man. From the information given, what conclusion do you draw?

Smith and Jones, let us suppose, were seated at the next table eavesdropping on the attorney's conversation, and each drew a separate conclusion. Smith concluded that the tall man is a bank robber and Jones concluded that the attorney is a bank robber. Smith and Jones, and you, reasoned. Each of you reasoned from one thing to another; you each concluded one thing on the basis of another thing.

More specifically, Smith and Jones reasoned from the same two premises, namely:

1. The first person the attorney defended in court had robbed a bank.

and

2. The tall man was the first person the attorney ever defended in court.

However, Smith reasoned to the conclusion that

3. The tall man robbed a bank.

while Jones reasoned to the conclusion that

4. The attorney robbed a bank.

Two different lines of reasoning are displayed here. Smith's goes from statements (1) and (2) to statement (3), while Jones's goes from (1) and (2) straight to (4). That is, Smith concluded (3) on the basis of (1) and (2), while Jones concluded (4) on the basis of (1) and (2). Another way to put this is: Smith inferred (3) from (1) and (2), and Jones inferred (4) from (1) and (2). An *inference* is an act of reasoning that a person performs when, on the basis of a statement or group of statements, the person concludes some further statement. When someone makes an inference, we often say that the person has "drawn a conclusion." Thus, to make an inference is to think through a line of reasoning to a conclusion.

CORRECT VERSUS INCORRECT REASONING

Consider again the inferences drawn by Smith and Jones. Is one of these inferences more reasonable than the other? Wouldn't you agree that Smith's inference—from (1) and (2) to (3)—is obviously correct? And isn't Jones's inference—from (1) and (2) to (4)—just as obviously incorrect? Thus, not all instances of reasoning are equal, for in some cases the reasoning is correct and in other cases it is not.

Now, what about Smith's reasoning makes it an example of correct reasoning? And why is Jones's reasoning faulty? Logic offers answers to these questions. Earlier, logic was characterized simply as the study of reasoning. Our subject matter may now be defined more precisely: Logic is the study of the principles of *correct* reasoning. Principles of correct reasoning, if established, would guide the way people ought to reason, just as the principles of arithmetic guide the way people ought to add and subtract.

Upon hearing that logic is the study of the principles of correct reasoning, principles that people ought to follow if they wish to reason well, some people ask the following: "Doesn't each person have his or her own individual principles of reasoning? Why lay down principles—absolutes—that everyone ought to follow?" These are interesting questions. However, before one can intelligently debate the merits of the principles of logic, one must first understand those principles that logicians have developed over the centuries and across many cultures. Then, reflection upon the validity of those principles may properly begin. Incidentally, questions such as the two stated above are taken up in the branch of philosophy known as *epistemology*—the study of the nature of knowledge. In Chapter 6, when we touch on the epistemology of logic, we shall briefly take up the question of how, if at all, we know the principles of logic.

ARGUMENTS

Let's go back to the story involving Smith, Jones, and the tall man. Suppose a few days later Smith is standing in line at the bank, sees the tall man waiting in the next line, and says to a woman standing nearby, "That tall man over there robbed a bank." The woman, let us suppose, is the tall man's wife, and this is the first time she has ever heard this. The natural thing to do—the rational thing to do—when presented with a questionable claim is to ask for the reasoning or evidence supporting the claim. "Why do you say that?" she asks. In reply, suppose Smith offers the following to support his claim:

1. I know an attorney, and the first person she ever defended had robbed a bank.
2. The tall man is the first person she ever defended.
Therefore,
3. The tall man robbed a bank.

Within logic there is a special term for what Smith has done. Smith has presented an *argument*. An argument may be defined as a sequence of statements, one of which, the conclusion, is claimed to follow from the other or others, which are called premises. The premise or premises are offered as evidence or reasons in support of the conclusion. Smith's argument consists of just two premises, statements (1) and (2), plus the conclusion, statement (3), together with the (implicit) claim that the conclusion follows from or is justified by the premises.

This definition of the term 'argument' may not fit everyone's usage of the word. For instance, when we say that two people are "having an argument," we sometimes mean that they are exchanging angry words, perhaps even yelling at each other. Often, neither side is giving reasons in support of a conclusion. However, in logic, we shall not be concerned with the emotional aspects of such disputes; logic concerns itself just with the evaluation of the lines of reasoning, if any, that are involved.

From what has been said so far, it follows that for every inference there is a corresponding argument representing that inference. If an inference goes from statements A and B to statement C, then there is an argument that has A and B as its premises and C as its conclusion. To go back to Smith's and Jones's reasoning, we earlier noted that the inference from statements (1) and (2) to (3) seemed obviously correct but that the inference from (1) and (2) to (4) seemed obviously faulty. We may now express the same point by saying that the argument from premises (1) and (2) to the conclusion (3) is a good argument but the argument from premises (1) and (2) to the conclusion (4) is faulty.

Here is another way to think about the concept of an argument. We often think just to ourselves about things. Sometimes, we are reasoning about something, trying to reach a conclusion about some matter. Occasionally, we wish to share our reasoning with someone, perhaps because we wish to reason with that person; that is, we want that person to draw the same conclusion we've drawn. We must then express our inner act of reasoning in a public form, in a manner that is accessible to another human being. We do this, of course, by putting our thoughts into the words and sentences of a language. That is one of the things language is for. When it is an inner act of reasoning we put into linguistic form, we are presenting an argument. An argument, then, is an episode of reasoning expressed within the sentences of a language.

Our study of reason will naturally focus on arguments, for arguments are the public expressions of various acts of reasoning, and our object of study must, of course, be a public object.

2. DEDUCTION

●— DEDUCTIVELY INVALID ARGUMENTS

Let us now press our investigation of reasoning further by asking: What is it that makes an argument a good argument? In other words, what are we looking for in an argument? In order to approach this question, let's first examine an obviously incorrect argument. Suppose someone argued as follows:

Premise one:	Pat is a cousin of Chris.
Premise two:	Jan is a cousin of Chris.
Conclusion:	Pat is a cousin of Jan.

Now, what is wrong with this argument? To begin with, suppose you were to reason from the premises. What conclusion should be drawn from them? On the basis of the information presented in the premises, should you draw the argument's conclusion? Obviously, you should not, for it is possible that the premises are true and the conclusion false. So, even if we assume that the premises are true, the possibility remains that the conclusion is nevertheless false. Therefore, even if you were to know for sure that the premises are true, you would not, on the basis of that, be justified in concluding that the conclusion is true as well.

An argument such as this, in which there is any possibility—no matter how far-fetched or unlikely—that the premises are true and the conclusion is false, is said to be a *deductively invalid* argument.

Here are some examples of deductively invalid arguments. For convenience, each argument will be presented in a standardized form by listing the premises first, the conclusion last, with the conclusion introduced by words that typically serve, in the context of an argument, to introduce the conclusion of the argument, words such as 'therefore,' 'so,' 'thus,' and 'consequently.'

— Examples of Deductively Invalid Arguments —•

Argument A
1. Whenever it rains over the roof, the roof gets wet.
2. The roof is wet.
3. Therefore, it is raining over the roof.

Argument B
1. All of Pat's friends are artists.
2. Some artists are musicians.
3. Thus, some of Pat's friends are musicians.

Argument C
1. If you drink that, you'll get sick.
2. So, if you don't drink that, you won't get sick.

Argument D
1. According to the polls, 70 percent plan to vote for Bloggs.
2. Therefore, Bloggs will be elected.

If you will reflect upon each of these four arguments, you will see that in each case, there are possible circumstances in which the premises are true but the conclusion is false. In each case, even if we were to be certain that the premises are true, we would not thereby be guaranteed that the conclusion is also true. In other words, with these arguments, the truth of the premises does not guarantee the truth of the conclusion.

●— DEDUCTIVELY VALID ARGUMENTS

Let's turn next to what is obviously a correct argument:

> *Argument E*
> 1. Whenever it rains over the roof, the roof gets wet.
> 2. It is raining over the roof.
> 3. So, the roof is getting wet.

The problem with a deductively invalid argument, as we saw, is that there's a possibility of true premises and a false conclusion in the case of such an argument. Reflect carefully on argument E. Is there any possibility that it has true premises and yet a false conclusion? The answer is no. In order to see this, you will have to think about the argument in a hypothetical way. Suppose hypothetically or "for the sake of argument" that the premises are true. Whether they are actually true or not doesn't matter at the moment. Basing your judgment just upon the supposition that the premises are true, what follows? Clearly, the conclusion must be true, for there is no possibility of the premises being true and the conclusion false. If the premises are true, then the conclusion must be true as well.

An argument such as this one, in which there is no possibility that the premises are true and the conclusion is false, is said to be a *deductively valid* argument. In the case of a deductively valid argument, if the premises are true, then the conclusion must be true. That is, the conclusion of a deductively valid argument cannot possibly be false if the premises are true.

At this point, some people think they can see a way in which the premises of argument E could be true while its conclusion is false. They suggest this: Suppose a big tarp covers the roof and protects it from rain. They conclude from this that it is possible for the premises of E to be true while its conclusion is false. However, this would not be a case in which the premises of the argument are true while its conclusion is false, for if a tarp were to cover the roof, then the first premise of the argument would be false.

We saw above that it is not possible that the premises of argument E are true and its conclusion is false. The definition of deductive validity involves a special kind of possibility, called *logical possibility*, to distinguish it from other kinds of possibility. In everyday life, when we say that a thing is possible, we sometimes mean just that it is likely. For example, because the probability of a write-in candidate getting elected President is astronomically small, someone might say, "It is not possible to be elected President if you run as a write-in candidate." Other times, when we say that something is possible, we mean that on the basis of what we presently know, it's possible. For example, someone might say, "We've inspected every inch of Loch Ness, and it's not possible a Cadborosaurus lives there." In contrast, the concept of possibility employed within logic is broader than either of these two ordinary concepts of possibility. Nothing will count as logically impossible except that which constitutes or implies a self-contradiction. So, anything—no matter how unlikely or far-fetched—will qualify as logically possible as long as it does not constitute or imply a *self-contradiction.*

The notion of a self-contradiction will be explained in more detail in future chapters, but a preliminary definition is the following: A statement is explicitly self-contradictory when it consists of a declarative statement conjoined to the negation of the very same statement. We typically conjoin two statements by placing the word 'and' between them, and we may form the negation of a statement by prefixing 'It is not the case that' to it. Thus, each of the following statements is explicitly self-contradictory:

    i. The final exam will be on Tuesday and it is not the case that the final exam will be on Tuesday.

    ii. The computer has a hard drive and it is not the case that the computer has a hard drive.

In order for a statement to count as an explicit self-contradiction, the sentence component placed to the left of the 'and' must carry exactly the same meaning as the sentence component placed to the right of the 'it is not the case that'. Notice that with an explicit self-contradiction, the left side of the statement expresses an assertion and the right side retracts that very assertion, leaving absolutely nothing asserted overall. This is why (i) gives the students no information about the final exam and (ii) provides no information about the computer. Since a contradiction asserts nothing, it certainly doesn't describe anything, which is part of the reason that a contradiction does not describe or represent a genuine possibility.

Thus, it is logically impossible for an only child to have a sister, since the child would not be an only child if the child had a sister. In order to be an only child with a sister, a child would have to have no sisters at all and at the same time the child would have to have a sister, which would involve a self-contradiction. On the other hand, it is logically possible that a person lives for a thousand years, since there is nothing self-contradictory about such an extraordinarily long life.[1]

Now, back to the concept of deductive validity. Here are some examples of deductively valid arguments:

---

[1] Of course, we are assuming here that the words being used are given their standard meanings. This is something we ordinarily assume, whether at the grocery store, at the gas station, or in conversation with a neighbor. For instance, at the checkout line, when the clerk asks, "Would you like paper or plastic?", nobody ever replies, "Wait a minute, are you using those words with their standard meanings or are you giving those words special or secret meanings?" Effective communication depends upon the use of publicly agreed-upon standard meanings. So, when thinking about the examples of possibilities and impossibilities given in this section, give the words their standard meanings, just as you would when shopping at the grocery store.

Argument F
1. All whales are mammals.
2. All mammals have lungs.
3. Consequently, all whales have lungs.

Argument G
1. All snakes are reptiles.
2. Some snakes are blue.
3. Therefore, some reptiles are blue.

Argument H
1. It will either rain or snow.
2. But it won't snow.
3. So, it will rain.

Argument I
1. If it rains, then the roof will leak.
2. If the roof leaks, then the wall will get wet.
3. Thus, if it rains, then the wall will get wet.

If you reflect on each of these arguments, you will see that in each case, there is no possibility of the premises being true while the conclusion is false. Therefore in each case, if the premises are true, then the conclusion must be true. Thus, each is deductively valid.

Do you see the sharp difference between these deductively valid arguments and the deductively invalid arguments given previously? Consider the deductively invalid arguments, arguments A through D. In each case, there are possible circumstances in which the premises are true and the conclusion false. Therefore, with one of these arguments, if you rely just on the information presented by the premises, you cannot be certain that the conclusion is true. If the premises are true, the conclusion might be true, but then again it might not be. However, consider any of the five deductively *valid* arguments just given. In each case, there are no possible circumstances in which the premises of the argument are true while the conclusion is false. Consequently, if you rely just on the information presented by the premises, you can be certain that the conclusion is true, for if the premises are true, the conclusion must be true as well.

●— A TEST FOR VALIDITY

In order to test an argument for validity, ask yourself the following question: Are there any possible circumstances in which the premises are true and the conclusion is false? If the answer is "no," then the argument is valid. If the answer is "yes," then the argument is invalid. Essentially, there's nothing more to it than this.

Let's apply this test to two arguments. Is the following piece of reasoning deductively valid?

Argument J
1. Whenever Sue swims, Jim swims.
2. Sue won't swim tomorrow.
3. So, Jim won't swim tomorrow.

The question to ask is this: Are there any possible circumstances in which the premises are true while the conclusion is false? Well, consider the possibility in which Sue swims every other day and Jim swims every day. That is a possibility. If you are puzzled by this, notice that the first premise tells us that every time Sue swims, Jim swims, but it does *not* also tell us that every time Jim swims, Sue swims.

In this case, we found a possible circumstance in which the premises of the argument are true while the conclusion is nevertheless false. This shows that the argument is invalid. It is extremely important to observe that we evaluated this argument and we figured out that it is invalid, without knowing whether the premises are actually true or not. We proceeded by considering, abstractly, just the possibilities. Thus, there are two questions that were kept separate:

A. Are the premises true?
B. Is the argument valid or invalid?

We answered question (B) without having answered question (A). The information one might uncover by answering (A) would not contribute in any way to the answer to (B). And, of course, the reverse is true as well.

Consider next the following argument:

Argument K
1. Jim is Susan's father.
2. Fred is Jim's father.
3. Therefore, Fred is Susan's grandfather.

Is this argument valid or invalid? Well, is there any possibility of the premises being true while the conclusion is false? Clearly, there is no such possibility. The argument is therefore deductively valid. Incidentally, are the premises true? That's a completely separate issue. To resolve that question, we'd have to research Susan's family tree.

Once again, notice that we evaluated this argument, we figured out that it is valid, without knowing whether or not the premises are true. We looked only at the various possibilities, without any knowledge of the actual facts of the matter.

Counterexamples Suppose someone claims that the following argument is deductively valid:

Argument L
1. All the members of the club are Democrats.

2. Some Democrats are senior citizens.

3. So, some members of the club are senior citizens.

This argument is deductively invalid, of course. But suppose someone insists that the argument is valid. In saying that this argument is valid, this person is claiming that it is impossible for the premises to be true while the conclusion is false. How could you convince this person that the argument is invalid?

You could describe what is called a *counterexample* to the argument. A counterexample to an argument is a possible circumstance in which the premises of the argument are true and the conclusion is false. For example, consider the following possibility:

⋮ Perhaps the club is composed entirely of teenage Democrats.

This possibility constitutes a counterexample to argument L, for it is a possible circumstance in which the premises of L are true and its conclusion is false. When we describe for someone a counterexample to an argument, we help them see a way in which the premises of the argument could be true and the conclusion false. Thus, the presentation of a counterexample establishes that an argument is invalid.

Now, look back at argument K. Can you describe a counterexample to that argument? Try to describe a situation in which the premises of that argument are true while its conclusion is false. Of course, this can't be done, for the argument is valid. There is no such thing as a counterexample to a valid argument.

●— THE RELATIONSHIP BETWEEN VALIDITY and TRUTH

It is now time to demonstrate an exceedingly important point. If you will reflect on the definition of deductive validity, you will notice that the validity of an argument has nothing to do with whether or not the premises are actually true. When we say that a particular argument is deductively valid, we are *not* saying that its premises are actually true, nor are we saying that its conclusion is actually true. Rather, we are merely saying that it is not possible for the premises to be true and the conclusion false. Valid arguments do not always have true premises. With this in mind, let's look at the possible combinations of truth and falsity that may be found within various valid and invalid arguments.

First, some deductively valid arguments actually have true premises and a true conclusion. For example:

Argument M
1. All cats are mammals. (True)

2. All mammals are warm-blooded. (True)

3. Therefore, all cats are warm-blooded. (True)

However, in other deductively valid arguments, the premises and conclusion are actually false. For example:

> *Argument N*
> 1. All beer drinkers drink milk. (False)
> 2. All milk drinkers are poets. (False)
> 3. So, all beer drinkers are poets. (False)

Still other deductively valid arguments have false premises with a true conclusion:

> *Argument O*
> 1. All cats are birds. (False)
> 2. All birds are mammals. (False)
> 3. Thus, all cats are mammals. (True)

The only combination of truth and falsity that will never be displayed by any deductively valid argument is that of all true premises with a false conclusion. Can you explain why?

Here is another important point concerning deductive validity. The mere fact that the premises and the conclusion of a given argument are all true does not make the argument deductively valid. If this surprises you, remember that there are arguments having true premises and a true conclusion that are, nonetheless, deductively invalid. For example, consider the following argument:

> *Argument P*
> 1. There are more than 100 billion stars in the Milky Way galaxy. (True)
> 2. A galaxy is a large collection of stars. (True)
> 3. So, there are more than 100 billion galaxies. (True)

This argument has true premises and a true conclusion, yet it is nevertheless deductively invalid. True premises and a true conclusion do not, by themselves, make an argument valid.

Consider next deductive invalidity. When we say that an argument is deductively *invalid*, we are not saying that its premises are false, nor are we saying that its conclusion is false. Remember that some invalid arguments, argument P for example, have true premises and a true conclusion. Furthermore, the mere fact that an argument has false premises and a false conclusion does not mean that the argument is deductively invalid. Argument N, for example, has false premises and a false conclusion, yet is deductively valid.

With this in mind, let's look at some of the possible combinations of truth and falsity that may be found within deductively invalid arguments. Some invalid arguments have false premises with a false conclusion. For example:

Argument Q

1. The Earth is flat. (False)

2. The Moon is square. (False)

3. Therefore, the Sun is triangular. (False)

However, some invalid arguments have false premises with a true conclusion. For example:

Argument R

1. All fish are reptiles. (False)

2. All whales are fish. (False)

3. Consequently, all cats are mammals. (True)

Some invalid arguments have true premises with a false conclusion. For example:

Argument S

1. All cats have lungs. (True)

2. All dogs have lungs. (True)

3. So, all cats are dogs. (False)

Finally, as we've already seen in the case of argument P above, some invalid arguments have true premises and a true conclusion.

Upon being introduced to the terminology of logic, some people begin using the term 'true argument'. However, as the various terms of argument appraisal have been defined above, truth is *not* a property of an argument. There are true statements, true premises, true conclusions, but—as things have been defined—there is no such thing as a 'true argument'. There are, of course, valid arguments.

Deductively Sound Arguments   Suppose we know that a particular argument is deductively valid. Does it follow that the conclusion of the argument is true? As we have seen, the answer is no. To say that an argument is deductively valid is only to say that it is not possible for the premises to be true with the conclusion false. Typically, when we evaluate an argument, we want to know more than whether it is deductively valid or not. We also want to know whether the premises and conclusion are true. Truth is the ultimate goal of arguing, of using our capacity to reason. So, we are looking for more than deductive validity when we evaluate an argument—we are looking for truth. Suppose we know, of a particular argument, two things:

i. The argument is deductively valid.

ii. The premises are all true.

Such an argument is called a *deductively sound* argument.

Will the conclusion of a deductively sound argument be true or false? Since any sound deductive argument is deductively valid, it follows that if the premises are true, then the conclusion *must* be true. Furthermore, the premises of a deductively sound argument are true. The conclusion of a deductively sound argument must therefore be true as well. So, every deductively sound argument is valid, has true premises, and consequently has a true conclusion.

Since truth is, or should be, the ultimate goal of arguing, it is deductive soundness, rather than mere deductive validity, that we are ultimately interested in when we evaluate arguments such as the ones above. When we evaluate an argument, we hope to see valid reasoning plus true premises.

━━━━━━━━━━━ • EXERCISE 1.1 • ━━━━━━━━━━━

Which of the following arguments are valid and which are invalid? In which cases can you tell whether the argument is sound or unsound? In which of those cases is the argument sound?

1. Lawrence Welk is the leader of the Rolling Stones. Lawrence Welk started his first band during the 1930s. So the leader of the Rolling Stones started his first band in the 1930s.

2. If Arnold Schwarzenegger and Danny DeVito are twins, then Danny DeVito is not an only child. Danny DeVito is not an only child. So, Arnold Schwarzenegger and Danny DeVito are twins.

3. Since some Fords are purple and some Fords are trucks, it follows that some Fords are purple trucks.

4. No car is a fruit. All lemons are fruits. So, no car is a lemon.

5. Axl Rose is a Trappist monk. All Trappist monks live lives of quiet asceticism and devotion. So, Axl Rose lives a life of quiet asceticism and devotion.

6. Since the Monterey Pop Festival occurred before Woodstock, and Woodstock occurred after the 1968 Democratic National Convention, it follows that the Monterey Pop Festival occurred before the 1968 Democratic National Convention.

7. The United States House of Representatives has more members than there are days in a leap year. Therefore, at least two members of the U.S. House of Representatives have the same birthday.

8. Boys Town is located in Nebraska. Nebraska is located in the United States. So, Boys Town is located in the United States.

9. Since this room is perfectly circular, it follows that it has no corners.

10. Every member of the Progressive Labor Party must study the writings of Chairman Mao. Bob studied the writings of Chairman Mao. So, Bob is a member of the Progressive Labor Party.

11. All life forms require a substantial degree of orderly structure. There is no orderly structure in the middle of a supernova explosion. So, no life is possible in the middle of a supernova explosion.

12. Listening to loud music ruins your hearing. Joe is hard of hearing. Therefore, Joe has been listening to loud music.

13. The mind, by its very nature, has no mass, physical shape, or location in physical space. The brain does have a mass, a physical shape, and a location in physical space. If one thing possesses properties that another thing does not possess, then the two things are not the same thing. Therefore, the mind and the brain are not the same thing, since the brain possesses properties that the mind does not possess.

14. The Monterey Pop Festival (July 1967) featured some of the greatest bands of the 1960s. The Beatles did not perform at the Monterey Pop Festival. So, the Beatles were not among the greatest bands of the 1960s.

15. The Monterey Pop Festival featured some of the greatest bands of the 1960s. Big Brother and the Holding Company performed at the Monterey Pop Festival. So, Big Brother and the Holding Company was one of the greatest bands of the 1960s.

16. Only Hell's Angels live in the Brentwood apartments. Hank belongs to the Hell's Angels. Therefore, Hank lives in the Brentwood apartments.

17. All bikers are independent spirits. Joan is a biker. So, Joan is an independent spirit.

18. All Hell's Angels live in California. Pat lives in California. So, Pat belongs to the Hell's Angels Motorcycle Club.

● CONCEPTS RELATED TO DEDUCTIVE VALIDITY

There are a number of concepts logically related to the concept of deductive validity. This group of interrelated ideas includes the concepts of consistency, inconsistency, implication, equivalence, and necessity. If you read almost any work of contemporary philosophy, you will find reference after reference to these concepts. It is not too much of an exaggeration to say that the study of this family of ideas constitutes the heart of modern logical theory. This section uses ordinary English and everyday examples to introduce this group of important logical concepts. In future chapters, we will use symbolic logical languages to probe each of these concepts in greater depth. The understanding you will gain in this chapter will serve as a foundation for the logical studies you will undertake in subsequent chapters.

The first two concepts from this group are ones we often employ in everyday reasoning: the concepts of consistency and inconsistency.

Consistency and Inconsistency Within everyday discourse, there is no one precise definition of consistency. However, within logical theory, this concept is given a precise definition. A set of statements is said to be *logically consistent* if and only if there is any possibility—no matter how unlikely—that all the statements in the set are true together. If a set of statements is logically consistent, one may consistently assert all the members of the set. For example, suppose Detective Joe Friday is questioning a suspect and the suspect makes the following three statements:

1. On the night of the crime, I worked from 9 p.m. until 6 a.m.
2. I never left the building during my shift.
3. I took a dinner break from 1 to 2 a.m.

Since it is logically possible that all three of these statements are true, this set of statements is a *consistent* set. In other words, the suspect's story is consistent.

It is important that you understand the following point. When we say that a particular set of statements is consistent, we are not saying that all of the statements in the set are true. We are merely saying it is *possible* that all of the statements in the set are true. So, if Detective Friday reports that the suspect's story is consistent, he is not reporting that all of the suspect's statements are true. Rather, he is saying it is possible that all of the suspect's statements are true.

A set of statements is *logically inconsistent* if and only if it is not possible that all of the statements in the set are true together. It would be inconsistent to assert all the members of such a set of statements. For example, suppose Detective Friday interviews another suspect, and the suspect makes the following three statements:

1. On that night, I worked 9 p.m. to 6 a.m.
2. I never left the building during that time.
3. At midnight I went outside for a cup of coffee.

Since it is not possible that all the members of this set are true together, this set of statements is an inconsistent set of statements. In other words, this suspect's story is inconsistent.

When we say that a set of statements is inconsistent, we are not saying that all the statements in the set must be false. We are merely saying that there's no possibility that all of the statements in the set are true, which implies that at least one statement in the set must be false. So, if Detective Friday reports that the suspect's story is inconsistent, he is saying only that it is not possible that all of the statements in the story are true.

Consider next the relationship between deductive validity and the concepts of consistency and inconsistency. First, the *denial* or *negation* of a statement may be formed by prefixing 'It is not the case that' to the statement. If a statement is true, then its denial is false, and if a statement is false, then its denial is true. Now, if an argument is valid, then it is not possible that the premises are true and the conclusion is false. It follows that if an argument is valid, it is not possible that the premises are true and the *denial* of the conclusion is also true. Therefore, the set of statements consisting of the premises of a valid argument plus the denial of the argument's conclusion is an inconsistent set. Consequently, it would be inconsistent of someone to agree with the premises of a valid argument but then to deny the conclusion. That is, one cannot consistently accept the premises of a valid argument and then deny the conclusion of the argument. So, if someone presents to you a valid argument, and you agree with the premises, then you cannot consistently deny the conclusion.

Here is an example of the inconsistency that results when someone asserts the premises of a valid argument but denies the conclusion. Suppose Smith argues:

1. All cats are mammals.
2. All mammals are warm-blooded.
3. Therefore, all cats are warm-blooded.

Imagine Jones replying:

> I accept your premises. I agree that all cats are mammals and that all mammals are warm-blooded. However, I deny your conclusion. It is not true that all cats are warm-blooded.

Do you see the inconsistency in Jones's reply?

Valid arguments are valuable pieces of reasoning. If you accept the premises of a valid argument, you must also accept the conclusion . . . unless you want to be inconsistent.

Implication The concept of deductive validity is logically related to another concept, the concept of *logical implication*. Implication is a concept used in everyday reasoning. We naturally say things such as "Are you implying that . . ." or "One of the implications of what she said was. . . ." Within everyday reasoning, there is no one precise definition of implication. However, within logical theory, the concept is precisely defined: One statement implies a second statement if and only if there is no possibility that the first statement is true and the second is false.

In each of the following examples, the first statement implies the second:

1. Jan is exactly 18 years old.
2. Jan is a teenager.

1. Pat is exactly 90 years old.
2. Pat is over 65 years old.

1. Chris is exactly 50 years old.
2. Chris is not 30 years old.

Does the second statement imply the first in any of the above examples?

In each of the following examples, the first statement does not imply the second:

1. Jan is religious.
   Jan attends church every Sunday.

2. Pat is Hindu.
   Pat lives in India.

3. Chris is an Olympic athlete.
   Chris is a good swimmer.

A *set* of statements implies a given statement if and only if it is not possible that all the statements within the set are true and the given statement is false. For example, the set of statements consisting of the following two statements:

1. All persons are rational.

2. Socrates is a person.

implies the statement

3. Socrates is rational.

The set consisting of (1) and (2) implies statement (3) because there's no possibility that (1) and (2) are true but (3) is false.

The relationship between validity and implication may now be put this way. If an argument is valid, then it is not possible that its premises are true and its conclusion is false. Therefore, the set of statements consisting of the premises of the argument implies the conclusion of the argument. In other words, in the case of a valid argument, the premises, considered as a set, imply the conclusion.

Equivalence Another concept related to deductive validity is the concept of *logical equivalence*. Two statements are logically equivalent just in case it is not possible for the two to differ in terms of truth or falsity. In other words, if two statements are equivalent, then in any possible circumstance, if one is true, the other will also be true and if one is false, the other will also be false. Two equivalent statements will always match in terms of truth and falsity. Each of the following pairs of statements is a pair of equivalent statements:

1. Sue is married to Ron.
   Ron is married to Sue.

2. Sue is taller than Ron.
   Ron is shorter than Sue.

3. Sue is exactly as old as Ron.

   Ron is exactly as old as Sue.

4. The charge on a returned check is twice the value of the check, or $100, whichever is larger.

   The charge on a returned check is twice the value of the check with a $100 minimum charge.

If you will reflect on the concepts of equivalence and implication, you will see that if two statements are equivalent, then they must imply each other. In each of the four examples above, the first statement implies the second and the second implies the first.

In everyday discourse, we often use the concept of equivalence in accord with the logical definition given above. For instance, if Pat says, "Joe is taller than Bob," and Chris snaps back as if in disagreement, "Bob is shorter than Joe," it would be natural for Pat to reply, "The two statements are equivalent."

◆ EXERCISES ◆

Exercise 1.2

In each of the following cases, are the two sentences consistent or inconsistent? When trying to understand what each sentence means, give words their standard meanings.

1. a. The President is 16.
   b. The President is a teenager.

2. a. Sue is 36.
   b. Sue's age is between 10 and 40.

3. a. The room contains 10 people.
   b. The room contains more than 30 people.

4. a. Joan is taller than Joe.
   b. Joe is taller than Joan.

5. a. Pat is 40 and Pat is not 40.
   b. Rita is 50.

6. a. Joan is at least as tall as Joe.
   b. Joe is at least as tall as Joan.

7. a. Pat is taller than Chris.
   b. Chris is shorter than Pat.

8. a. Some journalists are science fiction buffs.
   b. Some science fiction buffs are journalists.

9. a. Some journalists are science fiction buffs.
    b. Some science fiction buffs are not journalists.

10. a. Some astronomers are not stamp collectors.
    b. Some stamp collectors are not astronomers.

11. a. An all-loving, all-knowing, all-powerful God created the world and watches over it.
    b. The world contains large quantities of suffering.

12. a. The ring contains five huge diamonds.
    b. The ring's price is 10 cents.

13. a. Michael Jackson is President of the United States.
    b. John Candy is Vice President of the United States.

14. a. No horologist is a member of the Peace and Freedom Party.
    b. One member of the Peace and Freedom Party is a horologist.

15. a. If you drink that liquid, then you will get sick.
    b. If you don't drink that liquid, then you won't get sick.

16. a. Ann and Bob are not both home.
    b. Either Ann is not home or Bob is not home.

17. a. Either Ann or Bob is not home.
    b. Ann and Bob are not both home.

18. a. Ann and Bob are both not home.
    b. Ann and Bob are not both home.

19. a. Neither Ann nor Bob is home.
    b. Both Ann and Bob are not home.

20. a. Ann and Bob are both not home.
    b. Either Ann or Bob is home.

21. a. The Vice President of the United States is a female.
    b. The President of the United States is a female.

22. a. Someone is 60.
    b. Someone is not 60.

Exercise 1.3

In each of the following cases, does p imply q? Does q imply p?

1. p: Pat is rich.
    q: Pat is materialistic.

2. p: Pat is 16.
    q: Pat is a teenager.

3. p: Pat is over 39.
    q: Pat is under 60.

4. p: Pat is under 21.
    q: Pat is under 100.

5. p: Some people earn less money than others.
    q: Social injustice exists.

6. p: Chris couldn't possibly have done otherwise.
    q: Chris is not morally responsible for what she did.

7. p: The figure is a square.
    q: The figure has four equal sides.

8. p: Pat just won the state lottery.
    q: Pat is a millionaire.

9. p: The world cannot possibly be infinitely old and it cannot possibly have come into being out of absolutely nothing.
    q: The world was created by a creator.

10. p: Some old neighborhoods have many new buildings.
    q: Some neighborhoods with many new buildings are old neighborhoods.

11. p: Every human action is determined by a prior state of the world, which is itself determined by a prior state of the world, and so on back billions of years.
    q: Nobody has free will.

12. p: If you drink the stuff in that cup, then you will get ill.
    q: If you don't drink the stuff in that cup, then you won't get ill.

13. p: The physical world is structured according to an enormously complex set of natural laws.
    q: The physical world was designed by an intelligent being.

14. p: Muhammad Ali is Cassius Clay, and Cassius Clay fought in the 1960 Olympics.
    q: Muhammad Ali fought in the 1960 Olympics.

15. p: All mammals are warm-blooded.
    q: All warm-blooded creatures are mammals.

16. p: If the Beatles perform at the event, then tickets will be sold out.
    q: So, if the Beatles don't perform at the event, then tickets won't be sold out.

Exercise 1.4

1. Refer back to Exercise 1.2. In which of those cases is the pair of statements a pair of equivalent statements?

2. Refer back to Exercise 1.3. In which of those cases is the pair of statements a pair of equivalent statements?

Necessity   Three more logical concepts are related to the concept of deductive validity. In everyday reasoning, when we say that something is "necessarily" a particular way, we mean that it could not possibly have been otherwise. In logic, an individual statement is said to be *necessarily true* if and only if it could not possibly have been false. In other words, a statement is necessarily true just in case there's no possible circumstance in which it is false. Consider the following statements:

a. Nothing is red all over and green all over at the same time.

b. The Space Needle either is 100 years old or is not 100 years old.

c. If today is Tuesday, then today is not Monday.

Notice that in each of these three cases, there's no possibility that the statement—given its standard meaning—is false. Therefore, each statement—if given its standard meaning—expresses a necessary truth.

A statement is said to be *necessarily false* if and only if there's no possible circumstance in which it is true. Consider the following examples:

d. Jan and Pat are each older than the other.

e. Pat is 30, and it's not the case that Pat is 30.

f. Some circles are square.

Notice that in each of these three cases, there's no possibility that the statement—given its standard meaning—is true. That is, each statement—if given its standard meaning—expresses a necessary falsehood.

In everyday discourse, we generally use the term 'contingent' to mean 'dependent'. For example, the sale of a house is sometimes said to be "contingent" upon the sale of the buyer's house. A statement that is neither necessarily true nor necessarily false is going to be true in some possible circumstances and false in others. That is, its truth or falsity will vary from circumstance to circumstance. Such a statement is called a *contingent* statement, since whether it is true or false depends on the circumstance. Here are some examples of contingent sentences:

g. Red Delicious apples are 59 cents a pound.

h. Socrates had a beard.

i. Rita works for the Department of Parking Enforcement.

In future chapters, we will investigate the concepts of necessity and contingency and their relationship to deductive validity.

3. INDUCTION

●— INDUCTIVELY STRONG ARGUMENTS

Consider the following argument:

> *Argument T*
> 1. A scientific poll was conducted in Smith's district, and 97 percent of the voters polled said they planned to vote for Smith.
> 2. Therefore, Smith will win the election.

Does this piece of reasoning qualify as a deductively valid argument? It is clearly invalid, for it is possible that the premise is true and the conclusion false. After all, it's at least possible that the voters will change their minds before election day.

Nevertheless, this argument seems to be pretty persuasive. Although it does not satisfy the standard of deductive validity, there is something right about it, isn't there? What is "good" about this argument? Suppose—just for the sake of argument—that the premise is true. Now, given that the premise is true, what follows? It follows that the conclusion is *probably* true. That is, the truth of the premise would make the conclusion highly probable. An argument such as this, in which the truth of the premise or premises makes the conclusion highly probable, is called an *inductively strong* argument.

Now consider another argument. Is the following piece of reasoning inductively strong?

> *Argument U*
> 1. We polled 50 people over at the nearest church, and 30 said they planned to vote for Smith.
> 2. Smith's district has 10,000 registered voters.
> 3. Therefore, Smith will be elected.

Obviously, the premises of this argument do not make the conclusion probable at all. That is, if the premises are true, the conclusion is not, on the basis of that, probable. Call such an argument an *inductively weak* argument.

● EXERCISE 1.5 ●

In each case below, do you find the argument inductively strong or weak?

1. There is a pulsar star 30,000 light years away with a peculiar wobble. For five years, scientists have been studying the pulsed radio signals emitted by this star and have found that the wobble can best be explained by the gravitational tug of a nearby planet. Therefore, a planet orbits this pulsar.

2. The surviving Beatles will hold a reunion concert next August. Tickets will cost $1. There are tens of millions of fans all over the world who would pay $1 to attend a Beatles reunion. Therefore, the stadium where the concert will be held will sell at least five tickets.

3. I've known four of the forty members of the school chess club, and all four were Star Trek fans. Pat belongs to the chess club. So, Pat is a Star Trek fan.

4. Most religions contain a belief in a supreme deity. Therefore, the religious beliefs of the Navaho Indians include a belief in a supreme deity.

5. Eighty percent of the hamburgers ordered at Dick's Drive-In are cheeseburgers. Sue is about to order a burger at Dick's. Sue will order a cheeseburger.

6. Fay and Barbara each own nearly identical 1969 Dodge Darts. Barbara's car is making the very same noise Fay's made just before it broke down. So, Barbara's car is about to break down as well.

7. Someone used up the last of the hamburger meat last night. It wasn't Sue, since she's a vegetarian. It wasn't Randy, since he's on a strict diet. It wasn't the dog, since it can't open the refrigerator. Pat is the only other person in the house. The best explanation is that it was Pat. So, Pat did it.

●— THE RELATIONSHIP BETWEEN INDUCTIVE STRENGTH and TRUTH

The relationship between inductive strength and truth is analogous to the relationship between deductive validity and truth. When we say that an argument is inductively strong, we are not saying that its premises are true, and we are not saying that its conclusion is true. We are merely saying that *if* the premises are true, then the conclusion is probably true. Thus, there are inductively strong arguments—such as the second argument in the exercise above—that nevertheless have false premises.[2]

[2] Within statistics, probabilities are measured on a scale ranging from 0 (which represents absolute impossibility) to 1 (which represents absolute inevitability or certainty). Thus, the probability of a contradiction is 0, the probability of a necessary truth is 1, and the probability that the sun will come up tomorrow is so high that it is very close to 1. In everyday matters, when we say that something is "highly probable," we normally mean that the probability is significantly above .5 but less than 1. Things with a probability of 1—things that are absolutely inevitable or certain—are not ordinarily called "highly probable." This allows a quick answer to a question that often arises at this point. Are valid arguments also strong arguments? One might think so at first, for in the case of a valid argument, if the premises are true, then based on that the probability of the conclusion is 1. However, given the assumption that by "highly probable" we mean "with a probability that is high but less than 1," it follows that no valid argument ever qualifies as a strong argument.

If an argument is a strong inductive argument, and in addition it has true premises, we say that it is an *inductively sound* argument. Such an argument is also sometimes called a *cogent* argument.

CONTRASTING DEDUCTIVE VALIDITY and INDUCTIVE STRENGTH

There are a number of differences between deductively valid arguments and inductively strong arguments. One difference that deserves to be highlighted is this: Deductive validity does not come in degrees, whereas inductive strength does come in degrees. Thus, an argument either is deductively valid or it is not. There is no such thing as an argument that is partly deductively valid or partly deductively invalid, and one deductively valid argument cannot be more deductively valid than another. On the other hand, some inductively strong arguments are stronger than other inductively strong arguments. You can verify this point by reflecting upon the examples of deductively valid and inductively strong arguments we have examined so far.

Deductive validity and inductive strength differ in another significant respect: If any premises are added to a deductively valid argument, the resulting expanded argument will remain valid no matter what premises were added. Even if contradictory premises are added to a valid argument, the resulting expansion of the argument will remain valid. You can see this for yourself by picking a valid argument, adding contradictory premises, and then verifying that the argument does indeed remain valid. However, premises can be added to an inductively strong argument in such a way that the expanded argument turns into a weak argument. For instance, a pile of circumstantial evidence might make it look extremely likely that the new cashier has been embezzling money from the firm. However, one new piece of evidence added to the pile might suddenly make it look extremely unlikely that the cashier is the embezzler. In this case, the addition of one premise makes the conclusion unlikely.

EXERCISES

Exercise 1.6

1. Construct an inductively strong argument and then specify additional premises that would turn it into a weak argument.

2. Construct a deductively valid argument and then add any additional premises. Notice that the argument remains valid. Explain why this is so.

Exercise 1.7

Which of the following are true and which are false?

1. If the premises of an argument are all true, then the argument must be valid.

2. If the premises of an argument are false, the argument must be invalid.

3. If the conclusion of an argument is true, then the argument must be valid.

4. If an argument is valid, then it is also deductively sound.

5. If an argument is deductively sound, then it is valid.

6. If the premises and conclusion are all true, then the argument must be valid.

7. If an argument is deductively sound, then all its premises are true.

8. If an argument has a false premise, then it must be deductively unsound.

9. If an argument is deductively unsound, then it must be invalid.

10. If an argument is deductively unsound, then it has at least one false premise.

11. All invalid arguments have false premises.

12. All valid arguments have true conclusions.

13. If an argument is inductively strong, then it has true premises.

●— DISTINGUISHING DEDUCTIVE and INDUCTIVE ARGUMENTS

Some logicians classify arguments as either *deductive* or *inductive*. The distinction between these two categories is roughly this: If an argument claims—explicitly or implicitly—that there's no possibility of true premises and a false conclusion, then the argument is called a *deductive* argument. If an argument claims—explicitly or implicitly—that the premises make the conclusion probable, then the argument is called an *inductive* argument.

If this terminology is adopted, then a valid deductive argument may be defined as a deductive argument whose claim is correct, and strong inductive argument may be defined as an inductive argument whose claim is correct. Of course, using this terminology, a deductive argument is deductively invalid if and only if its claim is incorrect and an inductive argument is inductively weak if and only if its claim is incorrect.

This terminology allows us to classify arguments without evaluating them. That is, we can classify an argument as deductive or inductive without deciding whether it is valid or invalid, strong or weak.

If you want to claim explicitly that your argument is deductive, you should introduce your conclusion with a word or phrase that makes your deductive claim clear. Indicators that typically signal a deductive claim within an argument include such phrases as 'it necessarily follows that', 'it must follow that', and 'it is therefore certain that'. If you want to claim in an explicit way that your argument is inductive, you should introduce your conclusion with words indicating an inductive claim. Phrases that typically indicate an inductive claim include 'it's therefore probably true that', 'it's therefore likely that', and 'it's therefore plausible to suppose that'.

Textbooks within other academic fields sometimes define induction and deduction differently from the way they are defined here. In classes outside of the field of philosophy, students are sometimes told that an inductive argument is an argument that proceeds from the particular to the general and a deductive argument is one that proceeds from the general to the particular. A statement is a "particular" statement if it makes a claim about a particular member of a group of things, and a statement is a "general" statement if it makes a claim about all the members of some group. This way of distinguishing deductive and inductive arguments does fit some cases. However, it does not fit all cases. Here is an example of a deductively valid argument that proceeds from the particular to the general:

> Susan is a Buddhist.
> James is a Buddhist.
> Susan and James are the only adults in this room.
> Therefore, all the adults in this room are Buddhists.

And here's a deductively valid argument that goes from the particular to the particular:

> Susan owns a microscope.
> Susan owns a telescope.
> So, Susan owns a microscope and a telescope.

Finally, here's an example of an inductively strong argument that proceeds from the general to the particular:

> All the crows I've ever seen are black.
> So, the next crow I'll see will be black.

After reflecting on these examples, it should be clear that it is simply not correct to define or characterize a deductive argument as one that proceeds from the general to the particular. Neither will it do to define or characterize an inductive argument as one that proceeds from the particular to the general.

●— SOME COMMON PATTERNS OF INDUCTIVE REASONING

In order to further clarify the idea of an inductive argument, let us briefly examine some common patterns of inductive reasoning.

Inference to the Best Explanation   An inference to the best explanation is an inductive argument that fits the following format.

1. The argument cites one or more purported facts that, it is claimed, are in need of explanation.

2. Possible explanations are considered.

3. It is argued that one particular explanation is the best or most reasonable explanation of the phenomena under consideration.

4. The conclusion, which is claimed to be probably true if the premises are, is that the explanation singled out in (3) is the correct explanation.

For example, suppose some jewels have been stolen from the Gotrocks's residence. It is known that the thief was seven feet tall, had long red hair and a pointy red beard, and spoke with a strong Chicago accent. On top of all that, the burglar was bitten by the Gotrocks's pet poodle, Charles, as the burglar ran out of the house. Who stole the jewels?

Suppose the police learn that Bruno, the Gotrocks's butler, is seven feet tall, has long red hair and a pointy red beard, and speaks with a thick Chicago accent. Coincidence, you say? Perhaps a seven-foot tall burglar who looks and talks just like Bruno happened to be working the neighborhood? Suppose further that Bruno has what looks like a poodle bite on his left leg. What's the most reasonable explanation of the theft? Isn't the best explanation the hypothesis that Bruno stole the jewels? Doesn't the evidence given make it *probable* that the butler did it?

━━━━━━━━━━━━━ • EXERCISE 1.8 • ━━━━━━━━━━━━━

Think of other possible explanations of the theft of the jewels. Compare these to the hypothesis that the butler did it. What makes the hypothesis that the butler did it the better explanation? What criteria do you use when you decide which is the most reasonable explanation?

Arguments given in support of various scientific theories are often inferences to the best explanation. For example, at the beginning of the twentieth century, there were a number of atomic phenomena that physicists simply couldn't explain with Newtonian mechanics or other scientific theories. After Einstein put forward the Special Theory of Relativity, it eventually became clear to physicists that Einstein's theory offered the best explanation of the phenomena. The fact that the Special Theory provided the best explanation of the phenomena gave physicists a good reason to suppose that Einstein's theory was correct. The argument that Darwin presented in the *Origin of Species* in favor of his theory of evolution was also an inference to the best explanation. Essentially, Darwin cited phenomena that, he argued, were best explained by his theory of evolution. From this, he concluded that his theory is probably true.

Generalization from a Sample   A *generalization from a sample* is an inductive argument that begins with a claim about a sample of a group of things and from this concludes something about the whole group. For example, a product tester at the

Ace Widget Company might select a random sample of 500 widgets from the factory's assembly line. After finding these widgets to be of high quality, the tester might conclude that Ace widgets are high-quality widgets. If the information about the sample is correct, then this makes the tester's conclusion probable.

Whether a generalization from a sample is an inductively strong or weak argument depends upon several things. Generally, the larger the sample in relation to the group as a whole and the more random the selection of the sample, the stronger the argument. Various types of racial or sexual stereotypes are based on generalizations from a sample. In these cases, a sample of a group is observed, and various characteristics are attributed to the individuals in the sample. A conclusion is then drawn about all the members of the group. If the selected sample is not representative of the group, then the stereotype is based on a *weak* argument. If the characteristics attributed to the sample are not actually possessed by the members of the sample, then the stereotype is based on an argument with a false premise.

Argument from Analogy Scientists sometimes test medicines on laboratory animals before testing them on human beings. If a medicine effectively cures a disease in monkeys, for example, scientists reason that since a monkey's physiology is analogous to a human's, the medicine will probably cure the disease in humans as well. This reasoning is called an *argument from analogy* or an *analogical* argument.

Analogical arguments, which are inductive in nature, typically fit the following general format:

1. Entities x and y are similar in respect to features a, b, c, and d.
2. Entity x has feature e.
3. Entity y is not known to lack feature e.
4. These considerations make it probable that y—like x—also has feature e.

Supporting the inference to the conclusion is the following consideration: The fact that two things are alike in many ways makes it probable that they are alike in other relevant ways as well. Of course, the more features x and y have in common, the stronger the analogy and therefore the stronger the argument.

One way to critique an argument of this form is to establish a *disanalogy* between the things being compared. A disanalogy between two things consists of respects in which the two are different. The more ways x and y are shown to be different from each other, the weaker the analogy, the stronger the disanalogy, and the weaker the analogical argument.

• EXERCISES •

Exercise 1.9

In philosophy, one of the important arguments for the existence of a supreme being is the *argument from design*. David Hume, an eighteenth-century Scottish philosopher, discussed the argument in his *Dialogues on Natural Religion*. The following is a paraphrase of Hume's version of the argument:

> Look at the world of nature around you. It is very similar to a giant machine. Machines typically have many parts and the parts are organized systematically so as to form a functioning whole. The natural world is like this, as the scientific study of nature reveals. But in addition to having parts that fit together into a functioning whole, machines have another feature: machines are typically the product of an intelligent designer. Typically, it takes an intelligent mind to design and assemble a complicated device consisting of parts organized into a functioning whole. Since the natural world is similar to a machine in terms of the arrangement of parts to whole, it is probable that it is similar to a machine also in terms of being the product of intelligent design. These considerations make it probable that the universe was designed and put together by an intelligent being.

Some Questions

1. Explain why this argument counts as an analogical argument.

2. In what ways is the natural world similar to a machine? In what ways does it differ from a machine? Is the analogy between the structure of the natural world and the structure of a machine a strong or a weak analogy? Support your answer with reasons.

3. Based on your answer to question 3, is the argument from design presented above a strong or a weak argument?

4. Rewrite the argument from design and put it in the form of an inference to the best explanation.

Argument from Authority   An *argument from authority* is a type of inductive argument that generally begins by citing a statement made by an authority on some subject. A conclusion is then drawn based on the authority's statement. For example, in the 1970s, a number of prominent scientists announced that the Earth is entering a new ice age. People then argued that because some eminent scientists believe we are about to enter a global ice age, it follows that we are about to enter a global ice age. That was an argument from authority. However, since scientists are sometimes mistaken, the fact that eminent scientists claim something does not guarantee that the claim is true. Rather, the testimony of an authority merely makes the conclusion of the argument probable to some degree.

Exercise 1.10

1. Critique the following argument:

   St. Thomas Aquinas said it is wrong to argue from authority. Therefore, an argument from authority is not a good argument.

2.   Take the argument from authority given above and the generalization from a sample given earlier and rewrite each in the form of an inference to the best explanation.

Exercise 1.11

In each case below, is the argument best interpreted as a deductive argument or as an inductive argument? If you think the argument is deductive, evaluate it as valid or invalid. If you think the argument is inductive, evaluate it as strong or weak. Record any indicator words that support your interpretation.

1.   Since Theresa is the mother of Holly and the sister of Alfred, Alfred must be Holly's uncle.

2.   The sign in front of the building says the building was built in 1708. Therefore, it must have been built in 1708.

3.   Since the sun has risen every morning for thousands of years, it is likely that it will rise tomorrow morning.

4.   Tomorrow's concert in the park will probably be canceled. There's a 98 percent chance of rain showers and the musicians won't perform in the rain.

5.   Alice and Jane are cousins. Jane and Robert are cousins. Therefore, Alice and Robert must be cousins as well.

6.   Joe accidentally cooked his hamburger in the microwave for an hour. In all likelihood, the burger is no longer fit for human consumption.

7.   Since New Year's Day is always one week after Christmas, Christmas and New Year's Day are necessarily going to be on the same day of the week during any one holiday season.

4. ARGUMENTS

● EVALUATING ARGUMENTS

The logical evaluation of an argument may be divided into two completely independent tasks. Either of the two tasks may be performed first; neither is more or less important than the other. First, determine whether the argument is deductively valid or invalid, or inductively strong or weak. This involves evaluating just the logical relation between the premises and the conclusion. It is important to remember that this part of the evaluation can be accomplished without knowing whether the premises are actually true. The second task involves determining whether the premises are actually true. For example, recall Smith's argument for the claim that the tall man had robbed a bank. In the case of Smith's argument, this second task would involve determining whether or not the tall man actually was the attorney's first client, and whether her first client did or did not rob a bank.

This second task—determining whether or not the premises are actually true—does not fall within the special province of logic. Any number of fields of study may be relevant when determining the truth or falsity of the premises. For instance, if a premise makes a claim about the charge on the electron, we would turn to a physicist for help in determining whether the premise is true or not. If a premise makes a claim about the size of Napoleon's army, we would turn to a historian, and so on.

When you evaluate an argument, it is best to perform these two tasks one at a time. You might begin simply by asking: Is the argument valid? If it is not valid, is it strong? Then, once that part of the evaluation process is complete, the next question is: Are the premises true?

●— ARGUMENTS IN EVERYDAY LANGUAGE

The arguments we've examined so far have had simple structures. In each case, we listed the premises first and placed the conclusion last. Let us say that an argument is in *basic form* when the premises are listed one after the other and the conclusion is listed last. However, when arguments are presented in the usual circumstances—in books, speeches, newspapers, conversations, and so on—they are seldom organized so simply. Often, people will put the conclusion first with the premises afterward. Sometimes, premises will be given, a conclusion will be drawn, and then more premises will be added. The conclusion might be spread across two sentences, or it might appear as a clause inside a larger sentence. In contrast to the short, neatly ordered arguments we've been looking at, arguments in everyday contexts can appear all jumbled up.

Often, it is easier to understand an argument if the argument is put into basic form. This is not always a simple task, for in many cases, when someone is presenting an argument, it is not obvious to the reader or listener which of the statements is supposed to be the conclusion, which are supposed to be the premises, and which are neither premises nor conclusion. Fortunately, the English language has a variety of expressions that may be used to indicate to the reader or listener which statements are offered as premises and which is offered as the conclusion. *Conclusion indicators* are words or expressions that are used in the context of an argument to signal the location of the conclusion.

SOME TYPICAL CONCLUSION INDICATORS

therefore	shows that
hence	consequently
so	it follows that
implies that	in conclusion
thus	accordingly

Conclusion indicator words are generally prefixed to the sentence that expresses the conclusion. For example, if someone is giving an argument, and after presenting a number of statements says "consequently," or perhaps "therefore," you naturally expect the conclusion to follow.

Premise indicators are words or expressions that are used in the context of an argument to identify the premises.

SOME TYPICAL PREMISE INDICATORS

because	since
for	may be inferred from
for the reason that	as shown by
given that	on the assumption that

For example, if someone makes a questionable claim and then says "because" or "since," you naturally expect the next statement to serve as a premise for the questionable claim.

Suppose that an argument presented to you contains no indicator words whatsoever. What then? You will have to *interpret* the argument as best you can. It is helpful to begin by asking: What is the author trying to prove? When you answer this question, you will have identified the conclusion of the argument. Next, you should ask: What evidence is being offered in support of the conclusion? What is the conclusion supposed to be based on? In answering this, you will be identifying the premises.

———————————• **EXERCISES** •———————————

Exercise 1.12

1. Find a newspaper or magazine editorial containing an interesting argument. Summarize that argument in your own words and write an evaluation of it. (Your college newspaper probably has some interesting arguments in it.)

2. Listen to a radio or TV talk show that has on it a controversial guest. Summarize one of the arguments given and write an evaluation of the argument.

3. Find a commercial or a political advertisement containing an interesting argument. Summarize the argument and then write an evaluation of it.

Exercise 1.13

Each of the following passages contains or constitutes one or more arguments. In each passage, identify the premises and conclusion of each argument.

1. Santa is a Red. He represents the Red Peril. And he's got a beard, therefore he's a hippie. I'd be a red-bearded Santa Claus because I got a red beard. And what does Santa smoke in that pipe? Grass! Santa gives everything away free. He's a commie. Santa is a stoned commie.

   —Jerry Rubin, *Do It*

2. Many of the most important chemical processes of industry and of daily life depend upon reactions in which the valence of elements changes. Such reactions are called oxidation-reduction reactions. The burning of fuels, the smelting of metals, the operation of dry-cells and of storage batteries, the rusting and tarnishing of metals, and the growth of plants and animals are intimately connected with oxidation-reduction reactions.

   —Lionel Porter Home Chemistry Lab Manual, 1951

3. Ideas about causation were at the focus of attention in the early days of Greek philosophy, and it occurred to some to wonder whether all physical events are caused or determined by the sum total of all prior events. If they are—if, as we say, *determinism* is true—then our actions, as physical events, must themselves be determined. If determinism is true, then our every deed and decision is the inexorable outcome, it seems, of the sum of physical forces acting at the moment, which in turn is the inexorable outcome of the forces acting an instant before, and so on, to the beginning of time. How then could we be free?

   —Daniel Dennet, *Elbow Room*

4. Sulfur is one of the world's most useful elements. You rarely see it in everyday life, but it enters directly or indirectly into the manufacture of almost everything you use. Rubber, paper, metal, concrete, pigments, matches, medicines, and glass have all come into contact with sulfur or sulfur compounds at some stage of their manufacture.

   —Lionel Porter Home Chemistry Lab Manual, 1951

5. If happiness is the highest good of a being endowed with reason and if that which can by any means be snatched away is not the highest good (since that which cannot be taken away is always better), it is plain that the gifts of Fortune by their instability cannot ever lead to happiness.

   —Boethius, *The Consolation of Philosophy*

6. Astronomers may have finally, and inadvertently, found where most of the mass in the universe is hiding: in small groups of galaxies like the one to which our Micky Way belongs. [Researchers] used the orbiting Rosat observatory to search for X-rays from hot gas in a trio of galaxies called the NGC 2300 group. . . . On January 4th . . . the researchers announced that Rosat detected X-rays coming from a cloud 1.3 million light-years across, heated to 10 million degrees Kelvin. But the

combined gravitational attraction of the three visible galaxies is much too weak to hold the gas in place. The balance of the force apparently is exerted by 230 trillion solar masses of unseen, or dark, matter.

—*Sky and Telescope* magazine, March 1993

●— DIALECTIC

Have you participated lately in a good discussion? One in which individuals with different points of view reasoned with one another? If so, you took part in a *dialectical* process of thought. A dialectical discussion is one in which people with different points of view reason with each other.

Have you ever noticed that your understanding of something has deepened after you discussed it with someone who holds a different view? Reasoning with others generally leads to increased understanding. Since understanding generally leads to tolerance and cooperation, and since tolerance and cooperation are key ingredients of any viable human society, it seems that dialectical discussion has a valuable role to play in human affairs.

The Greek Dialectic   Human beings have always asked "why" questions. Why does the Sun move across the sky every day? Why does its path across the sky change every day? And why do these changes make up a pattern that recurs yearly? And why is the sky blue? Why does water always flow downhill?

From the beginning of human thought, the natural reaction to such perplexity has been to seek explanations. The ancient myths were the earliest systematic attempts at explaining our world and its regularities. It is important to note that ancient myths were not irrational speculation. Every myth had to hang together in a logically consistent way, and its many parts had to fit together into an intelligible whole. The ancient myths were affairs of human reason.

In the sixth century B.C., a group of thinkers associated with the Greek seaport of Miletus constructed the first philosophical explanations of the world. The Milesians, as these philosophers are now called, were simply doing what human beings had always done—they were trying to understand the world. However, their ideas were philosophical in at least two respects. First of all, the Milesian thinkers offered *evidence* in support of their theories. That is, they gave reasons in support of their theories. And second, the advocates of various theories offered reasons against the theories of others. As a result, a dialectical discussion developed in which different views were put forward, discussed, criticized, and defended. Theories were then either abandoned or modified in the light of argumentation. As one would expect, out of this dialectic emerged better and better explanations and theories.

Indeed, this was the beginning of a dialectical process of thought that has transformed the entire world. For out of these early attempts at reasoning—these attempts to explain the world—emerged such subjects as physics, chemistry, biology, mathematics, history, literature, and theology. When human beings reason together in a dialectical context, the result is generally an increased understanding.

●— THE ETHICS OF ARGUMENTATION

There are ideals associated with each type of human activity. Let's reflect for a moment upon the activity of rational discussion and thought. When you have intellectual discussions with others, what ideals should you try to live up to? This is a question you can think about on your own. Perhaps you might begin by asking yourself, "When engaging in discussions, how do I want others to treat *me*?" The following two ideals of intellectual interaction are offered merely as suggestions to get you thinking about the question.

Intellectual Fairness   Ideally, when you are trying to make your mind up on an issue, you ought to try to understand the main reasons for and against each of the different points of view on the matter. Someone who tries to understand and evaluate all sides of an issue before drawing conclusions may be called an *intellectually fair* individual. For example, an intellectually fair individual, when trying to decide which theory of social justice is the best theory, ideally tries to understand the reasoning for and against Marxist theories, classical liberal theories, democratic socialist theories, democratic capitalist theories, market socialist theories, welfare state theories, and anarchist theories. Each alternative point of view, such a person will tell you, is worthy of serious consideration.

Intellectual Kindness   Some people, when engaged in reasoned discussion, can become angry, mean-spirited, and even abusive. In the heat of argument, they sometimes treat opposing points of view with contempt. In short, some people are just no fun to argue with. In contrast, an *intellectually kind* individual treats each point of view with equal respect. This individual is trying to understand the valuable points put forward by others. If someone is having difficulty expressing an idea clearly or fully, the intellectually kind individual won't make fun of that person, or cut the person off, or make the person feel incompetent. Rather, he or she will help the person clarify and express the thought.

●— THE FOCUS OF OUR STUDY

Deductive logic may be defined as the branch of logic concerned with the study of deductive validity and associated concepts. *Inductive logic* may be defined as the branch of logic concerned with the study of inductive strength and related concepts. Within inductive logic, theorists develop methods to measure the degree of probability or rational support conferred by the premises upon the conclusion. As you might suspect, a substantial amount of mathematics (probability theory) goes into the measurement of degrees of probability. Within deductive logic, techniques are developed to help us determine whether or not an argument is deductively valid.

This text is concerned primarily with the presentation of the deductive branch of logical theory. Deductive logic has been an active field of thought for well over 2,000 years. Studies in this field were initiated by Aristotle (384–322 B.C.). They were carried further by the Stoic philosophers, who wrote from the third century B.C. to the second century A.D. In addition, important contributions to deductive logical theory were made by a number of scholastic philosophers during the Middle Ages. During the twentieth century, a number of significant advances have occurred through the work of such philosophers as Gottlob Frege, Bertrand Russell, C. I. Lewis, Kurt Godel, Alonzo Church, W. V. O. Quine, Saul Kripke, Alvin Plantinga, and David Lewis.

A Look Ahead   In this chapter, we tried to decide whether various arguments were valid or invalid, whether various groups of statements were consistent or inconsistent, and so on. In each case, we did this simply by mulling over the various possibilities. That is, we evaluated arguments and sentences *intuitively*. In many cases, it was difficult to decide on a definite answer. In Chapter 2, you will learn how to *compute*—in a precise and systematic way—definite answers to logical problems.

If computational methods give precise answers to the sorts of questions we have been asking, why haven't we used those methods in this chapter? The answer is that before you can understand those methods, you first must understand the conceptual foundation underlying those methods. That foundation has been laid, using fairly plain English, in this opening chapter. Now that you understand the fundamental concepts of this chapter, you are ready to study the computational methods of Chapter 2.

EXERCISE 1.14

I.   To strengthen your understanding of the concept of validity, here are some more arguments to evaluate. In each case, is the argument valid or invalid?

1.   Ann and Bob won't both be home. Ann will be home. So, Bob won't be home.

2.   Ann and Bob both won't be home. So, Bob won't be home.

3.   If Ann is home, then Bob will be home. Ann won't be home. So, Bob won't be home.

4.   If Ann goes swimming today, then Bob will go swimming today. Bob won't go swimming today. So, Ann won't go swimming today.

5.   If Ann is home, then Bob will be home. Bob will be home. So, Ann will be home.

6.   Either Ann or Bob won't go on the picnic. Bob will go on the picnic. So, Ann won't go on the picnic.

7.   Either the school will buy a new snowplow or the campus will be closed tomorrow. So, if the school doesn't buy a new snowplow, the campus will be closed tomorrow.

8.   Rita sees the moon rising. Therefore, the moon is rising.

9.   Rita believes she sees the moon rising. Therefore, the moon is rising.

10. The car won't run unless we put in a new carburetor. We will put in a new carburetor. So, the car will run.

11. The picnic will be canceled unless the sun comes out. So, if the sun comes out, then the picnic won't be canceled.

12. The concert will be held if and only if the permit is approved. The permit won't be approved. So, the concert won't be held.

13. Everyone who has read Aquinas's *Summa Theologica* believes that Aquinas was a deep thinker. Professor Smith believes that Aquinas was not a deep thinker. Therefore, Professor Smith hasn't read Aquinas's *Summa Theologica*.

14. Socrates taught Plato. Plato taught Aristotle. Aristotle taught Alexander the Great. Therefore, the teacher of the teacher of Alexander the Great was taught by Socrates.

15. All who love logic love poetry. Susan loves poetry. Therefore, Susan loves logic.

16. Rodney Dangerfield gets no respect from anyone. Everyone respects someone who has won a Nobel Peace Prize. So, Rodney Dangerfield hasn't won a Nobel Peace Prize.

17. Jones will pass the logic test if Jones joins a study group. So, if Jones joins a study group, Jones will pass the logic test.

18. Nothing created by God can be all bad. Each person was created by God. So, nobody is all bad.

19. If the doctrine known as skepticism is true, then nothing can be known for sure. If nothing can be known for sure, then skepticism cannot be known for sure. Therefore, if skepticism is true, it cannot be known for sure.

20. If God changes, God changes for the better or for the worse. If God changes for the better, God isn't perfect. If God changes for the worse, God isn't perfect. God is perfect. Therefore, God is changeless.

21. If time had no beginning, then the present moment is the endpoint of an infinite series of preceding moments. It is not possible that an infinite series of moments have an endpoint. So, it is not possible that time had no beginning. Therefore, time had a beginning.

22. The following is a short version of an argument against the existence of God. The argument is known as the "argument from evil" because the suffering to which it refers is a bad thing or an "evil."

A wholly good being would be opposed to any instance of suffering. An all-knowing being would know about any instance of suffering. An all-powerful being would have the power to prevent any instance of suffering from occurring. So an all-knowing, all-powerful, wholly good being would not allow any suffering to exist. Therefore, it is not possible that (a) an all-knowing, all-powerful, wholly good God exists and at the same time (b) suffering exists. The world does contain instances of suffering. Therefore, an all-knowing, all-powerful, wholly good God does not exist.

23.   The following is a short version of an argument for the existence of God. The argument is known as the "cosmological argument" because it is concerned with the nature of the universe—the cosmos—as a whole:

Nothing can possibly exist without a sufficient reason why it exists. The universe exists so there must be a sufficient reason why it exists. Every entity is either a contingent being (a being that exists but that might not have existed) or a necessary being (a being that exists and could not possibly have failed to exist). In the case of a contingent entity, the sufficient reason for its existence must lie in something prior that brought the entity into existence. A necessary being must carry within its inner nature the sufficient reason for its existence. If each thing in existence is contingent, then there is no explanation constituting a sufficient reason for the universe's existence. There must therefore exist a necessary being. The universe might not have existed so it is a contingent entity. Therefore, there must exist a necessary being that is other than the universe. Only a necessary being could constitute a sufficient reason for the universe's existence. So, there must exist a necessary being that is itself other than the universe and that is the sufficient reason for the existence of the universe.

II.   Write an argument of your own which is:

    a.   valid and has false premises.

    b   invalid and has true premises.

    c.   inductively strong and has false premises.

    d.   inductively weak and has true premises.

III.   For each of the invalid arguments in Part I above, describe a counterexample.

TRUTH-FUNCTIONAL LOGIC

1. INTRODUCTION

In many fields of thought, special symbolic languages have been developed. You are probably familiar with some if not all of the languages in the following examples:

Mathematics: $$x = \frac{-b + \sqrt{b^2 - 4ac}}{2a}$$

Music:

Chemistry: $$2Na\ (s) + Cl_2\ (g) \rightarrow 2NaCl\ (s)$$

Economics: $$XPx + YPy = I$$

The symbolic languages in the examples above are called *artificial* languages and may be contrasted with *natural* languages such as English, French, and Spanish. A natural language is typically learned during the first few years of life and serves as a general tool of communication. Natural languages are not designed by a person or a committee. Rather, they evolve without central direction slowly and spontaneously within a social group over long periods of time. In contrast, an artificial language is typically designed for the expression of a restricted and specialized body of ideas and not for general communication. Artificial languages are learned after one has first learned a natural language and are learned with the aid of a natural language. Although artificial languages also evolve over time, their development involves a significant degree of central direction and conscious planning.

Artificial symbolic languages serve an important purpose: they enable us to simplify the expression of complex ideas. Consider, for example, the symbolic language of musical notation. Beethoven could have written his compositions entirely in German—the natural language he grew up with. By spelling out in words every note to be played, along with the associated instructions for each note, he could have avoided using any musical notation. However, one composition would fill up hundreds of pages of text, and that text would be extremely difficult to read or play an instrument from. By using the symbols of musical notation, Beethoven was able to express his ideas in a more compact and easier to read form.

Similarly, a calculus book could be written entirely within a natural language, without the use of a single mathematical symbol. The ideas normally expressed in symbols would have to be put into words and sentences. However, when compared to an ordinary calculus text, a natural language calculus book would be very lengthy and far more difficult to read.

In addition to the fact that they simplify the expression of complex ideas, artificial symbolic languages serve another important purpose. The words of a natural language typically have many meanings and often evoke emotions as well. This may be good for certain types of communication such as poetry or drama, but it generally leads to confusion and error in the context of a technical field of thought such as mathematics, computer science, or logic. In a technical field of thought, ideas must be expressed in precise and unambiguous terms or else communication breaks down. The symbols of an artificial language can be given the precise meanings required for the communication of technical ideas.

This chapter introduces a symbolic logical language that is designed to represent the logical properties of an important type of argument. Future chapters will develop additional logical languages representing the logical properties of additional types of argumentation. Once our first symbolic language is developed, we will translate arguments out of English and into that special logical language. Procedures will then be developed that will allow us to solve a variety of logical problems. The main reason we will translate arguments out of English and into an artificial language is that it is only with such a language that certain absolute, exceptionless rules of argument evaluation can be formulated and applied. Using these rules, you will be able to evaluate arguments in a very precise and systematic way, similar to the way a mathematician solves an algebra problem. The first step will be to learn how to translate English sentences into the symbolism of our logical language.

●— SIMPLE and COMPOUND SENTENCES

Consider the following sentence:

    1.   Birds have wings, and fish have scales.

Sentence (1) is an example of a *compound* sentence. Any sentence that contains one or more simpler sentences qualifies as a compound sentence. Embedded within sentence (1) are two simpler sentences, namely:

    a.   Birds have wings

and

    b.   Fish have scales.

The two sentences within sentence (1), sentences (a) and (b), may be called *component* or *embedded* sentences because they are part of or embedded within sentence (1). Here are some further examples of compound sentences, with the component sentences underlined:

2. <u>Susan is a doctor</u> or <u>Susan is an engineer.</u>
3. It is not the case that <u>Susan is an engineer.</u>

A sentence that is not compound shall be termed a *simple* sentence. Each of the component sentences embedded within sentences (1), (2), and (3) is a simple sentence.

2. SENTENCE OPERATORS and TRUTH FUNCTIONS

In general, when components are joined to form a compound, something must bind or link the components together to form the compound. For example, in the chemical realm, atomic bonds join atoms together to form molecules. In the construction industry, cement is used to join sand and rocks together to form concrete. The English language contains certain words and expressions, called *sentence operators* or, alternatively, *sentence connectives*, that are used to form compound sentences out of component sentences.

For example, suppose we start with a sentence such as:

4. The Moon has mountains.

If we prefix to this sentence the expression 'it is not the case that', we produce the sentence:

5. It is not the case that the Moon has mountains.

When you prefix 'it is not the case that' to a sentence, you produce the *negation* of that sentence. Thus, sentence (5) is the negation of sentence (4).

Sentence (5) qualifies as a compound sentence because within (5) you will find a component sentence, namely, the string of words 'the moon has mountains', and any sentence containing one or more component sentences qualifies as a compound sentence. Notice that (5) contains just *one* component sentence, namely, 'the Moon has mountains'. The expression 'it is not the case that', which also lies within (5), does not qualify as a component sentence, for it is not a sentence in its own right. Rather, it is used within (5) as a sentence operator, since it is used in this case to form a compound sentence out of a simpler sentence.

The word 'and' may also be used as a sentence operator to form a compound sentence. For example, suppose we begin with the two sentences:

    a.   Carbon is an element.

    b.   Oxygen is an element.

We may bind these two sentences together to generate a compound sentence by placing 'and' between them:

    6.   Carbon is an element, and oxygen is an element.

Of course, by letting the two subjects share a common verb phrase, we could have *abbreviated* sentence (6) as:

    7.   Carbon and oxygen are elements.

    Likewise, the word 'or' may be used as a sentence operator. For example, beginning with the following two sentences

    a.   Pat is 21.

    b.   Randy is 21.

we may write:

    8.   Pat is 21 or Randy is 21.

If we let the two subjects share the same verb, then (8) may be abbreviated as:

    9.   Pat or Randy is 21.

    In sum, a sentence operator or connective is a word or expression that may be attached to one or more sentences so as to generate a new sentence, a compound sentence, out of that to which it is attached. The words 'not', 'and', and 'or' are called "sentence operators" because they "operate" on sentences to form compound sentences. They are also called "sentence connectives" because they "connect" sentences to form compounds. Let us now look at each of these three operators in more detail.

●— THE NEGATION OPERATOR

The expression 'It is not the case that' may be called a *negation operator* because when you attach it to a given sentence, the compound formed is the *negation* of the original sentence. An operator that attaches to just one sentence is called a *monadic* operator. Since 'It is not the case that' attaches to one sentence at a time, it is a monadic operator. Of course, the sentence that this operator attaches to may itself be either compound or simple. Notice that the operator 'it is not the case that' has, in a sense, a "hook" on its right side by which it attaches to a sentence to form a compound.

Alternatives to 'it is not the case that'   There are many different but equivalent ways to express the negation of a given sentence. For instance, if we wish to negate 'Mars is a planet', we may write any of the following:

> *It is not the case that* Mars is a planet.
>
> *It is false that* Mars is a planet.
>
> *It is not true that* Mars is a planet.
>
> Mars is *not* a planet.

Each of these sentences should be viewed as a different but equivalent way of expressing the same thing, namely, the negation of 'Mars is a planet'.

Negation and Truth   From here on, we will frequently be attributing truth and falsity to things. Consequently, it will be helpful if we have some abbreviated ways of doing so. Two terminological conventions will be used. First, if a sentence in a given context expresses a truth, we will sometimes simplify things by saying that the sentence itself is true. However it should be understood that when we simplify things by calling a sentence "true," this is just a shorthand way of saying that the sentence *expresses* a truth. Similarly, if a sentence in a given context expresses a falsehood, we'll simplify things by saying that the sentence itself is false. In Chapter 5, we will see why some philosophers hold that, strictly speaking, a sentence expresses the truth but is not itself that which is true.

Second, instead of saying that a sentence "expresses" the truth, or that the sentence "is" true, it will occasionally be convenient to say that the sentence *has the truth-value true*. Similarly, instead of saying that a sentence "expresses" a falsehood, or that the sentence "is" false, it will sometimes be more convenient to say that the sentence *has the truth-value false*. The truth-values truth and falsity will often be abbreviated **T** and **F** respectively. In this chapter, we shall only speak of two truth-values, truth and falsity. However, in the Philosophical Interlude, the possibility of a third truth-value will be considered briefly.

Let us now consider the relationship between the negation operator and truth and falsity. Suppose the following sentence is true:

> **10.**   Ann owns a pet aardvark.

It follows that the *negation* of (10), namely:

> **11.**   It is not the case that Ann owns a pet aardvark

is false.

Next, suppose the following sentence expresses a falsehood:

> **12.**   The Earth has six moons.

It follows that the *negation* of (12), namely:

13. It is not the case that the Earth has six moons

expresses a truth.

In the logical language we are building, capital letters will be used to abbreviate specified English sentences. Thus, instead of writing out the sentence

Ann owns a pet aardvark

we'll abbreviate this by writing

A

with the understanding that 'A' abbreviates or stands for the entire sentence 'Ann owns a pet aardvark'. In symbolic languages, a *constant* is a symbol that stands for one specified thing. When capital letters are used to abbreviate specified sentences, those letters are being used as *sentence constants*.

If a sentence contains a negation operator, we'll abbreviate the operator with the symbol '~', called the *tilde* (pronounced "*till*-da"). Thus,

It is not the case that Ann owns a pet aardvark

will be symbolized in our logical language as:

~A

with the understanding that the tilde abbreviates 'It is not the case that' and 'A' abbreviates 'Ann owns a pet aardvark'.

Now we can state the relation between the truth-value of the sentence 'Ann owns a pet aardvark' and the truth-value of its negation 'It is not the case that Ann owns a pet aardvark' as follows:

If 'A' is true then '~A' is false and if 'A' is false then '~A' is true.

This relationship may also be displayed on what is called a *truth-table*:

A	~A
T	F
F	T

The first row of the table (running horizontally across the table under the horizontal line) indicates that when 'A' is true, '~A' is false. The second row of the table indicates that when 'A' is false, '~A' is true.

Generalizing This   The truth-table above expresses the relationship between the truth-value of the specified sentence 'A' and its negation. However, the table expresses something that is true not only of sentence 'A' but of any sentence, namely:

> *If any sentence in a given context is true, then the negation of that sentence is false; and if any sentence in a given context is false, then the negation of that sentence is true.*

We need to represent this general claim in a symbolic way. How shall we proceed? At first, one might think we could simply pick a capital letter such as 'B', stipulate that the letter 'B' stands for any sentence, and then write:

If 'B' is true then '~B' is false and if 'B' is false then '~B' is true.

However, this won't work, for in the symbolic language we are building, a regular capital letter will be used only to abbreviate a specified sentence. Thus, the sentence displayed immediately above could only be used to say something about a specified sentence, a sentence abbreviated 'B'. We therefore can't use capital letters such as 'B' to make a general claim about all sentences.

The solution is to use symbols called *variables*. Whereas a constant is a symbol that stands for one specified thing, a variable is a symbol that is said to *range over* or stand for anything from a group of things. Mathematicians often make general claims about numbers using *numerical* variables. For example, the *commutative law of addition*

$$x + y = y + x$$

contains two variables, 'x' and 'y', that are said to range over numbers. The commutative law specifies that for any number x and for any number y, the result of adding x plus y equals the result of adding y plus x. The variables we'll use, called *metalinguistic* variables, will range over sentences. (The reason for the term 'metalinguistic' will be explained after certain preliminaries have been covered.)

To distinguish our variables from our constants, we'll use bold-faced capital letters from the middle of the alphabet, **P, Q, R, S**, as metalinguistic variables. If a situation requires the use of more than four variables, we'll add subscripts to our variables to generate additional variables.

The general relationship between negation and truth-value may now be put as follows:

> *Where **P** is any sentence, if **P** is true, ~**P** is false and if **P** is false, ~**P** is true.*

This is merely a shorthand way of saying:

> *If any arbitrarily selected sentence happens to be true, then the negation of that sentence will be false, and if that arbitrarily selected sentence is false then its negation will be true.*

Our statement is now about sentences in general and is not about one specified sentence.

This general claim may also be expressed on what is termed the *truth-table for negation*. With the understanding that **P** is any sentence, the table is the following:

P	~P
T	F
F	T

Truth-Table for Negation

The first row of the table, running horizontally under the horizontal line, specifies that in any circumstance in which **P** is true, the compound ~**P** is false. The second row specifies that in any circumstance in which **P** is false, the compound ~**P** is true.

Since the table specifies the conditions under which a negation is true and the conditions under which it is false, the table specifies the *truth-conditions* for negation.

●— THE CONJUNCTION OPERATOR

Suppose we begin with two separate sentences, namely:

    a. Sergeant Pepper taught the band to play.

    b. The band is guaranteed to raise a smile.

Next, suppose we wish to assert just that both conjuncts express the truth. Then we may join these two sentences together by placing the word 'and' between them:

    14. Sergeant Pepper taught the band to play, and the band is guaranteed to raise a smile.

The component sentence to the left of the 'and' is called the *left conjunct*, while the component sentence to the right is called the *right conjunct*. The compound as a whole is called a *conjunction*. The word 'and' is being used in (14) as a sentence operator. Since the sentence formed is a conjunction, 'and' is being used here as a *conjunction operator*.

Any operator that joins two sentences into a compound is called a *dyadic* operator. Thus, the word 'and' has been used here as a dyadic operator, for it binds two sentences into a compound sentence. We can use 'and' to join two simple sentences into a compound, as we did above, or we can use 'and' to join together compound sentences.

Alternatives to 'and' The English language contains many other words and expressions that may, in various contexts, be used in place of 'and' to form conjunctions. For instance, instead of saying "Pat grew up in Liverpool during the 1960s, *and* he has never heard of the Beatles," one might say one of the following instead:

Pat grew up in Liverpool during the 1960s, *but* he has never heard of the Beatles.

Pat grew up in Liverpool during the 1960s, *yet* he has never heard of the Beatles.

Pat grew up in Liverpool during the 1960s, *although* he has never heard of the Beatles.

Pat grew up in Liverpool during the 1960s, *however*, he has never heard of the Beatles.

We typically use words such as 'but', 'yet', 'although', and 'however' in place of 'and' when a conjunction is surprising in some way—as in the above examples.

Notice that a conjunction operator has, in a sense, two "hooks"—one on each side—by which it links two component sentences to form a compound sentence.

Conjunction and Truth   Let us now examine the relation between conjunction and truth. Consider again sentence (14):

14.   Sergeant Pepper taught the band to play, and the band is guaranteed to raise a smile.

Suppose both conjuncts are true. That is, suppose

a.   Sergeant Pepper taught the band to play

is true, and

b.   The band is guaranteed to raise a smile

is also true.

Given that (a) and (b) are both true, what's the truth-value of the conjunction of these two sentences? Clearly, the conjunction, (14), which was used to assert just that both conjuncts are true, is true, isn't it?

Next, suppose that just one conjunct is true while the other is false. For example, suppose that Sergeant Pepper really did teach the band to play but the band is not guaranteed to raise a smile. What's the truth-value of the conjunction in this case? The conjunction as a whole is clearly false.

Next, suppose that both component sentences are false. Sergeant Pepper never taught the band to play, and the band is not guaranteed to raise a smile. What's the truth-value of the conjunction in this case? The conjunction is false in this case.

Instead of writing the word 'and' between the components, let us use, as an abbreviation, the ampersand, '&'. This symbol will simply be used as an abbreviation for 'and' when that operator is used to assert just that both conjuncts are true.

If we abbreviate 'Sergeant Pepper taught the band to play' with 'S' and 'The band is guaranteed to raise a smile' with 'B', then we may symbolize sentence (14) as:

S & B

The truth-table for this conjunction follows:

S	B	S & B
T	T	T
T	F	F
F	T	F
F	F	F

The first row, which contains three **T**'s, indicates that in any circumstance in which 'S' is true and 'B' is true, the conjunction as a whole is true. The second row indicates that when 'S' is true and 'B' is false, the conjunction is false. The third row indicates that when 'S' is false and 'B' is true, the conjunction is false. And the fourth row indicates that when both 'S' and 'B' are false, the conjunction is false.

Generalizing This   In general, a conjunction that asserts just that both conjuncts are true is true if and only if *both* conjuncts are true. So, if one or both conjuncts are false, then the conjunction as a whole is false as well. This general claim may be expressed on the following truth-table, with the understanding that **P** is any sentence and **Q** is any sentence:

P	Q	P & Q
T	T	T
T	F	F
F	T	F
F	F	F

Truth-Table for Conjunction

The first row indicates that in any circumstance in which **P** is true and **Q** is true, the conjunction of the two is true. The second row indicates that in any circumstance in which **P** is true and **Q** is false, the conjunction as a whole is false. The third row indicates that in any circumstance in which **P** is false and **Q** is true, the conjunction as a whole is false. Finally, the fourth row specifies that in any circumstance in which **P** and **Q** are both false, the conjunction as a whole is false.

The four rows represent every combination of truth-values that the component sentences could possibly have. The table specifies the truth-value of the compound for each combination of truth-values the components might have. Since the table specifies the conditions under which a conjunction is true and the conditions under which it is false, the table specifies the truth-conditions for conjunction.

●— THE DISJUNCTION OPERATOR

When two sentences are linked together by 'or', the resulting compound sentence is called a *disjunction*. Thus, if we take

15.   The starter motor isn't working

and

16. The battery is dead

and join them together with the word 'or', we get the disjunction

17. The starter motor isn't working or the battery is dead.

which is itself a compound sentence.

Sentence (15) constitutes the *left disjunct* of the disjunction, while sentence (16) constitutes the *right disjunct*. Since the 'or' is used to form a disjunction, it is used as a sentence operator. Since the sentence formed is a disjunction, 'or' is being used as a *disjunction operator*. Notice that 'or' functions as a dyadic operator, for it joins two sentences into a disjunction.

Disjunction and Truth   There are two different ways to use the word 'or' in English. When we use the word 'or' in its *exclusive sense*, we mean to assert that one or the other of the two disjuncts is true but not both. For example, the menu at Billy Bob's Brontosaurus Burger Bar might tell you that Billy Bob's famous brontosaurus burger combo plate comes with stegosaurus soup *or* with fossilized fries. If Billy Bob's place is an ordinary restaurant, this means that the burger comes with the soup, or with the fries, but not with both. The restaurant 'or' is typically an exclusive 'or'.

When we use the word 'or' in its *inclusive sense*, we mean to assert that one, or the other, or both, of the disjuncts is true. For example, when Father Flanagan started Boys Town in 1917, he offered a home to boys who were either homeless *or* in trouble with the law. Did he also accept boys who were *both* homeless and in trouble with the law? Of course. Father Flanagan helped many boys who were both homeless *and* in trouble with the law.[1] Father Flanagan's 'or' was an inclusive 'or'.

⎯⎯⎯⎯⎯• SOME QUESTIONS •⎯⎯⎯⎯⎯

In the following sentences, is the 'or' an exclusive or an inclusive 'or'?

   i.   If an argument is unsound, then it is either invalid or has it least one false premise.

   ii. Her favorite recording is either Liberace's version of "As Time Goes By" or Dean Martin's "Everybody Loves Somebody."

   iii. I can't remember exactly who did it, but "Light My Fire" was recorded in 1967 by either The Doors or Jackie Gleason.

---

[1]   Incidentally, most of the young people Father Flanagan helped grew up to lead productive lives. A wonderful account of Boys Town is the book *Father Flanagan of Boys Town* by Fulton and Will Ousler (Doubleday, 1956).

Let us use the symbol 'v', called the *wedge*, to abbreviate the word 'or' as that word is used in its inclusive sense. If we abbreviate 'Father Flanagan helped homeless boys' with 'H' and 'Father Flanagan helped boys who were in trouble with the law' with 'L', then the disjunction of the two, namely, 'Father Flanagan helped homeless boys or Father Flanagan helped boys who were in trouble with the law' may be symbolized

H v L

When we use 'or' in its exclusive sense, the disjunction as a whole is true if and only if exactly one of the two disjuncts is true. However, when we use 'or' in its inclusive sense, the disjunction is true just in case one or both of the disjuncts is true, which is to say that it is false only when both disjuncts are false. This general claim may be expressed on a truth-table. If **P** is any sentence and **Q** is any sentence, the truth-table for the inclusive disjunction of the two is:

P	Q	P v Q
T	T	T
T	F	T
F	T	T
F	F	F

Truth-Table for Inclusive Disjunction

Since the table specifies the conditions under which an inclusive disjunction is true and the conditions under which it is false, the table specifies the *truth-conditions* for inclusive disjunction.

We do not need to build into our logical language symbols for both the inclusive and exclusive disjunction operators, for as long as one type of 'or' is symbolized, the other type of 'or' can be expressed using a combination of other symbols. This will be demonstrated later. It is customary, in introductory logic, to provide a symbol for inclusive disjunction, and then to symbolize exclusive disjunctions using a combination of symbols. Consequently, from here on, let the term 'disjunction' refer to *inclusive* disjunction, and unless an 'or' is specifically identified as an exclusive disjunction operator, let us treat it as an inclusive disjunction operator and symbolize it with the wedge.

●— TRUTH-FUNCTIONS

If you will examine the truth-tables for negation, conjunction, and disjunction, you will notice that in each case the truth-value of the compound sentence as a whole is determined precisely by the truth-values of the component sentences. Compound sentences such as these are said to be *truth-functional* compound sentences since the truth-value of the compound as a whole is a function of the truth-values of the components. In general, a compound sentence is truth-functional if and only if the truth-value of the compound as a whole is a function of the truth-value of the component or components. Any operator that, when attached to one or more component sentences, forms a truth-functional compound sentence will be called a *truth-functional sentence operator*. It will be helpful if we spend a few minutes on these two new concepts, the concepts of a function and of a truth-function, since they play such important roles in logical theory.

It is best to begin with the more fundamental notion of a function. Essentially, a function is a rule that correlates one set of values with another set of values in such a way that each member of one of the sets, the set called the *domain* of the function, is paired with exactly one member of the other set, the set called the *range* of the function.

For example, suppose that every time Wimpy eats hamburgers, he eats ten french fries with each burger. (Wimpy was, in his day, one of the truly great connoisseurs of the hamburger.) Wimpy's dining habits may be described by the following function:

$$y = 10x$$

where 'y' is a variable standing for the number of french fries Wimpy eats and 'x' is a variable standing for the number of hamburgers he eats. This function may be called a *numerical* function, since it correlates one set of numerical values with another set of numerical values. That correlation may be represented by the following table of values:

x	y
1	10
2	20
3	30
4	40

Compare this table with a truth-table. The table above pairs one set of values with another set of values. Each truth-table also pairs one set of values with another set of values. Thus, each truth-table also defines a function. The conjunction truth-table, for example, specifies a rule that relates the truth-value of a conjunction to the truth-values of its components. Call this function the *conjunction function*. Since this function correlates one set of truth-values with another set of truth-values, it is called a *truth-function*. Since the truth-value of the conjunction is a function of the truth-values of the components, this type of conjunction is called a *truth-functional conjunction*, and the conjunction operator that forms such a sentence is called a *truth-functional conjunction operator*.

The negation table also specifies a truth-function, for the truth-value of the compound in this case is a function of the truth-value of the component. Call this function the *negation function*. The negation operator, when it is used to form a sentence whose truth-conditions are specified by the negation truth-table, qualifies as a truth-functional operator, since it is used to form a truth-functional compound. A sentence so formed is a *truth-functional negation*. Similarly, the disjunction truth-table

the compound as a whole. Thus, we are unable to fill in even the first row of a truth-table for the operator 'because':

PQ	P because Q
TT	?

This shows that 'because' is not being used in these examples in a truth-functional way.

Here is another example of a nontruth-functional use of an operator. An expression such as 'Franklin D. Roosevelt believed that' may be used as a sentence operator. If we prefix this expression to a sentence, we form a compound sentence and thereby attribute a belief to a former President:

> Franklin D. Roosevelt believed that Winston Churchill was the Prime Minister of Britain.

A compound sentence formed in this way is not truth-functional, for the truth-value of the whole compound is not determined by—it is not a function of—the truth-value of the component sentence. For instance, in the example just given, the belief operator is attached to a true sentence that also happens to be a sentence stating what Roosevelt believed, and the resulting compound sentence is true. However, if we attach this belief operator to a true sentence that happens to be a sentence Roosevelt did not believe, the component sentence will be true while the compound as a whole is false. The following is an example:

> Franklin D. Roosevelt believed that protons and neutrons are composed of smaller particles called "quarks."

The component sentence within this sentence is true. However, since quarks weren't discovered until 1964, the compound as a whole is surely false. If we change the component to a sentence that is false but nevertheless happens to be one Roosevelt believed, we will have a case in which the compound formed is true while the component is false.

This demonstrates that there exists no systematic or functional relationship between the truth-value of the embedded component and the truth-value of the compound as a whole in this case. Therefore, the operator 'Franklin D. Roosevelt believed that' is not used here truth-functionally. This is why no truth-table exists for this operator.

Some Additional Terminology We pointed out a few paragraphs ago that 'and', 'or', and 'not' may be used in truth-functional and in nontruth-functional ways. In order to make sure that we do not confuse truth-functional with nontruth-functional uses of 'and', 'or', and 'not', we shall from here on presuppose the following definitions. A *truth-functional conjunction* is a compound sentence that is true if both components are true and that is false otherwise. A conjunction operator that produces a truth-functional conjunction is a *truth-functional conjunction operator*. A *truth-functional negation* is a compound sentence that is true if its component is false and that is false if its component is true. Any operator that produces a truth-functional negation when combined with a sentence is a *truth-functional negation operator*. A *truth-functional disjunction* is a compound sentence that is true if one or more of its components are true and that is false if both of its components are false. Any operator forming such a sentence is a *truth-functional disjunction operator*. Let us further stipulate that the terms 'conjunction', 'disjunction', and 'negation', will be used as abbreviations of 'truth-functional conjunction', 'truth-functional disjunction', and 'truth-functional negation' respectively, unless otherwise specified.

●— ON THE IMPORTANCE OF PRECISE DEFINITIONS

Why does logic have so many definitions? And why do those definitions have to be so "wordy" and technical? In the sixteenth century, as modern science was first taking shape, the philosopher Francis Bacon (1561–1626) wrote:

> Although we think we govern our words . . . certain it is that words, as a tartar's bow, do shoot back upon the understanding of the wisest, and mightily entangle and pervert the judgment. So that it is almost necessary, in all controversies and disputations, to imitate the wisdom of the mathematicians, in setting down in the very beginning the definitions of our words and terms, that others may know how we accept and understand them, and whether they concur with us or no. For it cometh to pass, for want of this, that we are sure to end there where we ought to have begun, which is in questions and differences about words.[2]

If we try to communicate using undefined terms with imprecise meanings, the result will be confusion. Instead of getting our point across, we will end up going back to explain exactly what our undefined terms mean. Definitions and terminology are crucial to every field of thought, which is why the early part of just about any course of study involves the introduction of terminology.

————————● EXERCISE 2.1 ●————————

1. On the assumption that God is omniscient, the expression 'God believes that' does serve as a truth-functional operator. Can you explain why?

2. In the sentence 'Lemon and lime taste good together', the 'and' is not being used in a truth-functional way. Explain why.

3. Find other examples of nontruth-functional uses of sentence operators.

---

[2] This passage is quoted in Hacking [1988], p. 5.

3. SYMBOLIZATION

In order to put the truth-tables to work solving various logical problems, we shall first have to learn to translate English sentences into logical symbolism. As noted earlier, we shall use sentence constants to abbreviate specified English sentences. For example, instead of writing:

> Gertrude is happy

we may write simply:

> G

with the understanding that 'G' abbreviates or stands for the entire sentence 'Gertrude is happy'. Likewise, instead of writing:

> It is not the case that Gertrude is happy

we may simply write:

> ~G

with the understanding that 'G' abbreviates the sentence 'Gertrude is happy' while the tilde abbreviates the negation operator 'It is not the case that'. When we use a constant to abbreviate a sentence, the constant is serving as a name of the sentence.

The compound sentence

> **23.**   Arnold is happy and Betty is happy.

may be symbolized as:

> A & B

with the understanding that 'A' abbreviates the left conjunct 'Arnold is happy' and 'B' abbreviates the right conjunct 'Betty is happy'. The ampersand, of course, abbreviates the conjunction operator 'and'.

Of course, sentence (23) could be simplified to

> **24.**   Arnold and Betty are happy.

However, (24) would still be symbolized 'A&B', for (24) is merely an abbreviation of (23).

Similarly, we may abbreviate

> **25.**   Pat loves peaches smothered in ketchup or Chris loves waffles covered with mustard.

as simply

P v C

It is important that you understand that 'P' does not abbreviate just 'Pat'. Rather, 'P' abbreviates the entire component sentence 'Pat loves peaches smothered in ketchup'. Likewise, 'C' abbreviates the full component sentence 'Chris loves waffles covered with mustard'.

It is a relatively simple matter to symbolize a sentence when it contains only one sentential operator. However, sentences containing two or more operators present a whole new type of problem. In order to see this, consider the following. Suppose you have been invited to a party, and the invitation includes the following:

> **26.** Michael Jackson will be there and Stevie Wonder will be there or Bruce Springsteen will be there.

What does this sentence tell you? If you examine it closely, you will see that this sentence is ambiguous—there are at least two ways to interpret it. If you read it to yourself and emphasize the 'or' with your voice, you will get one interpretation, and if you read it and emphasize the 'and' you will get a different interpretation.

> **Interpretation One:** Michael Jackson will be there and Stevie Wonder will be there *or* Bruce Springsteen will be there.
>
> **Interpretation Two:** Michael Jackson will be there *and* Stevie Wonder will be there or Bruce Springsteen will be there.

Do you see the difference between these two interpretations? One of the two is more specific than the other. That is, on one of the interpretations, you are being assured of the presence of a specific individual, while on the other interpretation, no one specific individual is guaranteed to be there. The original sentence, (26), as written, does not indicate which interpretation is intended.

It is interpretation two, of course, that tells us something definite about one individual. Interpretation two tells us that Michael Jackson will definitely be there and then adds that one or the other or both of Stevie Wonder and Bruce Springsteen will also be there. Interpretation one, however, is less specific. On the basis of interpretation one, we don't know if the guest list will include Michael Jackson and Stevie Wonder, or just Bruce Springsteen, or Stevie Wonder and Bruce Springsteen, or Michael Jackson and Bruce Springsteen, or all three.

How might we symbolize sentence (26)? If we let 'J' abbreviate 'Michael Jackson will be there', 'W' abbreviate 'Stevie Wonder will be there', and 'S' abbreviate 'Bruce Springsteen will be there', we could try this:

J & W v S

However, this formula would itself also be ambiguous. As it is written, is 'J' joined by the ampersand to 'W', or is 'W' joined by the wedge to 'S'? Compare this with a similar problem in arithmetic. Consider the unpunctuated expression

$$2 + 3 \times 4$$

In the absence of any prearranged conventions regarding punctuation, does '2' go with '3', or does '3' go with '4'? As this expression is written, two interpretations are possible:

Interpretation One: First add 2 and 3 and then multiply the result by 4.
Interpretation Two: Multiply 3 times 4 and then add the result to 2.

Mathematicians resolve this type of ambiguity by using parentheses to *disambiguate* formulas such as this. If we intend interpretation one, we write

$$(2 + 3) \times 4$$

and if we intend interpretation two, we write

$$2 + (3 \times 4).$$

In logic, we'll use parentheses in much the same way. The two interpretations of sentence (26) discussed above may be expressed symbolically as:

Interpretation one: (J & W) v S
Interpretation two: J & (W v S)

In the symbolization of interpretation one, the parentheses tell us that (i) the conjunction operator joins together 'J' with 'W'; and (ii) this compound—*as a whole*—is joined by the disjunction operator to 'S'. In the symbolization of interpretation two, the parentheses tell us that (i) the disjunction operator joins together 'W' with 'S', and (ii) this compound is joined by the conjunction operator to the 'J'.

In order to represent a compound English sentence with our logical symbols, one must first determine which parts of the sentence are linked by which operators. Fortunately, there are a number of English grammatical devices that indicate to the reader which components of a sentence are linked. When translating sentences from English into logical symbols, pay attention to the ways that these grammatical devices function. Here are some examples that refer back to sentence (26). In order to indicate that interpretation one is intended, that sentence may be punctuated in any of these ways:

    i.   Michael Jackson will be there and Stevie Wonder will be there, or Bruce Springsteen will be there.

    ii. Michael Jackson and Stevie Wonder will be there or Bruce Springsteen will be there.

iii. Both Michael Jackson and Stevie Wonder will be there or Bruce Springsteen will be there.

iv. Either Michael Jackson and Stevie Wonder will be there or Bruce Springsteen will be there.

In (i), notice that the comma divides the sentence and pairs the first two component sentences together. The comma is a very important grouping device. In (ii), two subject terms share a common verb phrase, and this serves to group two parts of the sentence together. In the remaining sentences, coordinate phrases such as 'both . . . and' and 'either . . . or' serve to group components of the complex sentence together. Thus, in (iii) the word 'both' helps group the 'Jackson' component and the 'Wonder' component together. And in (iv), the words 'either . . . or' group the 'Jackson' and 'Wonder' components together and also divide the sentence in half at the 'or'.

These grammatical grouping indicators may be rearranged so as to indicate that interpretation two is intended. Each of the following sentences expresses the second interpretation of sentence (26):

i. Michael Jackson will be there, and Stevie Wonder will be there or Bruce Springsteen will be there.

ii. Michael Jackson will be there and Stevie Wonder or Bruce Springsteen will be there.

iii. Michael Jackson will be there and either Stevie Wonder will be there or Bruce Springsteen will be there.

iv. Michael Jackson will be there and either Stevie Wonder or Bruce Springsteen will be there.

●— SYMBOLIZING SENTENCES THAT CONTAIN A NEGATION OPERATOR

In English the negation operator typically applies to that which is to its immediate right. Likewise, in a formula, the tilde applies to whatever is to its immediate right. Thus, in the formula '~A & B', the tilde applies only to 'A'. However, in '~(A & B)', the tilde applies to the entire conjunction '(A & B)', since the parenthesis forming the conjunction sits to the immediate right of the tilde.

Without punctuation, the following sentence is ambiguous:

27. It is not the case that Ann is home and Bob is home.

As the sentence is written, it is not clear whether the negation operator applies to the conjunction as a whole, as in

~(A & B)

or to the left conjunct only, 'Ann is home,' as in

~A & B

In order to specify, in English, the '~(A & B)' interpretation, we may write:

28.   It is not the case that both Ann and Bob are home.

And in order to specify, in English, the '~A & B' interpretation, we may write:

29.   Ann is not home but Bob is home.

Of course, we could also write, instead of (29):

30.   It is not the case that Ann is home, but Bob is home.

Notice that in (30), the comma confines the influence of the negation operator to the first conjunct.

Translating 'and' into 'or' and 'or' into 'and'   Now, consider sentence (28) again:

28.   It is not the case that both Ann and Bob are home.

That is,

It's not the case that: Ann is home and Bob is home.

In this case, the negation operator applies to the conjunction as a whole, and the sentence is symbolized

~(A & B)

Do you see how closely the logical symbols correspond to the English grammatical structure of the sentence?

Suppose that (28) is true. Then who is home? Does (28) tell us that Ann is not home and that Bob is not home? No, it does not. It only tells us that we won't find them both home together. Given this, might Ann be home? Perhaps. Might Bob be home? Possibly. Might neither be home? Maybe. Sentence (28) just tells us that we won't find both home. So, given (28), we might find one, or we might find the other, or we might find neither of them home.

Within the English language, there are often two ways to say the same thing. Notice that (28) may also be put this way:

31.   Either Ann is not home or Bob is not home

which is symbolized as:

~A v ~B

This is just another way of saying that Ann and Bob are not both home. So, these two symbolizations, namely '~(A & B)' and '~A v ~B', represent two different ways of saying the same thing.

Now consider this:

32.   Both Ann and Bob are not home.

This sentence tells us that Ann is not home and that Bob is not home. The symbolization is:

~A & ~B

Nobody is home in this case, assuming, of course, that nobody else lives at their house.

Notice that (32) could have been put this way:

33.   It's not the case that either Ann or Bob is home

which is abbreviated in English as

34.   Neither Ann nor Bob is at home.

Both (33) and (34) may be symbolized as

~(A v B)

So, (32), (33), and (34) are merely different ways of saying the same thing. Consequently, the two symbolizations, (~A & ~B) and ~(A v B), represent different ways of saying the same thing.

Here is a point that needs emphasis. Notice, in the examples above, that when 'both' operates on 'not', as in (32), the 'both' distributes the 'not' onto each of the subjects equally. If both are not home, then Ann is not home and Bob is not home. That is, neither is home. However, things are different when it is the 'not' that operates on the 'both', as in (28). In this case, the 'not' does not distribute onto each subject. If Ann and Bob are not both home, it does not follow that Ann is not home and Bob is not home.

This relation between 'not', 'and', and 'or' may be summarized as follows:

ENGLISH:	SYMBOLS:
1. Not both A and B	~(A & B) or equivalently ~A v ~B
2. Both A and B will not	~A & ~B or equivalently ~(A v B)
3. Neither A nor B	~(A v B) or equivalently ~A & ~B

Some General Hints on Symbolizing The words 'both' and 'either' are very important English punctuational devices. When symbolizing a sentence, it is important that you be able to use your understanding of elementary English grammar to spot which parts of the sentence these words apply to. For example, suppose you are symbolizing:

> **35.** Either both Ann will swim and Bob will swim or both Darla and Ed will swim.

In this example, the first 'both' joins the two components following it into a conjunction. The second 'both' joins the last two components, that is, 'Darla will swim' and 'Ed will swim,' into a conjunction. Finally, the word 'either' joins these two conjunctions into a disjunction:

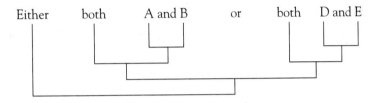

In symbols, (35) becomes:

$$(A \mathbin{\&} B) \mathbin{v} (D \mathbin{\&} E)$$

Punctuation marks are of crucial importance in written languages. In spoken languages, tone and inflection of voice serve as punctuation devices. In the symbolic logical language we are building, the only punctuation markers will be parenthetical devices of various kinds, including parentheses, brackets, and braces.

When symbolizing an English sentence, it is recommended that you follow this three-step procedure:

> Step 1: First, identify the simplest sentential components and replace these with constants.
> Step 2: Next, identify the sentence operators and replace these with their symbolic counterparts.
> Step 3: Finally, use the logical structure of the English sentence to determine which parts are to be grouped together, and then put parenthetical devices in the appropriate places.

For example, consider the following sentence:

> **36.** Either Jim swims and Pat swims or else Sam won't swim and Andy won't swim.

Following the three steps above we get:

Step 1: Either J and P or else not-S and not-A.
Step 2: J & P v ~S & ~A
Step 3: (J & P) v (~S & ~A)

Symbolizing Exclusive Disjunction   Suppose someone asserts an exclusive disjunction. For example, perhaps Chris says:

At noon, I'll be in Chicago or I'll be in New York.

Assuming that Chris does not have the power to bilocate, her claim is an exclusive disjunction and may be put more explicitly as:

I'll be in Chicago at noon or I'll be in New York at noon, but I won't be in both Chicago and New York at noon.

This exclusive disjunction is easily symbolized as:

(C v N) & ~(C & N)

• **EXERCISE 2.2** •

Symbolize the following sentences.

1. Either the Animals will perform and the Beatles will perform or the Doors will perform.

2. The Animals will perform, and the Beatles will perform or the Doors will perform.

3. The Young Rascals will perform and Bo Diddley will perform, or Santana will perform.

4. The Jefferson Airplane will perform or the Grateful Dead and the Byrds will perform.

5. Donovan will perform and either Fever Tree will perform or Otis Redding will perform.

6. Jimi Hendrix will perform and the Beach Boys and the Grateful Dead will perform.

7. Ike and Tina Turner won't both sing.

8. Grace Slick and Janis Joplin both won't perform at the Fillmore tonight.

9. Grace Slick and Janis Joplin won't both perform at the Fillmore tonight.

10. Marianne Faithful will not perform but the Isley Brothers will.

11. Both the Stones and Procol Harem will be there but the Move will not.

12. Either the Monkees or the Supremes will not perform.

13. Either Canned Heat performed or Moby Grape performed or The Group with No Name and Buffalo Springfield performed.

14. Neither the Wailers nor the Sonics will perform, but the Bards and the Natural Gas Company will perform.

15. Either Tom Jones or Tiny Tim did not sing or Engelbert Humperdinck and Dusty Springfield both did not sing.

16. Although Boxcar Willie and Zamfir both won't be available, either Slim Whitman or Locomotive Louie will be available.

4. THE LANGUAGE TL

A language can be specified in terms of its *syntax* and its *semantics*. The syntax of a language consists of its vocabulary and its rules of grammar. The semantics is the theory of meaning for the language. Essentially, the vocabulary gives us the elements we can use to construct expressions of the language, the grammar tells us how to construct properly formed expressions, and the semantics supplies the meanings for the various expressions.

At the beginning of this chapter, the distinction between artificial and natural languages was discussed. Our artificial logical language will be named TL (for 'truth-functional logic'). While the syntax of a natural language such as English has hundreds of rules, has an extremely complex structure, and takes years to learn, the syntax of TL can be specified precisely by a set of three simple rules and can be learned in a few hours.

The Vocabulary for TL

    i. Sentence constants with or without subscripts:

$$A, B, \ldots Z, A_1, A_2, \ldots B_1, B_2, \ldots \text{etc.}$$

    ii. Sentence operators:

$$\sim \lor \ \&$$

iii. Parenthetical devices:

Parentheses: (   )

Brackets: [   ]

Braces: {   }

The Grammar for TL

G1. Any sentence constant, with or without a subscript, is a sentence of TL.

G2. If **P** is a sentence of TL, then **~P** is a sentence of TL.

G3. If **P** and **Q** are sentences of TL, then **(P v Q)** and **(P & Q)** are sentences of TL.

Any expression containing only items drawn from the vocabulary of TL and that can be constructed by a finite number of applications of the rules of grammar G1 through G3 is a *sentence* of TL. Nothing else counts as a sentence of TL. Sentences of TL will also be called *well-formed formulas* or *formulas* of TL.

The sentences specified by rule (G1) are called *atomic* sentences of TL. An atomic sentence contains no operators. Sentences constructed out of atomic sentences according to rules (G2) or (G3) are *molecular* sentences of TL. The atomic sentences out of which a molecular sentence is built are the *atomic components* of the molecular sentence.

We can establish that a molecular sentence of TL is indeed a sentence of TL by beginning with the atomic components of the sentence and applying the rules of grammar until the molecular sentence is constructed. For example, consider the expression

$$\sim(\sim P \text{ v} \sim Q)$$

We can show that this is a well-formed formula of TL (or 'wff' for short) with the following line of reasoning. First, 'P' is a wff according to rule G1. Next, 'Q' is a wff according to rule G1. Consequently, '~P' is a wff according to rule G2 and '~Q' is also a wff according to G2. Therefore '(~P v ~Q)' is a wff according to rule G3. And finally, it follows from this that '~(~P v ~Q)' is a wff according to rule G2.

Longer wff's can get very complicated and difficult to read. Consequently, the adoption of the following rule will help simplify things:

> One may omit the outer pair of parenthetical devices on a wff if that pair of parentheses does not have a tilde applied to it.

As you will see shortly, outer parentheses add no new information to a formula and so may be dropped without any loss of meaning. When outer parentheses are dropped, formulas become easier to read. For example, using this abbreviatory rule:

'(A v B)'	may be simplified to	'A v B'
'[(A v B) & C]'	may be simplified to	'(A v B) & C'
'[(G & J) v ~(E & ~S)]'	may be simplified to	'(G & J) v ~(E & ~S)'

Notice that while the outer parentheses in '[(A v B) & C]' may be dropped, you may not drop the outer parentheses in a formula such as '~(A & B)'.

It is important that you be able to tell the difference between an expression that is a wff and one that is not a properly formed expression. Here are some examples:

EXAMPLES OF WFFS:	EXAMPLES OF EXPRESSIONS THAT ARE NOT WFFS:	
~A v ~B	(A v B v C)	A & B ~G
~(A v B)	A ~B	~v
~(~A & ~B)	(AB) & C	A(B v C)
Av~(E v G)	H v B~	G vv E

• EXERCISE 2.3 •

Which of the following are wffs and which are not wffs?

1. (A & ~B) v (~E & ~H)
2. ~(A v S) & [(H v B) v S]
3. ~[~(~A v ~S)~(H v G)]
4. A v (B v C) v E
5. ~(~H & R)
6. R v S v (B & G)
7. H & ~(P & S)
8. ~(~S v ~G)
9. ~(A & B) v (v S v G)
10. A & (~& R v G)
11. P v (QA)
12. R & ~ v H
13. (~A v B) & ~(H v S)
14. A v & G
15. AB
16. ~[~(A v B) v ~E] v ~S

Object Language and Metalanguage When one language is used to talk about a second language, the first language is called the *metalanguage* and the language being talked about is called the *object* language. Thus, if we speak in Spanish about expressions of English, then in that context Spanish is the metalanguage and English is the object language. If we speak in English about expressions of Spanish, then in that context English is the metalanguage and Spanish is the object language.

We have used the English language to define TL. TL itself, our artificial logical language, is in this context the object language. The language we use to talk about the object language, namely, English, is the metalanguage. We'll be using both languages in this chapter. That is, most of our discussion will be in English about TL and expressions of TL will appear frequently as objects of analysis or discussion.

It is important that the reader distinguish the expressions of the object language from the expressions of the metalanguage. Consequently, the next section specifies some conventions that distinguish object language expressions from metalanguage expressions.

Use and Mention   We normally use linguistic expressions to talk about something other than the expressions themselves. For example, in the sentence

> **37.**   Nebraska is a state

the word 'Nebraska' is being used to refer to one of the fifty states.

Sometimes, however, we wish to talk about an expression itself rather than about what the expression refers to. In such a case, we normally place the expression within quotes, as in the following example:

> **38.**   'Nebraska' contains eight letters.

Here the word 'Nebraska' is not being used to refer to a state; rather, the word itself is the item being talked about. The word with quotes around it is serving in (38) as the name of the word within the quotes.

In the case of (37), we say that the word 'Nebraska' is being *used*. In the case of (38), we say that the word 'Nebraska' is being *mentioned*. In general, an expression is being used if its function is to refer to something else, and an expression is being mentioned if the expression itself—and not what it refers to—is being spoken of.

In the course of our logical studies, we will often have to mention rather than use various expressions of TL. That is, we will often have to talk about expressions of TL themselves rather than about what they refer to. Within logic, there are two standard ways to signal explicitly that an expression is being mentioned rather than used. First, we may mention an expression by placing it within quotes, as in the following examples:

> **i.**   '~A' is a wff of TL.
> **ii.**   'Hello' is an English word.

Notice that in this type of case, the expression together with its surrounding quotes serves as a name of the expression within the quotes. In (i), for instance, we are talking about the TL sentence that is enclosed in single quotes.

Second, we may mention an expression by placing it on display. For example, the sentence

> It's been a hard day's night

is a sentence Ringo Starr is said to have spoken and is mentioned in this case by being displayed. The TL sentence

A & B

is a conjunction and is in this case referred to by being put on display.

The names of expressions of TL are not themselves part of the object language TL, for TL itself contains no quote marks and no conventions for displaying expressions. Thus, the names of TL expressions belong to the metalanguage. For example, the expression

'(A v B)'

does not belong to TL although the expression named does belong to TL.

We have been using boldfaced capital letters 'P', 'Q', 'R', and 'S' as variables ranging over sentences. These variables belong to the metalanguage, for TL contains no variables. Since our variables belong to the metalanguage and since they range over linguistic entities—sentences—the variables we are using are called "metalinguistic" variables.

Note that an expression such as

P v **Q**

combines both object language and metalanguage elements because it is composed of variables from the metalanguage and an item—'v'—from the object language TL. Let us say that in this type of mixed sentence the operator—with no quotes around it—is serving as a name of itself.

We have been using metalinguistic variables to make general claims about TL sentences. For example, if we use a sentence constant and write:

If 'A' is true, then '~A' is false

we are making a claim about a specified sentence of TL, namely, the sentence 'A'. However, if we use a variable and write:

If **P** is true, then ~**P** is false

we are making a general claim about all sentences of TL. The sentence displayed immediately above is just a shorthand way of saying:

If any sentence of TL is true, then the negation of that sentence is false.

Notice that it would have been incorrect to have written:

If '**P**' is true, then '~**P**' is false

for this would be to make the incorrect claim that if the metavariable itself is true, then its negation is false. Metavariables are neither true nor false.

In contexts where the distinction between the English expressions and mentioned TL expressions should be clear, we will sometimes simplify things by omitting the quote marks around the TL expressions. For example, in the following sentence, the distinction between the mentioned TL expression and the English words should be clear:

We symbolized the sentence with ~A & (B v C).

Consequently, it is not necessary to write this as

We symbolized the sentence with '~A & (B v C)'.

5. TRUTH-TABLES

●— SCOPE and MAIN CONNECTIVE

In this section, we will need to use two new ideas. The *scope* of an operator occurring in a sentence of TL is the operator itself along with the formulas in the sentence (plus any parenthetical devices) that the operator applies to or links together. The *main operator* of a sentence of TL is the operator of largest scope. The main operator will be the operator whose scope spans the entire sentence and the truth-value determined by the main operator will always be the truth-value for the whole compound. This will become clearer as we work through several examples.

Suppose someone has symbolized an English sentence and the result is:

~A & B

Suppose further that the truth-value of the atomic component 'A' is false while the truth-value of the atomic component 'B' is true. To record this assignment of truth-values, we'll place **T** under 'B' and **F** under 'A':

~A & B

F   T

Given the truth-values of the components, what is the truth-value of the compound sentence as a whole? First, in English, 'it is not the case that' attaches only to that which is to its immediate right. Likewise, in TL, the negation sign applies only to that part of the compound to its immediate right. So, in this example the tilde applies only to the 'A' to its immediate right:

This diagram records the fact that when the negation operator is applied to a component that is false, the resulting compound is true.

We've evaluated the operator of smallest scope, the tilde. Next we evaluate the ampersand. The right conjunct of the ampersand is true while the left conjunct is also true. Now, according to the conjunction truth-table, when the ampersand joins two true components, the resulting compound is true:

The ampersand is the main operator, the operator of greatest scope. It determines the truth-value of the compound as a whole. So the truth-value of the compound as a whole is true. The diagram provides a visual representation of the way in which the truth-value of the whole is a function of the truth-values of the atomic components.

Let's look at a slightly more complex example. Suppose someone symbolizes an English sentence as follows:

$$\sim(A \& B)$$

where the component sentence 'A' has the truth-value false and the component sentence 'B' has the value true. In this case, the tilde applies to the parenthesis to its immediate right—not to 'A' alone. That is, the tilde applies to the conjunction '(A & B)' as a whole. The first step, then, is to calculate the value of '(A & B)'. After that step is completed, then we apply the tilde. Here are the two steps:

$$
\begin{array}{cc}
& \sim \;(A \;\&\; B) \\
(1) & F \qquad T \\
& F
\end{array}
$$

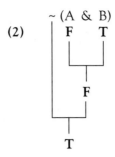

(2)

The value of the compound is thus true.

If we suppose that the component 'A' is true and the component 'B' is false, is ~[~(A & ~B) v ~(A v B)] true or false? We begin by first assigning truth-values to the atomic components:

~[~(A & ~B) v ~(A v B)]
    T   F   T F

Second, we evaluate the operators of smallest scope. The smallest scope possible is a tilde applied to a single atomic component with no intervening bracket:

~ [~ (A & ~B) v ~ (A v B)]
      T |F   T F
          T

Third, we evaluate the functions of next largest scope:

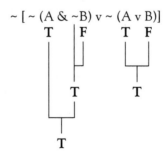

Fourth, the function of next largest scope:

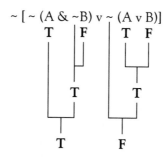

Fifth, the functions of next largest scope:

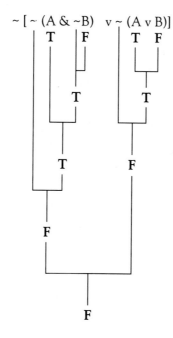

Finally, the remaining tilde—the sentence's main operator—is evaluated:

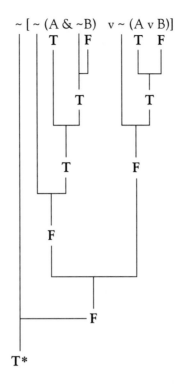

The final truth-value, which is starred, is true. Remember that the final value, the value for the compound as a whole, is always the value determined by the main operator. The main operator is always the last operator to be evaluated.

Notice the way in which the truth-values "flow" from the atomic components down through the structure and emerge at the end from the main operator. Notice also that as we calculate the truth-values, we move from the inside to the outside, starting with the functions of smallest scope and graduating in stages to larger functions, until we reach the function of greatest scope, the main operator.

● PAIRING PARENTHETICAL DEVICES

In the following formula, the pairs of parenthetical devices are marked with brackets:

$$\{ \, [\, A \lor \sim(A \lor \sim B)] \lor \sim (A \,\&\, B) \} \lor \sim(B \lor A)$$

If you are having difficulty seeing which parenthetical devices pair with which other parenthetical devices, here is a mechanical procedure that will allow you to match pairs accurately. Begin at the left end of the formula and proceed to the right until you come to the first right-hand parenthetical device. Pair this with the nearest left-hand device appearing before it. Continue this way through the formula, pairing each unmatched right-hand device with the closest unmatched left-hand device to the left. When you finish drawing in your brackets, no matching lines should cross and no parenthetical device should be unmatched. If any lines do cross, or if a parenthetical device is unmatched, either the formula is written incorrectly or you have matched things up incorrectly.

• EXERCISE 2.4 •

Identify the main connective in each of the following.

1. ~(A & S) & ~(G v B)

2. (A v B) v (H & ~S)

3. ~[~(A v B) v (E & F)]

4. ~A v B

5. ~(~A v ~B) & E

6. [(A v B) v C] v (~S & G)

7. A v ~(H v B)

Categorizing Compound Sentences   Truth-functional compound sentences may be categorized in terms of their main operators as follows. If the main operator of a sentence is a conjunction operator, the sentence is a *conjunction*. The following sentences are conjunctions:

(A v B) & (E v H)   [~(A & B) v (H v G)] & (E v G)   D & S

If the main operator is a disjunction operator, the sentence is a *disjunction*. The following are disjunctions:

(A & B) v ~(E & S)   ~(G v T) v ~(H & Y)   T v I

If the main operator is a negation operator, the sentence is a *negation*. Thus, the following are all negations:

~(A & H)   ~[(A v B) v (D & E)]   ~[~(G & S) & R]   ~J

• **EXERCISES** •

Exercise 2.5

Part I Let P abbreviate 'The Moon has mountains', let Q abbreviate 'The Earth is perfectly flat', and let R abbreviate 'Polar bears do not naturally live in tropical regions'. Determine whether each of the following is true or false.

1.	P & Q	13.	P & (Q & R)
2.	~(P v Q)	14.	(Q v P) v R
3.	~ P & ~Q	15.	(R v P) v Q
4.	~P & Q	16.	P v (P & Q)
5.	~(P & Q)	17.	(~P & ~Q) & ~R
6.	(P & Q) v R	18.	~[(P v Q) v R]
7.	~(P v Q) & R	19.	~[(P & Q) & R]
8.	P & (Q v R)	20.	(~P v ~Q) v ~R
9.	~P v ~Q	21.	(P & Q) v (~P & ~Q)
10.	(P v Q) v R	22.	(P v ~Q) v (~P & Q)
11.	P v (Q v R)	23.	~[(P & ~Q) v (~R & Q)]
12.	(P & Q) & R		

Part II Translate the above formulas into ordinary English sentences.

Exercise 2.6

Suppose you know that A and B are true and that C and D are false. However, suppose you do not know the truth-values of P and Q. Nevertheless, determine the truth-values of the following.

1.	P v A	8.	(D & P) & (P & Q)
2.	P & D	9.	~(P & Q) v P
3.	P & ~P	10.	P v (A v ~A)
4.	P v ~P	11.	P & (A & ~A)
5.	~(Q v B)	12.	~P & ~(A v ~A)
6.	~(C & P)	13.	~P v (~Q v P)
7.	(A v P) v (P & Q)	14.	(P & Q) v ~(P & Q)

15. ~B & P   17. ~(C & P) v ~(~P & Q)

16. ~P v (Q v P)   18. (P v Q) & ~(Q v P)

●— TRUTH-TABLES FOR FORMULAS

The English sentence 'It's not the case that Annette is both 40 years old and not 40 years old' is symbolized as follows:

$$\sim(A \mathbin{\&} \sim A)$$

What's the truth-value of the compound? Well, that depends on the truth-values assigned to the constant 'A'. Let's consider all the possible combinations of truth-values. There are only *two*. If 'A' has the truth-value true, then:

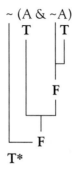

If 'A' has the truth-value false, then:

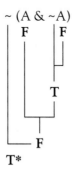

Notice that in both cases, the final truth-value—the truth-value of the whole—is true.

Both of these possibilities may be compactly displayed on a truth-table as follows. First, on top of a table write the formula we are evaluating and fill in all possible truth-values the atomic components might be assigned:

A	~(A & ~A)
T	
F	

Next, the truth-value of ~(A & ~A) is calculated for each assignment of truth-values:

A	~(A & ~A)
T	T T F FT
F	T F F TF
	*

On each row of this table, the truth-values were calculated just as they were in the diagrams above, except that truth-values were placed on the table and the long lines (drawn above) were not drawn in.

The column of truth-values under the main operator is called the *final column*. In the table above, a star has been placed under the final column. The table indicates that if 'A' is assigned **T**, then ~(A & ~A) is true and if 'A' is assigned **F** then ~(A &~A) is also true.

Next, consider a slightly more complex example:

$$\sim[(A \ \& \ B) \ \& \ \sim(A \ \& \ B)]$$

What's the truth-value of this formula? Again, we'll have to consider all the different combinations of truth-values that could possibly be assigned to the atomic components. There are only four possible combinations of truth-values that two components could be assigned, and so there are only four truth-value assignments to consider in this case:

1. Perhaps 'A' is true and 'B' is true:

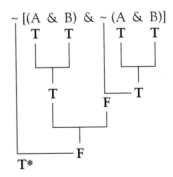

2. Or, perhaps 'A' is true and 'B' is false:

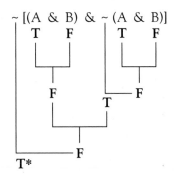

3. Perhaps 'A' is false and 'B' is true:

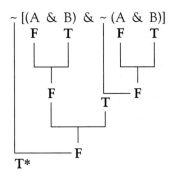

4. Or perhaps 'A' is false and 'B' is false:

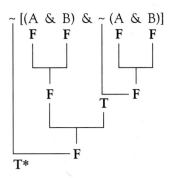

These four possibilities may be displayed more compactly on a truth-table:

A	B		~[(A	&	B)	&	~(A	&	B)]
T	T	T	T	T	T	F	F	T	T	T
T	F	T	T	F	F	F	T	T	F	F
F	T	T	F	F	T	F	T	F	F	T
F	F	T	F	F	F	F	T	F	F	F
		*								

This truth-table simply displays all combinations of truth-values the atomic components could possibly be assigned and shows how each combination of truth-values determines the truth-value of the compound that is a truth-function of those sentences.

Each row of a truth-table represents a different truth-value assignment. That is, each row of a table represents a different possible combination of truth-values that could be assigned to the atomic components of the formula or formulas on top of the table. Between them, the rows of a table collectively represent all possible combinations of truth-values that could be assigned to the atomic components of the formula or formulas on top of the table.

Eight-Row Tables Suppose we have a formula containing three different atomic components:

$$[(A \lor {\sim}B) \& {\sim}(B \& C)] \lor {\sim}B$$

How many different truth-value assignments are there for this formula? The formula contains three atomic components, namely 'A', 'B', 'C'. A fairly simple process of trial and error reveals that there are only eight possible combinations of truth-values for three components considered three at a time:

	A	B	C
1.	T	T	T
2.	T	T	F
3.	T	F	T
4.	T	F	F
5.	F	T	T
6.	F	T	F
7.	F	F	T
8.	F	F	F

So, the truth-table for the formula above begins with the following truth-value assignments:

A	B	C	[(A v ~B) & ~(B & C)] v ~B
T	T	T	
T	T	F	
T	F	T	
T	F	F	
F	T	T	
F	T	F	
F	F	T	
F	F	F	

The table now displays all the different combinations of truth-values that could be assigned to the atomic components.

Once the truth-value assignments have been filled in, the rows of the table may be calculated:

A	B	C	[(A v ~B) & ~(B & C)] v ~B
T	T	T	T T FT F FT T T F FT
T	T	F	T T FT T TT F F T FT
T	F	T	T T TF T TF F T T TF
T	F	F	T T TF T TF F F T TF
F	T	T	F F FT F FT T T F FT
F	T	F	F F FT F TT F F F FT
F	F	T	F T TF T TF F T T TF
F	F	F	F T TF T TF F F T TF
			*

The Number of Rows on a Truth-Table   In general, if you place a sentence containing **n** atomic components on a truth-table, the resulting table will have 2^n rows. For example:

IF THE SENTENCE HAS:	THE TABLE WILL HAVE:
1 atomic component	2 rows
2 atomic components	4 rows
3 atomic components	8 rows
4 atomic components	16 rows
5 atomic components	32 rows

INTRODUCING TWO MORE TRUTH-FUNCTIONAL OPERATORS

Conditional Sentences   Consider the following two sentences:

39.   If you eat hamburgers and french fries for lunch and dinner every day for a year, then your body will not have a shortage of cholesterol.

40.   If Chris is 80 years old, then Chris is a senior citizen.

These two sentences are *conditional sentences*, or *conditionals* for short. In each, the component sentence following the word 'if' is called the *antecedent* of the conditional, and the component following the word 'then' is called the *consequent* of the conditional. In the above examples, the expression 'if, then' functions as a dyadic sentential operator, for it links two component sentences into a compound sentence. Since it forms a conditional sentence, 'if, then' is being used here as a *conditional operator*.

Since conditionals play an important role in argumentation, it would be desirable to have a way of representing the 'if, then' operator within truth-functional logic. First, let the symbol '⊃', called the *horseshoe*, represent the 'if, then' operator. If we abbreviate 'Chris is 80 years old' with 'C' and 'Chris is a senior citizen' with 'S', then (40) may be symbolized as

$$C \supset S$$

In this formula, 'C' abbreviates the antecedent of the conditional—sentence (40)—and 'S' abbreviates the consequent. The antecedent will always be placed in front of the horseshoe, and the consequent will always be placed after the horseshoe. For another example:

41.   If it's sunny, then we will go on a hike.

will go into symbols as

$$S \supset H$$

where 'S' abbreviates the antecedent 'It's sunny' and 'H' abbreviates the consequent 'We will go on a hike'.

A *Truth-Function for the Conditional Operator*   The next step in the process of representing 'if, then' within truth-functional logic is not as simple. We must associate the horseshoe operator with a truth-function. Unfortunately for our purposes, 'if, then' is rarely used in everyday discourse in a purely truth-functional way. That is, the conditional sentences we construct using 'if, then' almost never display a functional relationship between the truth-value of the compound and the truth-values of the components. How then can a truth-function be associated with the 'if, then' operator?

Our strategy will be as follows. First, we'll isolate an element of meaning common to all uses of an 'if, then' operator. This core element constitutes part of what we normally mean when we write or utter a conditional sentence. Next we will specify a truth-function that represents this common element. Then we will associate this function with the horseshoe symbol. Finally, we'll go on to develop truth-functional logic in such a way that the representation of 'if, then' is based only on this core element of meaning.

Let us now try to isolate a core meaning that is common to all conditional sentences. Anytime anyone asserts an English conditional, and we find that the antecedent is true and that the consequent is false, we automatically say that the conditional as a whole is false. Thus, if someone asserts a conditional sentence **P ⊃ Q** where **P** is the antecedent and **Q** is the consequent, at least *part* of what the speaker means to assert is:

> **42.** It is not the case that: **P** is true while **Q** is false.

Let's support this by looking at some specific examples. Suppose someone says:

> **43.** If Wimpy eats one more hamburger, then he will get sick.

This is an example of a *causal* conditional. A causal conditional is a conditional that asserts the existence of a causal relationship between that which the antecedent refers to and that which the consequent refers to. If Wimpy does eat one more hamburger but does not get sick, then (43) is false. So, one who asserts (43) is asserting at least:

> It is not the case that: Wimpy will eat one more hamburger and yet not get sick.

Notice that this fits the general pattern constituted by sentence (42).
Here's another example. Suppose someone says:

> **44.** If it's a square, then it has four sides.

This is an example of an *implicative* conditional. This type of conditional asserts a relation of implication between the antecedent and the consequent. The claim in an implicative conditional is that the antecedent implies the consequent. Sentence (44) at least means:

> It is not the case that: it is a square but it does not have four sides.

Notice that this also fits the general pattern expressed in sentence (42).
Thus, it seems that we have isolated or abstracted at least part of what is asserted when someone asserts a conditional sentence. The core meaning that has been isolated is this: To say 'If **P** then **Q**' is at least to say 'It's not the case that: **P** but not-**Q**'. And this part of the meaning of a conditional may be represented by:

$$\sim(P \ \& \sim Q)$$

The corresponding truth-table for this formula is:

P	Q	~(P & ~Q)
T	T	T
T	F	F
F	T	T
F	F	T

In modern logic, this truth-function is associated with the horseshoe. Therefore, the truth-table for the horseshoe is the following:

P	Q	P ⊃ Q
T	T	T
T	F	F
F	T	T
F	F	T

Truth-Table for the Horseshoe

This truth-function—specified in terms of this configuration of **T**'s and **F**'s—at least represents part of what we mean when we assert a conditional sentence. This core meaning happens to be truth-functional in nature.

The horseshoe is called the *material conditional* operator. A sentence whose main operator is the horseshoe and whose truth-conditions are given by the truth-table for '⊃' is called a *material conditional sentence*. So, when we take an English conditional and translate it into a material conditional sentence, we have two different sentences: the English conditional with its (complex) meaning, and the associated material conditional with its truth-table specified meaning. However, the material conditional represents a core of meaning found in every English conditional.

Upon first learning the truth-table for the horseshoe operator, it is natural to feel a little puzzled. The table doesn't seem to represent exactly what we mean when we utter a conditional sentence. That is, the truth-conditions specified by the horseshoe's table don't seem quite "right." This is at least partly because when we write or utter an ordinary English conditional, we almost never mean to assert only what the corresponding material conditional asserts; we normally assert more than this. Remember, the horseshoe table only captures part of the meaning of 'if, then'.

Why then, do we translate English conditionals in logic as material conditionals? The answer is that there is a certain relationship between the truth-conditions of an English conditional and those of a material conditional. This relationship, it turns out, suffices for our purposes in truth-functional logic. Let's briefly examine that relationship.

Consider any conditional sentence, written in English, that is true. For example:

45. If the figure is square, then it has four sides.

Now consider the associated material conditional, abbreviated:

$$F \supset S$$

where 'F' abbreviates 'the figure is square' and 'S' abbreviates 'it has four sides'. Clearly, if the English conditional sentence is true—sentence (45)—then so is the material conditional. For, if the English conditional is true, then it is not the case that the antecedent is true but the consequent is false. That implies that the associated material conditional is also true. Since this holds for any conditional sentence, we may say: If an English conditional is true, then its associated material conditional is also true.

Consider next a material conditional that is false. If a material conditional is false, then the antecedent is true while the consequent is false. But then the associated English language conditional will also be false. Thus: If the material conditional is false, then the associated English conditional is also false.

The truth-conditions for an ordinary English conditional and the material conditional are thus intimately related. It must be admitted, however, that this relationship is not perfect. There are situations in which the truth-value of an English conditional is unknown and yet the corresponding material conditional is true; and there are cases where the English conditional will be false and yet the corresponding material conditional will be true. For example, suppose someone says:

46. If Joe quits, then Sue will quit.

Now, if it is true that Joe quits, and it is also that true Sue quits, this alone doesn't make the English conditional true. Perhaps the reason Sue quit had nothing to do with the fact that Joe quit. However, in this case, if we symbolize the sentence as a material conditional and place it on a truth-table, the sentence comes out true.

Or, suppose someone says:

47. If you push that button, then the door will open.

If nobody pushes the button, and the door does not open, then the antecedent and consequent are both false. From this alone, it is not clear whether the English conditional is true or false. However, in this situation, the corresponding material conditional is, according to its truth-table, true. Thus, the truth-conditions of an ordinary English conditional and of the material conditional do not match perfectly.

Although the horseshoe function does not represent all of what we mean when we utter a conditional sentence, it does represent a truth-functional core of meaning common to all uses of 'if, then'. Incidentally, no logician has yet captured the full range of meanings associated with the English 'if, then' operator in one precisely defined logical symbol. Perhaps there is no possibility of such a symbol.

When we translate an English conditional into our logical symbolism using the horseshoe, keep in mind that we are abstracting from the full English meaning of the conditional and focusing only on the truth-functional part of that meaning. The truth-table for the horseshoe thus represents only the relevant truth-functional properties of the English conditional. This is sufficient for our purposes within truth-functional logic.

●— BICONDITIONAL SENTENCES

The last truth-functional operator we shall introduce is called the *material biconditional*. Suppose that a club has the following rule:

48. The vote may be taken *if and only if* at least ten official members are present.

In place of the operator 'if and only if', called the *material biconditional* operator, we shall use the symbol '≡', called 'triple bar'. Thus, we may symbolize sentence (48) as:

$$A \equiv B$$

where 'A' represents 'The vote may be taken' and 'B' represents 'At least ten official members are present'. In ordinary discourse, the operator 'if and only if' is rarely used. However, it is used in several technical fields, including mathematics, law, physics, and philosophy. In these fields, 'if and only if' is typically used in a way that is closely related to the ordinary conditional operator. Typically, to say, 'The vote may be taken if and only if at least ten official members are present' is just to say 'If the vote may be taken then at least ten official members are present and if at least ten official members are present then the vote may be taken'. In general, something of the form 'P if and only if Q' is equivalent to something of the form 'If P then Q *and* if Q then P'. The truth-table for 'If P then Q *and* if Q then P' is:

P	Q	$(P \supset Q)$ & $(Q \supset P)$
T	T	T
T	F	F
F	T	F
F	F	T

We shall therefore assign this configuration of T's and F's to the triple bar's truth-table:

P	Q	$P \equiv Q$
T	T	T
T	F	F
F	T	F
F	F	T

The Truth-Table for Triple Bar

●— SYNTAX and CATEGORIZATION AGAIN

Earlier, we specified the syntax for our formal logical language TL. In order to incorporate the horseshoe and triple bar operators into TL, we must make two additions to TL's syntax. First, we must add the horseshoe and triple bar to the list of operators specified in TL's vocabulary. Thus, TL's amended list of sentence operators is:

> ii. Sentence operators: ~ v & ⊃ ≡

Second, we must amend TL's third rule of grammar to:

> G3. If **P** and **Q** are sentences of TL, then (**P** v **Q**), (**P** & **Q**), (**P** ⊃ **Q**), and (**P** ≡ **Q**) are sentences of TL.

Earlier, we categorized sentences in terms of their main operators. Let us continue this by saying that a *conditional* is any sentence whose main operator is a horseshoe. Thus, each of the following sentences of TL is a conditional:

> A ⊃ B   (A & H) ⊃ (F v S)   ~D ⊃ (~G v S)

And if the main operator is the triple bar, the sentence will count as a biconditional. For example:

> A ≡ B   ~(A v D) ≡ ~(E & S)   ~S ≡ (F v T)

────────────────● **EXERCISES** ●────────────────

Exercise 2.7

Part I Let P abbreviate 'Pat wins the lottery', let Q abbreviate 'Pat takes a trip to Europe', and let R abbreviate 'Chris will be happy'. Assume P, Q, and R are each true, and determine truth-values of the following.

1. P ⊃ Q	9. ~P ⊃ (~Q & ~R)
2. Q ⊃ P	10. ~R ⊃ ~(P ⊃ Q)
3. ~Q ⊃ ~P	11. (P & Q) ⊃ R
4. (P ⊃ Q) ⊃ R	12. ~(P & Q) ⊃ ~ R
5. P ⊃ (Q ⊃ R)	13. (P & Q) ≡ (P & R)
6. P ≡ (Q & R)	14. P ≡ R
7. P ⊃ (Q ≡ R)	15. R ≡ P
8. (P ⊃ Q) ⊃ (Q ⊃ P)	16. (~P & ~Q) ⊃ ~R

17. $Q \supset (P \supset R)$ 19. $(P \equiv Q) \vee (P \equiv R)$

18. $P \supset [(Q \vee R) \vee (Q \& R)]$

Part II   Rework the above problems assuming P is true, Q is false, and R is true.

Exercise 2.8

Suppose A and B are true, C and D are false, but the truth-values of P and Q are unknown. Determine the truth-values of the following.

1. $C \supset P$ 10. $A \equiv (C \& P)$

2. $P \supset A$ 11. $A \supset (P \vee \sim D)$

3. $\sim A \supset Q$ 12. $\sim (P \equiv \sim P)$

4. $Q \supset \sim C$ 13. $(Q \supset A) \supset C$

5. $P \supset (Q \supset P)$ 14. $(C \supset P) \supset (D \supset Q)$

6. $D \supset (P \supset Q)$ 15. $(P \& C) \supset Q$

7. $(P \vee Q) \supset A$ 16. $[P \supset (Q \supset P)] \supset D$

8. $Q \supset (P \supset A)$ 17. $(P \& A) \supset B$

9. $P \supset (P \supset P)$ 18. $\sim [P \supset (\sim B \supset Q)]$

●─ SYMBOLIZING SENTENCES USING THE HORSESHOE and TRIPLE BAR

The sentence

   49.   If it's sunny, then we will run.

translates easily into our logical symbolism as

$$S \supset R$$

However, consider the next sentence:

   50.   We will run if it's sunny.

It won't work to symbolize this as

$$R \supset S$$

for 'R ⊃ S' symbolizes in this context

   51.   If we run, then it will be sunny.

Sentence (51) states that our running will in some way make the sun come out, which is absurd. Sentence (50) does not mean what sentence (51) means.

Consider again sentences (49) and (50). If you will carefully compare them, you will see that both say exactly the same thing. In English, we often give the consequent first, as in (50). In such cases, the antecedent is placed second. However, in every case, the antecedent is always introduced with the word 'if'. Since (49) and (50) amount to the same claim, both should be symbolized with 'S ⊃ R'.

When translating conditionals, two points need to be kept in mind:

   i. In English, the antecedent is typically introduced by the word 'if';

   ii. In our logical symbolism, the abbreviated antecedent is always placed to the left of the horseshoe.

Suppose we wish to represent the following sentence in our logical symbolism:

52. If Ann wins and Bob wins then either Ed wins or George wins.

The first thing to notice is that everything in this sentence from 'if' to 'then' forms a logical unit, namely the antecedent of the whole sentence. Second, everything from 'then' to the end forms another unit, the consequent of the sentence. The sentence may be symbolized in the following three steps:

   1. If A and B then either E or G.
   2. If (A & B) then (E v G)
   3. (A & B) ⊃ (E v G)

Here are several interesting examples of conditional sentences. Suppose an insurance company has the following policy:

53. If you have an accident and a ticket in any one-year period, then your policy will be canceled.

This may be symbolized as:

54. (A & T) ⊃ C

Sentence (54) implies that:

55. *If* you have an accident, then *if* you get a ticket then your policy will be canceled.

In TL, this becomes:

56. A ⊃ (T ⊃ C)

Suppose you invite a friend to drive across the country with you. Your friend says, "If I go with you, I have to take Ralph with me." Ralph is your friend's 200-pound Saint Bernard, a dog that could easily win the world championship in the dog slobbering category of next year's Dog Olympics. You reply, "If that's the case, then forget it." Your reply may be put this way:

57.   If it's the case that if you go then Ralph goes, then you're not going.

And this goes into symbols as

58. $(Y \supset R) \supset \sim Y$

Compare sentences (55) and (57). Do you see the difference in logical structure? That difference is revealed by the symbolism in formulas (56) and (58). Notice that (55) is a conditional whose consequent is itself a conditional, while (57) is a conditional whose antecedent is a conditional.

• A QUESTION •

Does (57) imply that Ralph is not going on the trip?

Suppose a country has the following law:

A person has the right to vote if and only if the person is an adult and a citizen.

It follows that:

Bart has the right to vote if and only if Bart is an adult and a citizen.

In symbols, this is:

$$B \equiv (A \ \& \ C)$$

• EXERCISE 2.9 •

Symbolize the following.

1. If Fats Domino performs then either Jerry Lee Lewis or the Big Bopper will perform.

2. If both Buddy Holly and Little Richard perform, then Chuck Berry will perform.

3. Either Cab Calloway will perform at the Cotton Club tonight or if Duke Ellington performs then Ella Fitzgerald will perform.

4. If Ethel Waters performs and either Louis Armstrong or Lena Horne performs, then Mel Torme will sing.

5. If it's not the case that Bill "Bojangles" Robinson will be there, then either Cab Calloway or Duke Ellington will be there.

6. Either Benny Goodman's band or Glenn Miller's orchestra will perform if Lena Horne will not be available.

7. It is not the case that if Nat King Cole performs then ticket prices will be raised.

8. If Johnny Mathis sings, then it's not the case that either Ann or Bob will miss the show.

9. Dionne Warwick and Nancy Wilson will sing if and only if admission is free and the performance is not televised.

6. TRUTH-TABLE TESTS

Logic and Chemistry Sets If you have ever taken a chemistry course in school, or if you ever performed experiments with a home chemistry set, you know that chemists have developed tests for detecting the presence of various chemical properties. For instance, to test a solution for acidity, you dip a strip of blue litmus paper into the substance. If the strip turns red, the solution is acidic.[3]

Most of the remainder of this chapter will resemble a chemistry book in a way, for we are going to use truth-tables to test sentences and groups of sentences for the various logical properties introduced in Chapter 1. While the chemist tests physical substances for chemical properties, we'll be testing sentences and arguments for logical properties.

Logic resembles chemistry in another respect. Chemical tests often reveal chemical properties that aren't apparent to the naked eye. Thus, chemical analysis takes us beneath the surface appearance of things to an underlying chemical reality. Similarly, logical tests often reveal properties of an argument that aren't apparent at first glance. Thus, logical analysis also takes us, in a sense, beneath the surface of things to an underlying logical reality.

●— TESTING INDIVIDUAL SENTENCES

The Tautology Test Consider the sentence

59. Either today is Saturday or today is not Saturday.

---

3 Sheldon Glashow, a Nobel Prize winning physicist, has a wonderful discussion of the chemistry set that he had as a kid. See Glashow [1988] pp. 15–18.

When symbolized, this becomes:

$$S \text{ v} \sim S$$

where S abbreviates 'Today is Saturday'. The truth-table for this formula is:

S	S v ~S
T	T
F	T

Sentence (59) has a very interesting feature. Notice that on every row of its truth-table the formula computes true at the main operator. This indicates that the sentence has the truth-value true in every possible circumstance. In no circumstance will it ever be false. A sentence that has the truth-value true in every circumstance is called a *tautology*.

Recall that the final column on a truth-table is the column of truth-values determined by and placed directly underneath the main operator of the formula. Thus, we can say that a formula symbolizes a tautology if and only if the formula's final column shows all **T**'s.

The Contradiction Test   Consider the sentence

60.   Aluminum is an element and it's not the case that aluminum is an element.

The truth-table is:

A	A & ~A
T	F
F	F

Notice that the final column is all **F**'s, which indicates that the sentence has the truth-value false in every possible circumstance. In no possible circumstance is the sentence true. Such a sentence is said to be a *contradiction*. Here is another example of a contradiction:

A	B	(A v B) & ~(A v B)
T	T	F
T	F	F
F	T	F
F	F	F

The Contingency Test   A sentence is said to be *contingent* if the final column of its truth-table contains at least one **T** and at least one **F**. For example:

A	B	(A v ~B)
T	T	T
T	F	T
F	T	F
F	F	T

A contingent sentence is thus true in some circumstances and false in others.

These first three truth-table tests may be summarized as follows:

1. A sentence is a tautology if the final column on its truth-table contains all **T**'s.

2. A sentence is a contradiction if the final column on its truth-table contains all **F**'s.

3. A sentence is contingent if the final column on its truth-table contains at least one **T** and at least one **F**.

Logical Status and Some Alternative Terminology   Tautologies are also called *necessary truths* because they can't possibly be false, and contradictions are also called *necessary falsehoods* because they can't possibly be true. Necessary truth and falsity will be probed in greater detail in Chapter 6. If we specify whether a sentence is necessarily true, necessarily false, or contingent, we are specifying the sentence's *logical status*.

A variety of alternative terms may be found within logic. Tautologies are sometimes said to be "logically true" because they can be proven true using only the procedures of logical theory—without investigating the physical world. Tautologies are also said to be "truth-functionally true" because they can be proven true using just the procedures of truth-functional logic alone—without investigating the physical world. Similarly, contradictions are sometimes said to be "logically false" or "truth-functionally false" because they can be proven false using just the procedures of truth-functional logic alone—without investigating the physical world. Contingent sentences are sometimes called "logically indeterminate" or "truth-functionally indeterminate" sentences, since their truth-values are not constant but vary from situation to situation.

• **EXERCISES** •

Exercise 2.10

Use truth-tables to determine which of the following are tautological, which are contradictory, and which are contingent.

1. $(P \& Q) \supset (P \lor Q)$
2. $\sim(P \supset P)$
3. $[(P \supset P) \supset Q] \supset Q$
4. $P \equiv [(P \& Q) \lor (\sim P \& \sim Q)]$
5. $(P \supset Q) \supset (\sim P \lor Q)$
6. $(P \supset Q) \supset (\sim Q \supset \sim P)$
7. $(P \lor Q) \supset (P \& R)$
8. $P \supset (P \lor Q)$
9. $(P \& Q) \supset (R \supset \sim Q)$
10. $\sim(P \equiv P)$
11. $\sim P \supset P$
12. $P \supset \sim P$
13. $P \equiv (P \lor Q)$
14. $P \lor (Q \supset Q)$
15. $P \equiv \sim P$
16. $P \equiv (P \lor P)$
17. $(P \& Q) \supset Q$
18. $(P \& Q) \supset (\sim P \supset \sim Q)$
19. $P \supset [Q \supset (P \supset Q)]$
20. $[(P \supset Q) \& (Q \supset R)] \supset (P \supset R)$
21. $[(P \supset Q) \& \sim Q] \supset \sim P$
22. $[(P \supset Q) \& P] \supset Q$
23. $[(P \lor Q) \& \sim P] \supset Q$
24. $\sim[P \supset (\sim P \supset \sim Q)] \& (\sim P \& Q)$
25. $\sim[P \supset (P \& \sim P)]$
26. $(P \& \sim P) \supset Q$
27. $[(P \& Q) \& R] \supset [P \& (Q \& R)]$
28. $(P \supset Q) \equiv (\sim P \lor Q)$
29. $(P \supset Q) \equiv (\sim Q \supset \sim P)$
30. $(P \supset Q) \equiv (Q \supset P)$
31. $(P \& Q) \supset \sim(\sim P \lor \sim Q)$
32. $\sim(P \& Q) \supset (\sim P \lor \sim Q)$
33. $\sim(P \lor Q) \supset (\sim P \& \sim Q)$
34. $(P \supset Q) \supset (P \supset R)$
35. $\sim[P \supset (P \lor \sim P)]$
36. $[(P \supset Q) \& (R \supset Q)] \supset (P \supset R)$
37. $(P \supset Q) \supset [(P \supset Q) \lor R]$
38. $P \supset (\sim Q \supset P)$
39. $\sim P \supset (P \supset Q)$
40. $\sim Q \supset (P \supset Q)$
41. $\sim(P \& \sim P)$
42. $P \lor (P \& \sim P)$
43. $(P \& \sim P) \lor (Q \& \sim Q)$
44. $(P \& \sim P) \lor Q$
45. $[(\sim P \lor Q) \lor R] \lor (P \& \sim R)$

Exercise 2.11

Each of the following is true. In each case, explain why the statement is true. For the first three problems: If **P** is a sentence, then **~P** is the negation of **P**.

1. The negation of a tautology is a contradiction.

2. The negation of a contingent sentence is another contingent sentence.

3. If one disjunct of a disjunction is tautological, then the disjunction itself is tautological.

4. If one conjunct of a conjunction is a contradiction, then the conjunction itself is a contradiction.

5. If the antecedent of a conditional is a contradiction, then the conditional is a tautology.

6. If the consequent of a conditional is a tautology, then the conditional itself is a tautology.

7. The disjunction of a tautology and a contradiction is a tautology.

8. The conjunction of a tautology and a contradiction is a contradiction.

●― TESTING AN ARGUMENT FOR VALIDITY

Recall from Chapter 1 that an argument is valid if and only if there is no possibility that its premises are true and its conclusion is false. Suppose an English argument has been symbolized and the result is:

1. ~(A & B)
2. ___A___
3. ~B

The formulas above the line are the premises of the argument and the formula below the line is the conclusion of the argument. Is this argument valid? Perhaps you can see, intuitively, that it is. If A and B are not both true, and A is true, then it follows that B is not true. However, we can use a truth-table to decide the issue in a very precise way. The procedure involves the following three steps:

Step 1. Place the symbolized version of the argument on top of a truth-table and fill in the truth-value assignments:

		First premise:	Second premise:	Conclusion:
A	B	~(A & B)	A	~B
T	T			
T	F			
F	T			
F	F			

Step 2. Compute the truth-values of the premises and conclusion on each row. The final values are filled in below:

A	B	~(A & B)	A	~B
T	T	F	T	F
T	F	T	T	T
F	T	T	F	F
F	F	T	F	T

Step 3. Examine the table. If a row of the truth-table produces a **T** under the main operator of each premise but produces an **F** under the main operator of the conclusion, let us call that row an *invalidating row*. If the truth-table for an argument contains one or more invalidating rows, then the argument is invalid. If the table contains no invalidating row, then the argument is valid.

Here's the rationale behind this test for validity. An argument is valid if there are no possible circumstances in which the premises are true and the conclusion is false. Each row of the table represents a collection of possible circumstances. The first row on the table above, for instance, represents those circumstances in which A and B are assigned **T**; the second row represents those circumstances in which A is assigned **T** and B is assigned **F**, and so on. Between them, the rows of a table represent all possible circumstances. The truth-table allows us, then, to see how the truth-values of the premises and conclusion vary across all possible circumstances. An invalidating row represents a possibility of true premises and false conclusion. If a table contains an invalidating row, then it is possible for the argument represented on top of the table to have true premises and a false conclusion. If there is any possibility that the premises are true and the conclusion is false, then the argument is invalid. Thus, the existence of an invalidating row proves an argument invalid. If no row of a table shows all true premises with a false conclusion, then we know that in no possible circumstance are the premises true and the conclusion false. If it is not possible for the premises to be true and the conclusion false, then the argument is valid. Thus, if a table contains no invalidating row, the argument on top of the table is valid.

Is the following symbolized argument valid or is it invalid?

1. ~(A & B) v ~(C v B)
2. ~(~C & ~B)
3. ~A v ~B

The following truth-table establishes that it is valid:

A	B	C	~(A & B) v ~(C v B)	~(~C & ~B)	~A v ~B
T	T	T	F	T	F
T	T	F	F	T	F
T	F	T	T	T	T
T	F	F	T	F	T
F	T	T	T	T	T
F	T	F	T	T	T
F	F	T	T	T	T
F	F	F	T	F	T

Notice that the table contains no invalidating row. The argument is valid.

In this chapter, we are examining arguments whose validity or invalidity is due to the various truth-functional relationships connecting the premises to the conclusion. These truth-functional relationships are revealed in the configuration of T's and F's appearing on the argument's truth-table. In later chapters, we will study arguments whose validity is due *not* to truth-functional relationships but to logical relationships of other types. Consequently, in order to distinguish the valid arguments we have been examining from valid arguments of other types, we will say that a valid argument whose validity is established with the methods of truth-functional logic alone is a *truth-functionally valid* argument. An argument that is shown to be valid with a truth-table is sometimes said to be "truth-table valid."

• **EXERCISE 2.12** •

1. Any argument whose conclusion is a tautology is a valid argument, regardless of what its premises are. Explain why.

2. Any argument containing a contradiction as a premise is a valid argument, regardless of what its conclusion is. Explain why.

An Alternative Test for Validity A *corresponding conditional* for an argument is a conditional whose antecedent is a conjunction of the argument's premises and whose consequent is the argument's conclusion. In the following arguments, the premises are written above the line and the conclusion appears below the line.

ARGUMENT: CORRESPONDING CONDITIONAL:

1. A & B
2. C v D   [(A & B) & (C v D)] ⊃ (A & C)
3. (A & C)

1. (P v Q) ⊃ H
2. A & H   {[(P v Q) ⊃ H] & (A & H)} ⊃ ~(Q & A)
3. ~(Q & A)

In general, the corresponding conditional for an argument with premises **P**, **Q**, and **R**, and conclusion **S** is:

$$[(P \& Q) \& R] \supset S$$

Now, if we place the corresponding conditional of an argument on a table, and the final column—under the horseshoe—shows all **T**'s, then the argument is valid. Here's why. A horseshoe is assigned **F** just in case the antecedent is true and the consequent is false. In the case of an argument's corresponding conditional, the antecedent is constituted by the premises of the argument being tested and the consequent is the argument's conclusion. So, if each row computes **T** at the horseshoe, then on no row is the antecedent of the conditional true while the consequent is false. Since the antecedent is the conjunction of the corresponding argument's premises, and the consequent is the argument's conclusion, it follows that on no row are all the premise formulas true while the conclusion formula is false, which is to say that in no possible situation are the premises true while the conclusion is false. Thus:

> *If the corresponding conditional of an argument is a tautology, then the argument is valid.*

If we put the corresponding conditional on a table and the formula is not a tautology, the argument is invalid. (Can you explain why?)

EXERCISE 2.13

Part I In the case of each argument below, use a truth-table to test the argument for validity. In each problem, the premises are listed above the line and the conclusion follows the slash.

1. A ⊃ B
   ~B/~A

2. ~A v B
   ~A/B

3. A v B
   B v C/A v C

4. A v B
   ~A/~B

5. A ⊃ B/B ⊃ A

6. A ⊃ B/~A ⊃ ~B

7. A v B/A & B

8. A ⊃ B
   B ⊃ C/A ⊃ C

9. A ⊃ (B ⊃ C)/C ⊃ (B ⊃ A)

10. A ⊃ B
    B ⊃ E
    B v E/A v B

11. A ⊃ B/~B ⊃ ~A

12. A v B
    ~A/B

13. A ⊃ B
    A/B

14. (A & B) ⊃ C
    ~C/~A v ~B

15. A ⊃ B
    B/A

16. ~[(A & (B & C)]/~A & (~B & ~C)

17. ~(A & B) v C/~A & (B v C)

18. A ⊃ B
    ~A/~B

19. A & B/A

Part II Perform the "corresponding conditional" test for validity on the arguments above.

●— TESTING FOR CONSISTENCY and INCONSISTENCY

Recall from Chapter 1 that a set of sentences is *inconsistent* if and only if there's no possibility that all the members of the set are true, and a set of sentences is *consistent* if there is a possibility that all the members are true. Consider the following three sentences:

    i. Either Ann will apply or Bob will apply.

    ii. It's not the case that either Ann or Charlie will apply.

    iii. It's not the case that either Bob or Charlie will apply.

Are these three sentences consistent? The answer can be computed on a truth-table. First, let's symbolize the three sentences so that they can be placed on a table:

    i. A v B

   ii. ~(A v C)

  iii. ~(B v C)

Now, the following three-step procedure allows us to decide whether this set of three sentences is consistent or inconsistent:

Step 1. Place the symbolized sentences on top of a truth-table and fill in the truth-value assignments:

A	B	C	A v B	~(A v C)	~(B v C)
T	T	T			
T	T	F			
T	F	T			
T	F	F			
F	T	T			
F	T	F			
F	F	T			
F	F	F			

Step 2. Compute the truth-value of each formula:

A	B	C	A v B	~(A v C)		~(B v C)	
T	T	T	T	F	T	F	T
T	T	F	T	F	T	F	T
T	F	T	T	F	T	F	T
T	F	F	T	F	T	T	F
F	T	T	T	F	T	F	T
F	T	F	T	T	F	F	T
F	F	T	F	F	T	F	T
F	F	F	F	T	F	T	F
			*	*		*	

Step 3. Examine the final columns on the table, row by row. If on no row do all three formulas compute **T** together, the set is inconsistent. If there is at least one row of the table in which all the formulas compute **T**, the formulas are consistent.

Remember that each row of the table represents a collection of possible circumstances, and the rows of the table collectively represent all possible circumstances. If we find that on no row of the table do all the formulas compute **T**, then we know that there exists no possible circumstance in which the sentences represented by the formulas all come out true together. And this indicates that the set of sentences is inconsistent, for a set of sentences is inconsistent if there are no possible circumstances in which all the members of the set are true together. On the other hand, if we find one or more rows where all the members of the set compute **T**, then we know that there is at least one possible circumstance in which all the members of the set are true. Then the members of the set are consistent, for a set of sentences or formulas is consistent if there is at least one possible circumstance in which all the members of the set are true.

Notice how mechanical the truth-table test is. As we fill in the T's and F's, we do not concern ourselves with the meanings of the sentences that have been symbolized, nor with the context of utterance. We simply fill in truth-values in accord with the rules for the operators. The procedure is so mechanical that computers may be programmed to carry out the process.

●— TESTING FOR EQUIVALENCE and CONTRADICTION

Equivalence   Recall from Chapter 1 that two sentences are *equivalent* just in case they never differ as far as truth and falsity are concerned. Consider the following pair of formulas:

$$A \lor B \qquad \sim(\sim A \,\&\, \sim B)$$

Here are their truth-tables:

A	B	A v B
T	T	T
T	F	T
F	T	T
F	F	F

A	B	~(~A & ~B)
T	T	T
T	F	T
F	T	T
F	F	F

Notice that these two formulas have identical or matching final columns. That is, on each row, the final values match each other. Whenever, on a row, one formula computes true, then the other—on that row—also computes true; and whenever on a row one formula computes false, the other—on that row—computes false as well. When two formulas match in this way, we say they are *equivalent to each other*. Here are two more examples of pairs of equivalent formulas, along with their truth-tables:

~(A & B) is equivalent to ~A v ~B:

A	B	~(A & B)
T	T	F
T	F	T
F	T	T
F	F	T

A	B	~A v ~B
T	T	F
T	F	T
F	T	T
F	F	T

~(A v B) is equivalent to ~A & ~B:

A	B	~(A v B)
T	T	F
T	F	F
F	T	F
F	F	T

A	B	~A & ~B
T	T	F
T	F	F
F	T	F
F	F	T

When One Statement Contradicts Another   Consider the following two tables:

A	B	A ∨ B
T	T	T
T	F	T
F	T	T
F	F	F

A	B	~A & ~B
T	T	F
T	F	F
F	T	F
F	F	T

Notice that the final columns are exact opposites. Whenever, in one row, one formula is true, the other is false. In other words, the two formulas are always opposite in truth-value to each other. In such a case, we say that the two formulas are *contradictory to each other* or, more simply, the two formulas are *contradictories*. Thus, in the case of a pair of formulas, the test for contradiction is this: if two formulas have opposite final columns, the two formulas are contradictory to each other.

●— TESTING FOR IMPLICATION

Recall from Chapter 1 that one sentence implies a second sentence if and only if there's no possible circumstance in which the first is true and the second is false. The truth-table test for implication involves the following procedure:

Step 1. Place the symbolized sentences on top of a table.

Step 2. Complete the table.

Step 3. Examine the table. If there is no row on which the first is assigned **T** at its main operator while the second is assigned **F** at its main operator, then the first implies the second. If there is at least one row on which the first is assigned **T** at its main operator while the second is assigned **F** at its main operator, then the first does not imply the second.

For example, we can use a truth-table to show that

61.   Neither Ann nor Bob will be home

implies

62.   Bob will not be home.

First we abbreviate 'Neither Ann nor Bob is home' as ~(A ∨ B) and we abbreviate 'Bob is not home' as ~B. If we place these two formulas side by side on a truth-table, the result is:

A	B	~(A ∨ B)	~B
T	T	F	F
T	F	F	T
F	T	F	F
F	F	T	T
		*	*

Since no row shows sentence (61) true and sentence (62) false, this indicates that (61) implies (62). Incidentally, does (62) imply (61)?

In general, if we place two sentences on top of a table, and there is no row on which the first is true while the second is false, then the first *implies* the second. If the table does contain a row on which the first is true while the second is false, then the first does not imply the second.

EXERCISES

Exercise 2.14

Part I   In each case below, use a truth-table to determine whether the sentence on the left of the comma implies the sentence to the right of the comma.

1.   P v Q , P & Q
2.   P & Q , P v Q
3.   P ⊃ Q , Q ⊃ P
4.   P , ~~P
5.   P , P ⊃ P
6.   ~(P & Q) , (~P & ~Q)
7.   ~(P v Q) , (~P v ~Q)
8.   ~P , (P ⊃ Q)
9.   Q , (P ⊃ Q)
10.   (P & Q) , ~(~P v ~Q)
11.   (P & Q) , (P ⊃ Q)
12.   ~(P ⊃ Q) , ~(~P v Q)
13.   ~(P ⊃ Q) , ~P
14.   P ≡ Q , P ⊃ Q
15.   (P v Q) , ~(~P & ~Q)
16.   P ⊃ Q , ~Q ⊃ ~P
17.   [(P ⊃ Q) & P] , Q
18.   [(P v Q) & ~P] , Q
19.   [(P ⊃ Q) & ~Q) , P
20.   [(P ⊃ Q) & ~Q) , ~P
21.   P & Q , P
22.   P , P v Q
23.   P ≡ Q , Q ⊃ P
24.   P ⊃ Q , ~P v Q
25.   ~(P & Q) , ~P v ~Q
26.   ~(P v Q) , ~P v Q
27.   P v (Q & R) , (P v Q) & (P v R)
28.   (P ⊃ Q) & R , P ⊃ (Q & R)
29.   [(P ⊃ Q) & ~P] , ~Q
30.   [(P ⊃ Q) & Q] , P
31.   ~(P v Q) , P ⊃ Q

Part II   In which of the cases above are the two sentences consistent?

Exercise 2.15

Part I   Use truth-tables to determine, for each pair of sentences below, whether the sentences are equivalent, contradictory, or neither. Note that a comma separates the members of each pair.

1. ~(P v Q) , ~P & ~Q
2. ~~P , P
3. ~(P & Q) , ~P v ~Q
4. ~(P v Q) , ~P v ~Q
5. ~(P & Q) , ~P & ~Q
6. P v Q , ~(~P & ~Q)
7. ~(P & ~P) , (Q & ~Q)
8. P & (Q v R) , (P & Q) v (P & R)
9. P ⊃ Q , Q ⊃ P
10. P ⊃ Q , (~Q ⊃ ~P)
11. P ⊃ Q , ~P v Q
12. ~(~P v Q) , ~(P &~Q)
13. [~P ⊃ (P ⊃ Q)] , [Q ⊃ (P ⊃ P)]
14. ~(P ⊃ Q) , (~P ⊃ ~Q)
15. P ≡ P , P ⊃ P
16. P & P , P
17. P v P , P

18. P , P ⊃ P
19. P & Q , Q & P
20. P v Q , Q v P
21. P ≡ Q , [(P ⊃ Q) & (Q ⊃ P)]
22. P ≡ Q , [(P & Q) v (~P & ~Q)]
23. P ⊃ (Q ⊃ R) , (P ⊃ Q) ⊃ R
24. (P v Q) v R , P v (Q v R)
25. (P & Q) & R , P & (Q & R)
26. ~(P & Q) & R , ~P & (Q & R)
27. (~P v ~Q) v R , ~(P & Q) v R
28. (P ⊃ Q) v R , (~Q ⊃ ~P) v R
29. P ≡ (Q & R) , (Q & R) ≡ P
30. P ≡ (Q ≡ R) , (P ≡ Q) ≡ R
31. ~P ≡ ~Q , ~P ⊃ ~Q
32. P v (Q & R) , (P v Q) & (P v R)
33. P ⊃ (Q & R) , (Q & R) ⊃ P
34. (P & Q) v R , (P & Q) v R

Part II Use truth-tables to determine, for each pair of sentences above, whether the sentences are consistent or inconsistent.

●— PARTIAL TRUTH-TABLES

Is the following argument valid or invalid?

$$A ⊃ (B \text{ v } \sim G)$$
$$\underline{G ⊃ (E \& S)}/ S ⊃ A$$

We could construct a truth-table for this argument and determine the answer in a mechanical way. However, the table would have 32 rows because the argument contains five atomic components, and arguments containing five components require 32 rows. A far simpler way to test this argument is to construct a *partial truth-table*.

A partial truth-table for an argument is constructed as follows. First, on a line, write down the premises and conclusion of the argument under consideration. Next, assume—for the sake of the test—that the argument is invalid. In line with that assumption, assign **T** to the main operator of each premise and then assign **F** to the main operator of the conclusion. This assignment of truth-values represents a hypothesis that the premises are true and the conclusion is false. Next, try to fill in the values of the rest of the operators based on the values previously assigned to the main operators. If this task requires a contradictory assignment of truth-values to one or more components, then the hypothesis—that the premises are true and the conclusion is false—is contradictory or impossible. If this is the case, it is not possible for the premises to be true and the conclusion false, which proves the argument valid. A contradictory assignment of truth-values is an assignment that assigns **T** and **F** to the same atomic component.

However, if the truth-values can be assigned in such a way that the premises are true and the conclusion is false with no contradictory assignment of truth-values, then the hypothesis—that the premises are true and the conclusion is false—is a possibility. In this case, the argument must be invalid, for an argument is invalid if there is even a possibility it has true premises and a false conclusion.

For example, the first step in constructing a partial table for the argument above is to assign **T** to the main connective of each premise and **F** to the main connective of the conclusion:

$$A \supset (B \lor {\sim}G) \, / \, G \supset (E \,\&\, S) \, /\!/ \, S \supset A$$
$$ \quad T \qquad\qquad T \qquad\qquad F$$

Notice that the premises are separated by a slash and the conclusion is set off by a double slash.

Let us focus next on the conclusion. If S ⊃ A is false, then S must be assigned **T** and A must be assigned **F**, for a conditional is false only when its antecedent is true and its consequent is false:

$$A \supset (B \lor {\sim}G) \, / \, G \supset (E \,\&\, S) \, /\!/ \, S \supset A$$
$$ \quad T \qquad\qquad T \qquad\qquad T\,F\,F$$

If we are going to be consistent, we must assign **T** to each occurrence of S and **F** to each occurrence of A throughout the rest of the argument:

$$A \supset (B \lor {\sim}G) \, / \, G \supset (E \,\&\, S) \, /\!/ \, S \supset A$$
$$F \quad T \qquad\qquad T \qquad T \qquad T\,F\,F$$

From here, it is easy to assign values in such a way that the premises are true and the conclusion is false. If we add that B is true, G is true, and E is true, the premises all come out true while the conclusion is false:

$$A \supset (B \lor {\sim}G) \, / \, G \supset (E \,\&\, S) \, /\!/ \, S \supset A$$
$$F \; T \; T\,T\,F\,T \quad T\,T \; T\,T\,T \quad T\,F\,F$$

This partial table presents us with a possibility—the possibility of the argument's premises being true while its conclusion is false. In effect, this table reproduces an invalidating row from the argument's full truth-table. In this case, that row is the row on which A is assigned **F** and B, G, E, and S are each assigned **T**. Thus, the argument is proven invalid, for an argument is invalid if it is possible for it to have true premises and a false conclusion.

Here's another way to understand this example. Recall that a *counterexample* to an argument is a possible circumstance in which the argument's premises are true and its conclusion is false. This partial table presents a counterexample to the argument. If there is a counterexample to an argument, then the argument is invalid. Thus, this table, by presenting a counterexample, indicates that the argument is invalid.

In short, to test an argument for validity using a partial table, write down the premises and conclusion and then try to assign truth-values in such way that the premises are made true and the conclusion is made false. If there is no such assignment of truth-values, then the argument is valid. If there is such an assignment, then the argument is invalid.

• EXERCISE 2.16 •

Part I In the case of each argument below, symbolize the argument and then use a partial table to show that it is invalid:

1. If Sue has a hamburger, then Rita will have a hamburger if Joe will not have one. So, if Rita will have a hamburger, then Joe will not have one.

2. If Arnold quits his job, then Betty will quit her job. Arnold won't quit his job. So, neither Arnold nor Betty will quit.

3. If Pat goes swimming, then neither Quinn nor Rita will swim. If Rita doesn't go swimming, then Sue will swim, but Dolores won't swim. So, if Sue swims then Pat swims.

4. If Janet swims, then Bill will swim. Janet won't swim. So, Bill won't swim.

5. If Angie sings, then Randy will sing. If Randy sings, then either Chris or Pat will sing. If Pat sings, then Roberta will sing. Therefore, if Angie sings, then Roberta will sing.

6. Craig and Faith won't both order burritos. If Craig doesn't order a burrito, then Darla will order one. So, Darla and Faith will both order burritos.

7. Neither Ann nor Bob will order tacos but Chris and Darla will. If either Ann or Chris orders a taco, then Roberta will order a burrito. So, if Roberta orders a burrito, then Ann will order a taco.

8. Both Ann and Bob won't order french fries. If Bob doesn't order french fries, then either Chris or Darla will order fries. If Robert doesn't want to eat, then Chris won't order fries and neither will Darla. Either Robert will not want to eat or Chris will order fries.

9. If neither Ann nor Bubba orders the Bob's Brontosaurus Burger Bar Special Combo Plate, then Cherie and Zack will both be disappointed.

If Cherie is disappointed, then she will order four hamburgers. So, if Ann orders the Bob's Brontosaurus Burger Bar Special Combo Plate, then Cherie won't order four hamburgers.

Part II Use partial truth-tables to prove that each of the following symbolized arguments is invalid. In each case, the argument's symbolized premises are written above the line and symbolized conclusion appears after the slash mark.

1. P ⊃ (Q v R)
    Q ⊃ S/ P ⊃ S

2. P ⊃ Q
    R ⊃ S
    ~P v ~R/ ~Q v ~S

3. A v B
    P v Q
    (A v P) ⊃ (B & Q/ ~B

4. P v Q
    P/ ~Q

5. ~(P & Q)
    ~P/ Q

6. P ⊃ Q
    Q ⊃ R
    R ⊃ S
    S ⊃ H/ H ⊃ P

7. P ⊃ Q/ Q ⊃ P

8. ~(P & Q)/ ~P & ~Q

9. ~(P v Q)/ P & Q

10. ~(P & Q)/ ~P

11. ~(P ⊃ Q)/ ~P & ~Q

12. P v Q
    P/ Q

13. P ⊃ Q
    R ⊃ Q/ P ⊃ R

14. P ⊃ Q
    Q ⊃ R/ R ⊃ P

15. P ⊃ Q
    R ⊃ S
    Q v S/ P v R

16. P ⊃ Q
    Q/ P

17. P ⊃ Q
    ~P/ ~Q

18. ~(P & Q)/ ~P & ~Q

19. P ⊃ Q/ Q v P

20. P ⊃ Q
    Q ⊃ R
    S ⊃ R/ P ⊃ S

21. P ⊃ Q
    ~R ⊃ ~S
    ~Q ⊃ ~H/ ~H ⊃ ~S

22. P ⊃ Q
    R ⊃ S
    ~P v ~R/ ~Q v ~S

23. (P ⊃ ~Q)
    (~P ≡ Q)
    ~P/ ~Q

24. P ≡ Q
    Q v R / P ⊃ R

25. P ⊃ Q
    R ⊃ S
    Q v S/ P v R

26. (P v Q) ⊃ (Q v R)
    (Q ⊃ R) v P
    (Q ⊃ R) ⊃ (P v Q)
    ~P/~Q v ~R

27. P ⊃ Q
    ~P ⊃ ~R
    ~Q ⊃ ~S/~S ⊃ ~R

Shortcut Truth-Tables   When we place an argument on top of a table to test it for validity, we are looking for just one row: a row on which the premises are true and the conclusion is false. If we find one such row—an "invalidating" row—the argument is proven invalid. (Note: an invalidating row proves the argument invalid; the row itself is neither valid nor invalid.) If we complete the table and find no invalidating row, the argument is proven valid.

Perhaps it has occurred to you that, in searching for an invalidating row, we do not really need to fill in every spot on the table. If the conclusion is true on a particular row, then we know automatically that this particular row is not an invalidating row. And if a premise is false on a particular row, then we know automatically that this particular row is not an invalidating row. So as soon as we find a true conclusion or a false premise, we can ignore the rest of that row and proceed to the next row. This suggests the following shortcut method:

> *Begin by calculating the truth-value of the conclusion on row 1. If the truth-value is* **T** *, ignore the row and move down to the conclusion of the next row. If the truth-value of the conclusion on the first row is* **F** *, move to the left on the row and check the truth-value of the first premise. If the truth-value of the first premise is* **T** *, check the truth-value of the next premise. If the truth-value of the first premise is* **F** *, ignore the rest of the row and move down to the conclusion of the next row. Complete this process until you find an invalidating row or until you fill in the entire table. If no invalidating row is found, the argument is valid; as soon as one is found the argument is shown to be invalid.*

For example, consider the argument:

$$P \supset Q$$
$$\underline{\sim P \text{ v } \sim Q} / P$$

The shortcut table for this is:

P Q	P ⊃ Q	~P v ~Q	P
T T			T
T F			T
F T	T	T	F
F F			

The shortcut table contains an invalidating row, and so the argument is invalid.

• EXERCISE 2.17 •

Symbolize the following arguments and then test each for validity with a truth-table of your choice.

1. If Betsy takes Philosophy 110, then John will take Philosophy 110. John won't take Philosophy 110. So Betsy won't take Philosophy 110.

2. If Plektus IV invades Ruritania, then the Ruritania Senate will flee. Plektus IV will not invade Ruritania. Therefore, the Ruritania Senate will not flee.

3. Either Herman will be home or Lillian will be home. Herman won't be home, so therefore Lillian won't be home.

4. Gomez and Morticia won't both be home. But Morticia will be home. So, Gomez won't be home.

5. If either Granny or Jethro is home, then Mr. Drysdale will be happy. Mr. Drysdale won't be happy. So Granny won't be home and Jethro won't be home.

6. It is not the case that both Kirk and Spock will beam down. If Kirk doesn't beam down, then McCoy will. If Spock doesn't beam down, then McCoy will. So, McCoy will beam down.

7. If Ralph wins the game, then Alice will be happy. If Alice will be happy, then Norton will be happy. So, if Norton will be happy, then Alice will be happy.

8. If either Ralph or Alice is home, then Norton will come in and make himself comfortable. Therefore, either Ralph is not home or Norton won't come in and make himself comfortable.

9. If Moe slaps Larry up the side of the head, then Larry slaps Curly up the side of the head and Curly slaps Moe up the side of the head. Since Moe will not slap Larry up the side of the head, either Larry won't slap Curly up the side of the head or Curly won't slap Moe up the side of the head.

10. Either the sun will shine and there will be no wind, or the sun will shine and the race will be canceled. The sun will shine and the race will not be canceled. So there will be no wind.

11. If Moe slips then if Curly slips then Larry will slip. If Curly slips then if Moe slips then Larry will slip. However, neither Curly nor Moe will slip. So, Larry won't slip.

12. Dobie will be pleased if Maynard gets a job and saves some money. But if Maynard gets a job and doesn't like the job, then Dobie won't be pleased. Maynard will get a job only if he wants to. So, Maynard won't both get a job and save some money.

A Summary of the Truth-Table Tests We may summarize the truth-table tests and also introduce some new terminology with the following semantical principles. First, if truth-values are assigned to the atomic components of a sentence in such a way that the sentence is made true, then we shall say that the sentence is *true* *on that truth-value assignment.* If an assignment of truth-values makes a sentence false, then the sentence is *false on that truth-value assignment.* If 'P' and 'Q' are metavariables ranging over sentences of TL, then:

1. **P** is *truth-functionally true* if and only if **P** is true on every truth-value assignment.

2. **P** is *truth-functionally false* if and only if **P** is false on every truth-value assignment.

3. **P** is *truth-functionally contingent* if and only if **P** is true on one or more truth-value assignments and false on one or more truth-value assignments.

4. An argument is *truth-functionally valid* if and only if there is no truth-value assignment making the premises true and the conclusion false. An argument is *truth-functionally invalid* if and only if it is not truth-functionally valid.

5. **P** and **Q** are *truth-functionally inconsistent* if and only if there is no truth-value assignment upon which both are true.

6. **P** and **Q** are *truth-functionally consistent* if and only if there exists at least one truth-value assignment making both true.

7. **P** *truth-functionally implies* **Q** if and only if there is no truth-value assignment upon which **P** is true and **Q** is false.

8. **P** and **Q** are *truth-functionally equivalent* if and only if there is no truth-value assignment making one true and the other false.

7. THE CONCEPT OF LOGICAL FORM

●— SENTENCE FORMS

In this section, it is extremely important that you keep in mind the distinction between two very different kinds of symbols—variables and constants. Recall that a variable ranges over a prespecified group of things whereas a constant is a symbol that stands for one specified thing. For example, in the commutative law of addition,

$$x + y = y + x$$

x and y are variables that stand in, in each case, for any number whatsoever. Along with a variable goes the understanding that anything from a specified group of things may be named and the name inserted in place of the variable. In the case of the commutative law of addition, the variables serve as placeholders for names of numbers. In truth-functional logic, the metalinguistic variables we are using range over sentences of TL.

In ordinary English discourse, we typically use the word 'thing' to accomplish what a mathematician accomplishes with a variable such as 'x'. That is, the word 'thing' generally functions as a variable. Thus, when we want to refer to unnamed items, we naturally say things such as "Don't try to do too many things at once" or "I didn't hear a thing you said."

With the distinction between sentence variables and constants in mind, we may now turn to the concept of *form*. Consider the following three sentences:

    a. It will rain or it will snow.

    b. Jan is from Michigan or Jan is from Arkansas.

    c. It is Wednesday or it is Thursday.

Symbolized within TL, these become:

    d. R v S

    e. M v A

    f. W v T

What do these sentences of TL have in common? One thing they have in common is this: each consists of a left disjunct joined by a disjunction operator to a right disjunct. This feature common to all three sentences may be expressed by

$$P \ v \ Q$$

This expression represents a general pattern exhibited by sentences (d), (e), (f).

The expression **P v Q** is called the *form* of the three sentences (d), (e) and (f), and those three sentences are said to be *instantiations* or *substitution instances* of the form **P v Q**. A TL sentence will qualify as a substitution instance of a form if and only if the TL sentence can be generated from the form by replacing the variables in the form with TL sentences and, in addition, making any necessary parenthetical adjustments. This will become clearer after we work through several examples.

Consider the form **P v Q**. Suppose we abbreviate the English sentence 'Jean attended the Monterey Pop Festival in 1967' with 'J' and 'Chris attended the Monterey Pop Festival in 1967' with 'C'. If we begin with the form **P v Q** and then replace 'P' with 'J' and 'Q' with 'C', we generate the substitution instance:

$$J \ v \ C$$

To record the replacements, let us write:

$$P/J$$
$$Q/C$$

Here, 'P/J' indicates that each occurrence of the variable 'P' was replaced by 'J' and 'Q/C' indicates that every occurrence of the variable 'Q' was replaced by 'C'.

If we return to the form **P** v **Q** and replace '**P**' with the TL formula 'A', and if we replace '**Q**' with the TL formula 'B', we generate the substitution instance

A v B

To record the replacements, we write:

P/A

Q/B

If we replace '**P**' with '(A & B)' and if we replace '**Q**' with '(C v D)', we generate the substitution instance

(A & B) v (C v D)

To record the replacements, we write:

P/A & B

Q/C v D

Next, let us replace '**P**' with ~(A & ~B) and **Q** with ~(~A & E):

~(A & ~B) v ~(~A & E)

Again, we record the replacements by writing:

P/~(A & ~B)

Q/~(~A & E)

Consider next the form ~(**P** v **Q**). If we replace '**P**' with '(A & B)' and if we replace '**Q**' with '(C v D)', after we make the appropriate parenthetical adjustments we get the substitution instance

~[(A & B) v (C v D)]

If instead we replace '**P**' with ~(A & ~B) and '**Q**' with ~(~A & E) we get:

~[~(A & ~B) v ~(~A & E)]

Other TL substitution instances of this form include:

INSTANCE:	REPLACEMENTS:	
i. ~[(A & E) v ~(E v F)]	P/(A & E)	Q/~(E v F)
ii. ~(S v ~G)	P/S	Q/~G
iii. ~(H v ~R)	P/H	Q/~R

Notice that when you go from a form to one of its substitution instances, the arrangement of the form's operators—their relative positions and scopes—remains in place; the only change occurs when the form's variables are replaced. The arrangement of the form's operators and their associated scope-indicating devices gives the form its character, its own identity. It is this character, this arrangement of operators, that the form "shares with" or "passes on" to all of its instantiations. Thus, the arrangement of the form's operators constitutes the abstract pattern or structure that its instances have in common. Within the constraints of that abstract structure, the instances of a form may vary. A form therefore constitutes a sort of symbolic template, and substitution instances of a form are sentences that "fit" that template.

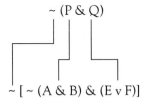

Notice how the arrangement of the form's operators carries over into the instance.

The relationship between a form and its instances is similar in some ways to the relationship between a cookie cutter and the cookies it is used to make. The cookie cutter shares with each cookie in the batch a common form, the shape of the cutter. However, within the constraints of that form, the individual cookies may vary. For instance, within the common form some might have chocolate chips, others might have cherries, and so on. Similarly, all the instances of a form share its arrangement of operators, but within that structure each instance of the form may have different sentences.

EXERCISES

Question

1. ~A is an instance of the form **P**, but A is not an instance of the form ~**P**. Can you explain why?

2. ~A v ~B is an instance of the form **P** v **Q**, but A v B is *not* an instance of ~**P** v ~**Q**. Can you explain why?

Exercise 2.18

For each sentence form, determine which of the numbered sentences is an instance of the form.

a. **P**   d. ~**P**

b. **P** v **Q**   e. ~(**P** & **Q**)

c. **P** & **Q**   f. ~(**P** v **Q**)

g. ~P & ~Q

h. P v P

i. ~P & Q

j. ~P v ~Q

k. P ⊃ Q

l. ~P ⊃ ~Q

m. (P v Q) & R

n. ~(P & Q) v R

o. P ⊃ (Q ⊃ R)

p. P ≡ Q

1. ~(A & B)

2. ~A v ~B

3. (A & B) ⊃ ~C

4. ~(H ≡ S) v ~(G & B)

5. (H v B) ⊃ [S ⊃ (R & H)]

6. ~A & ~B

7. [(R v H) v G] v [(R v H) v G]

8. ~[(E & S) v (F ≡ G)]

9. ~[~(A v B) & ~(H v S)]

10. ~A ≡ (S ⊃ H)

11. ~(A & B) v (H ⊃ Q)

12. [(H & B) v S] & S

13. ~(A & B) ⊃ ~(E v H)

14. ~(H v S) & (F ≡ G)

15. ~[(A v B) v C]

16. ~(A ≡ B)

17. (A & B) ≡ (C v E)

18. (H ⊃ S) ⊃ (F ⊃ R)

TAUTOLOGOUS and CONTRADICTORY FORMS

Consider the truth-table for the form **P v ~P**:

P	P v ~P
T	T
F	T

A form is a *tautological form* if and only if the final column of its truth-table displays all **T**'s. In this case, the table proves that **P v ~P** is a tautological form.

Consider next the truth-table for the form **P & ~P**:

P	P & ~P
T	F
F	F

A form is a *contradictory form* if and only if the final column of its truth-table displays all **F**'s. In this case, the table proves that **P & ~P** is a contradictory form.

If you reflect upon the relationship between a form and its instances, you will see that every substitution instance of a tautological form will be tautological and every instance of a contradictory form will be contradictory. This is because the final column of the form's truth-table is determined by the configuration of the form's operators. This configuration of operators carries over into each of the form's instances in such a way that the final column of each instance of a form will always match the form's final column. So if a form's final column is all **T**'s (or all **F**'s) then each of its instances will have a final column that is all **T**'s (or all **F**'s).

 • **Logic Joke**

What did the variable say to the constant?
"May the form be with you."

●— APPLICATION

Let us now put our understanding of forms to work. What is the logical status of the following sentence?

63. {[(P & ~Q) v ~(R & S)] v (T & U)} v ~{[(P & ~Q) v ~(R & S)] v (T & U)}

That is, is this sentence tautological, contradictory, or contingent? The truth-table method requires that we construct a table for the sentence and see whether the final column contains all **T**'s, all **F**'s, or a mixture of **T**'s and **F**'s. If we were to construct a table for (63), the table would have sixty-four rows. That's a lot of **T**'s and **F**'s. However, our understanding of forms makes possible a simpler solution. A careful inspection of (63) reveals that it is an instance of the form **P v ~P**. As we saw above, this form is tautological. Since we know that every instance of a tautological form is a tautology, it follows that (63) is tautological.

What is the logical status of the following?

64. {[(~P v ~Q) & (~R v ~S)] & (~T & ~U)} & ~{[(~P v ~Q) & (~R v ~S)] & (~T & ~U)}

If we were to construct a table to test this sentence, the table would again have sixty-four rows. However, inspection of (64) reveals that it is an instance of the form **P & ~P**. A little two-row table shows that this form is a contradictory form. Since (64) is an instance of a contradictory form, it follows that (64) is itself contradictory.

Now, for a somewhat confusing point. As noted above, any sentence that is an instance of a tautologous form will itself be a tautology, and any sentence that is an instance of a contradictory form will itself be contradictory. However, not all sentences that instantiate nontautologous forms are themselves nontautologous, for some tautologous sentences will instantiate nontautologous forms. And not all sentences that instantiate noncontradictory forms will themselves be noncontradictory, for some contradictory sentences will instantiate noncontradictory forms. For example, notice that the contradictory sentence

A & ~A

instantiates the noncontradictory form

P & Q

and the tautologous sentence

$$A \lor \sim A$$

instantiates the nontautologous form

$$P \lor Q.$$

Of course each of these sentences *also* instantiates the nontautologous form

$$P$$

This form, 'P', is a form every sentence instantiates, for any sentence can be generated from the form P by replacing P with the sentence to be generated.

Sentence forms that are neither tautologous nor contradictory are said to be *indeterminate* forms. The final column on the truth-table of an indeterminate form will have a mixture of T's and F's. Such forms are not called "contingent" forms because not every substitution instance of such a form will be a contingent sentence.

—————————————— • EXERCISES • ——————————————

Exercise 2.19

Without using truth-tables, determine in each case whether the sentence is a tautology or a contradiction.

1. {[(P v Q) & ~D] v B} v ~{[(P v Q) & ~D] v B}

2. {[(P v Q) & ~D] v B} & ~{[(P v Q) & ~D] v B}

3. ~{[(P v Q) & ~D] v ~[(P v Q) & ~D]} v {[(P v Q) & ~D] v ~[(P v Q) & ~D]}

Exercise 2.20

Which of the following forms are contradictory, which are tautologous, and which are indeterminate ?

1. $(P \supset Q) \equiv (Q \supset P)$
2. $P \supset (P \lor Q)$
3. $\sim(P \& \sim P)$
4. $\sim(P \lor \sim P)$
5. $(P \& Q) \supset P$
6. $\sim[P \supset (Q \supset P)]$
7. $(P \supset Q) \supset (\sim Q \supset \sim P)$
8. $[(P \supset Q) \& \sim Q] \supset \sim P$
9. $\sim P \lor (Q \lor P)$
10. $(P \equiv Q) \supset (P \supset Q)$
11. $(P \lor \sim P) \supset (Q \& \sim Q)$
12. $[(P \supset Q) \& P] \supset Q$
13. $[(P \lor Q) \& \sim P] \supset Q$
14. $[(P \supset Q) \& (Q \supset R)] \supset (P \supset R)$

15. [(P ⊃ Q) & Q] ⊃ P

16. (P ⊃ Q) ⊃ (Q ⊃ P)

17. P ⊃ ~P

18. P v (P & Q)

19. (P & Q) v (P & R)

20. ~[P ⊃ (P ⊃ P)]

21. P v (Q & P)

22. (~P ⊃ ~Q) ⊃ ~(P ⊃ Q)

23. (~P & ~Q) ⊃ ~(P & Q)

24. P ≡ (P v Q)

25. (P ≡ Q) ≡ (P v Q)

26. (P ⊃ P) ⊃ P

27. ~[(P v Q) v (P & Q)]

28. (P v Q) ⊃ P

29. (P ⊃ Q) ⊃ Q

30. Q ⊃ (P v ~P)

31. (P & ~P) ⊃ Q

32. [(P ⊃ Q) & ~Q] ⊃ ~Q

33. (P & Q) & (~P v ~Q)

●— ARGUMENT FORMS

The relationship between an *argument form* and a symbolized argument is similar to the relationship between a sentence form and a symbolized sentence. An argument form is a set of two or more sentence forms that can be transformed into a symbolized argument by replacing the variables with TL sentences. An argument formed in this way from an argument form is called a *substitution instance* of the argument form. For example, consider the argument:

1.   We'll either swim or jog.

2.   We won't swim.

3.   So, we'll jog.

In TL this is:

1.   S v J

2.   <u>~S</u>

3.   J

This TL argument is an instance of the following argument form:

1.   **P v Q**

2.   <u>**~P**</u>

3.   **Q**

for the TL argument can be generated from the form by making the following replacements:

P/S   Q/J

This argument form appears so often in everyday reasoning that it has been given a name: the *Disjunctive Syllogism* form. (A syllogism is a deductive argument. This form gets its name from the fact that its first step is disjunctive in nature.)

Another substitution instance of the Disjunctive Syllogism form is:

1. (A & B) v (E & F)
2. ~(A & B)
3. E & F

where the replacements are: P/A & B, Q/E & F. Each of the following is also a substitution instance of the Disjunctive Syllogism form:

E v G   ~(A v B) v (E ⊃ S)
~E   ~~(A v B)
G   (P/E, Q/G)   E ⊃ S   (P/~(A v B), Q/E ⊃ S)

Consider next this argument:

1. If it's raining, then the roof is wet.
2. It's raining.
3. The roof is wet.

In TL this is symbolized:

1. A ⊃ B
2. A
3. B

This argument is a substitution instance of the argument form:

1. P ⊃ Q
2. P
3. Q

This form, known as the *Modus Ponens* form, appears frequently in everyday reasoning. ('Modus Ponens' is Latin for 'method of affirmation.' This form gets its name from the fact that the second premise affirms the antecedent of the first premise.) Additional instances of the Modus Ponens form follow:

1. (A v B) ⊃ (E v F)
2. (A v B)
3. (E v F)   P/(A v B), Q/(E v F)

1. ~A ⊃ ~(H & G)

2. <u>~A</u>

3. ~(H & G)   **P**/~A, **Q**/~(H & G)

Notice that an argument form constitutes a general form or pattern of reasoning that many individual arguments may instantiate.

To test an argument form for validity, write the form across the top of a table and fill in the rows. Then, look for an invalidating row. If no invalidating row is found, the form is a valid argument form. If one or more invalidating rows are found, the form is an invalid argument form.

The Modus Ponens and Disjunctive Syllogism argument forms appearing in the two examples above are both valid argument forms, as the two tables below show:

PQ	P v Q	~P	Q		PQ	P ⊃ Q	~P	Q
TT	T	F	T		TT	T	T	T
TF	T	F	F		TF	F	T	F
FT	T	T	T		FT	T	F	T
FF	F	T	F		FF	T	F	F

Disjunctive Syllogism   Modus Ponens

Notice that neither table contains an invalidating row.

If an argument instantiates a valid argument form, then the argument itself must be valid. Every substitution instance of a valid argument form is a valid argument. Here is why. The arrangement of operators in the form carries over into every instance of the form. This arrangement of operators determines the final columns on the form's truth-table as well as on the truth-table of each instance. Thus, the final columns on the truth-table of any instance of the form will match the final columns on the form's truth-table. One consequence of this is that every instantiation of a valid form will itself be a valid argument.

In logic, we study valid argument forms because these serve as guides to help us reason properly. We can be sure our reasoning is correct if it instantiates a valid form of argument.

We must now pause to clarify an important but somewhat confusing point. Suppose we symbolize a particular English argument and find that it is an instance of a valid form. The argument 'fits' a valid form. This proves that the argument itself is valid. Any argument that fits a valid form will be valid; only valid arguments fit valid forms. However, the situation is different when it comes to invalid forms. Not every substitution instance of an invalid form will be an invalid argument, for in some cases a valid argument is an instance of an invalid form. That is, both valid and invalid arguments can instantiate invalid forms. Here are two examples.

i. The valid argument

A v B

<u>~A</u>

B

instantiates the invalid form:

P

Q

R   (P/A v B, Q/~A, R/B)

ii. The valid argument

A v (A & B)

(A & B)
_____

A

instantiates the invalid form

P v Q   P/A

Q   Q/(A & B)
_____

P

Therefore, if we symbolize an argument within TL and find that the argument is an instance of an invalid form, we should not conclude that the argument is invalid. It is possible for a valid argument to instantiate an invalid form. On the other hand, if we symbolize an argument within TL and find that the argument instantiates a valid form, the argument must be a valid argument, for every instantiation of a valid form is a valid argument.

• EXERCISES •

Exercise 2.21

Use truth tables to determine which of the following are valid argument forms and which are not.

1. P ⊃ Q

   Q
   _____

   P

2. P ⊃ Q

   Q ⊃ P

3. P ⊃ Q

   P
   _____

   Q

4. P ⊃ Q

   ~Q
   _____

   ~P

5. P v Q

   ~P
   _____

   Q

6. P v Q

   Q
   _____

   P

7. $P \supset Q$

   $\sim P$

   $\sim Q$

8. $P \supset Q$

   $\sim Q \supset \sim P$

9. $P \supset Q$

   $\sim P \vee Q$

10. $P \vee Q$

    $Q \vee P$

11. $P \supset Q$

    $Q \supset R$

    $P \supset R$

12. $P \& Q$

    P

13. $\sim(P \& Q)$

    P

    $\sim Q$

14. $(P \& Q) \supset R$

    $P \supset (Q \supset R)$

15. $\sim(P \vee Q)$

    $\sim P$

16. $P \supset Q$

    $P \supset R$

    $Q \supset R$

17. $P \supset Q$

    $Q \supset P$

    $P \equiv Q$

18. $P \supset Q$

    $P \supset R$

    $P \supset (Q \& R)$

19. $\sim P \vee \sim Q$

    $\sim(P \vee Q)$

20. $\sim P \& \sim Q$

    $\sim(P \& Q)$

21. $\sim (P \& Q)$

    $\sim P \& \sim Q$

22. $P \vee (Q \& R)$

    $(P \vee Q) \& (P \vee R)$

23. $P \supset R$

    $Q \supset R$

    $(P \vee Q) \supset R$

24. $\sim(P \& Q)$

    $\sim P \vee \sim Q$

25. $\sim(P \vee Q)$

    $\sim P \& \sim Q$

26. $P \& (Q \vee R)$

    $(P \vee Q) \& (P \vee R)$

27. $P \supset Q$

    $R \supset S$

    $P \vee R$

    $Q \vee S$

28. $(P \supset Q) \& (R \supset S)$

    $P \vee R$

    $Q \vee S$

29. $P \equiv Q$

    $(P \supset Q) \& (Q \supset P)$

30. $P \supset (Q \vee R)$

    $(Q \vee R) \supset P$

    $P \supset P$

31. $P \& \sim P$

    Q

32. Q

    $P \& \sim P$

33. $P \supset (Q \& R)$

    $(P \& Q) \supset R$

34. $P \supset Q$

    $\sim(Q \vee R)$

    $\sim P$

35. $P \vee Q$

    $Q \vee R$

    $P \vee R$

Exercise 2.22

Which of the arguments in Exercise 2.21 is an instance of the Modus Ponens form? Which is an instance of the Disjunctive Syllogism form?

The Value of Abstraction An argument form is an abstraction. It is a pattern *abstracted* from an argument. When we study an argument form after it has been abstracted from a specific natural language argument, we can see things about it we might not see if it were left embedded within that particular stretch of language. By studying abstract logical forms, modern logic—sometimes called "formal" logic—can teach us much about the inner structure of an argument that we might not learn otherwise.

When we abstract the form of an argument from a natural language argument, we are doing something significant. We are moving from the particular to the general. That is, we are shifting our analysis from the particular features of an individual argument, expressed within a particular context, to a general form that many arguments might share. The particular features of a natural language argument are those that distinguish one specific argument, expressed within a particular context, from other arguments. The general form of an argument is a feature many arguments may share. The process of going from the particular to the general is an important part of modern thought; it can be found in just about every subject. It is also an important part of modern formal logic.

●— SOME REFLECTIONS ON TRUTH-FUNCTIONAL LOGIC

In order to put this logical theory into perspective, let us reflect for a moment upon truth-functional logic as a whole.

Truth-Functional Logic Contains Decision Procedures A *decision procedure* or *algorithm* for a problem is a mechanical method, involving precisely specified steps, that is guaranteed to give a definite solution in a finite number of those steps. A procedure counts as a decision procedure if (a) using the procedure is purely a matter of following rules without the use of any ingenuity or creativity; and (b) the procedure is guaranteed to yield a definite answer in a finite number of steps. The rules for doing long division constitute a decision procedure that most people learn when they are in grade school.

The truth-table technique provides us with decision procedures for a number of logical problems, including the determination of validity, inconsistency, implication, equivalence, and so on. Sentences and arguments can be tested for the various logical properties by filling in the T's and F's on a truth-table according to definite rules. In this sense, truth-functional logic is 'mechanical'. However, the use of the truth-table, as a decision procedure, has its limits. There are certain types of arguments it can't be applied to. We will examine these types of arguments in future chapters.

Truth-Functional Logic Is Analytic The truth-table test for validity is an example of an *analytic method*. One way to understand something is to analyze it. This involves breaking something down into its constituent parts and then seeing how those parts fit together to constitute the whole. Imagine, for example, taking a watch apart and putting it back together to see how it works. An analytic method is one that involves the analysis of something. The analytic method has been successful in a number of fields of thought, especially the physical sciences where its application led to the discovery of photons, protons, neutrons, electrons, and various other subatomic particles.

Since logic is concerned with arguments, the analytic method in logic amounts to breaking arguments down into their component parts and studying how those parts fit together to produce a chain of reasoning. This is exactly what we are doing when we symbolize an argument and run a truth-table test on it. The truth-table procedure thus provides us with a way of analyzing an argument.

Truth-Functional Logic Is Formal Most everyday arguments are expressed within a natural language. That is, we normally put our reasoning into the words and sentences of the particular linguistic community we belong to. In the subject called *informal logic*, the student learns how to recognize and evaluate arguments as they appear within a natural language. Consequently, in an informal logic course few logical symbols are used. However, in *formal logic*, arguments are studied from a more abstract or general perspective, one that involves concentrating on argument forms rather than on specific natural language arguments. By studying forms, we can analyze patterns of reasoning common to many arguments across many natural languages. However, argument forms are very abstract things. In order to study them efficiently, we must use an artificial symbolic language. Thus, modern formal logic is sometimes called "symbolic" logic.

Before the nineteenth century, logicians used few symbols in their theorizing. Since logic is concerned with arguments, and since arguments are typically expressed within natural languages, logicians mainly studied arguments that were expressed within a natural language. However, in the late nineteenth century and during the first part of the twentieth century, Gottlob Frege and Bertrand Russell pioneered the development of symbolic languages for logical theory. The formal languages that were developed allowed logicians to simplify the expression of extremely complex ideas, and this led to enormous advances in logical theory. Incidentally, the study of symbolic logic also greatly advanced our understanding of language and laid the conceptual foundations of computer science. Thus, linguistics makes heavy use of symbolic logical languages, and computer science simply could not exist as a subject of study were it not for the development of symbolic logical languages.

In sum, truth-functional logic is mechanical, analytic, and formal, in the senses specified above.

APPENDIX 1. NECESSARY and SUFFICIENT CONDITIONS

The horseshoe is used within logical theory to symbolize statements about necessary and sufficient conditions. Philosophical arguments often refer to the distinction between a necessary condition and a sufficient condition. A brief discussion of this matter will help clarify the nature of both the horseshoe and the triple bar.

A condition s is called a *sufficient* condition for an event e just in case s is all that is required for e to occur. For example, jumping into a lake is sufficient for getting wet, for jumping into a lake is all that is required for getting wet. In contrast, a condition n is called a *necessary* condition for an event e just in case event e is *not possible without* n. For example, the presence of fuel is necessary for an engine to run, for an engine cannot possibly run without fuel.

Notice that while fuel is necessary for the operation of an engine, it is not sufficient. More is required for the operation of an engine than merely fuel. Oxygen, electricity, and parts in working order are also required. Notice also that while jumping into a lake is sufficient for getting wet, it is not necessary for getting wet. There are lots of ways to get wet that don't involve jumping into a lake.

Suppose we want to symbolize:

Gasoline is necessary for the engine's operation.

We could try representing the necessary condition in the antecedent:

$$G \supset R$$

where G abbreviates 'the engine has gasoline' and R abbreviates 'the engine is running'. It is obvious that this way of representing a necessary condition is incorrect. The mere presence of gasoline does not guarantee that an engine will run; engines often won't run even when their fuel tanks are full. However, if we represent the necessary condition in the consequent, the symbolism seems correct:

$$R \supset G$$

This indicates that if the engine is running, then it has gasoline in it. A statement of a necessary condition typically goes in the consequent and that for which the necessary condition is necessary is typically represented in the antecedent.

Next, suppose we want to symbolize:

Jumping in a lake is sufficient for getting wet.

We could try representing the sufficient condition in the consequent:

$$W \supset J$$

where W abbreviates 'You are wet' and J abbreviates 'You jumped in a lake'. This is obviously incorrect. The mere fact that someone is wet doesn't imply that she or he jumped into a lake. Clearly, the way to symbolize a claim about a sufficient condition is to represent the sufficient condition in the antecedent:

$$J \supset W$$

Next, consider the connective 'only if'. Suppose you claim something like

Human life is present only if oxygen is present.

The statement following 'only if' records a necessary condition for that which is referred to by the statement to the left of 'only'. Thus, we must symbolize this as:

$$H \supset O$$

where H abbreviates 'Human life is present' and O abbreviates 'Oxygen is present'. It would certainly be incorrect to symbolize this the other way around, that is, as $O \supset H$. The mere presence of oxygen does not guarantee the presence of human life. So, in general, a statement fitting the general form 'P only if Q' must be symbolized as:

$$P \supset Q$$

Earlier, it was noted that the word 'if' typically introduces the antecedent, which is always symbolized to the left of the horseshoe. A statement of the form 'P only if Q' is an exception to this general rule.

APPENDIX 2. EXPRESSIVE COMPLETENESS

●— INDIVIDUATION OF TRUTH-FUNCTIONS

Truth-functions may be *individuated* in terms of their truth-tables. Each truth-function may be defined in terms of a unique truth-table, a table displaying a configuration of **T**'s and **F**'s unlike that found on any other function's truth-table. It follows that different truth-functions cannot be represented by the same truth-table.

If two formulas of TL are equivalent, then the two formulas express the same truth-function. If two formulas of TL are not equivalent, then they express two different truth-functions. Thus, since '~(P v Q)' and '(~P & ~Q)' are equivalent, these two formulas express the same truth-function; namely, the function whose truth-table has the following configuration of **T**'s and **F**'s:

P	Q	~(P v Q)
T	T	F
T	F	F
F	T	F
F	F	T

In other words, '~(P v Q)' and '(~P & ~Q)' are merely different ways of expressing the same truth-function, namely, the function whose final column is:

$$
\begin{array}{c}
F \\
F \\
F \\
T
\end{array}
$$

However, the two formulas '(~P v ~Q)' and '~(P v Q)' are not equivalent—the final columns of their truth-tables differ—and so these two formulas express different truth-functions.

● **PROBLEM** ●

---

1. Show that '(P & Q)' and '~(~P v ~Q)' express the same truth-function.

2. Show that '~(~P v Q)' and '(~P & ~Q)' express different truth-functions.

●— AN INFINITY OF DIFFERENT TRUTH-FUNCTIONS

How many different truth-functions are there? We will approach the answer in stages. First, we need a way to count truth-functions. Call a truth-function a *two-row function* if its table contains just two rows. A two-row table contains a column on the left for just one atomic component:

P	
T	
F	

Next, call a truth-function a *four-row function* if its truth-table contains just four rows. A four-row table contains columns on the left for just two atomic components:

P	Q	
T	T	
T	F	
F	T	
F	F	

In general, an **n**-row truth-function is a function whose truth-table has **n** rows.

How many two-row truth-functions are there? A simple process of trial and error reveals that there are four, which we'll name $f_1(P)$, $f_2(P)$, $f_3(P)$, and $f_4(P)$:

P	$f_1(P)$	$f_2(P)$	$f_3(P)$	$f_4(P)$
T	T	F	T	F
F	T	F	F	T

Notice that each function displays a unique configuration of **T**'s and **F**'s in its final column. You are probably only familiar with one of these four, the function named 'f_4'. This function is the negation function and is normally expressed with an English negation operator. None of the other two-row functions is used in everyday speech. That is, there are no English operators associated with the three other two-row functions.

How many four-row functions are there? A process of trial and error reveals that there are sixteen four-row configurations of **T**'s and **F**'s:

P Q	$f_1(P,Q)$	$f_2(P,Q)$	$f_3(P,Q)$	$f_4(P,Q)$
T T	T	T	T	T
T F	T	T	T	F
F T	T	T	F	T
F F	T	F	T	T

P Q	$f_5(P,Q)$	$f_6(P,Q)$	$f_7(P,Q)$	$f_8(P,Q)$
T T	F	T	T	T
T F	T	T	F	F
F T	T	F	T	F
F F	T	F	F	T

P Q	$f_9(P,Q)$	$f_{10}(P,Q)$	$f_{11}(P,Q)$	$f_{12}(P,Q)$
T T	F	F	F	F
T F	T	T	F	F
F T	T	F	T	F
F F	F	T	T	T

P Q	$f_{13}(P,Q)$	$f_{14}(P,Q)$	$f_{15}(P,Q)$	$f_{16}(P,Q)$
T T	F	F	T	F
T F	F	T	F	F
F T	T	F	F	F
F F	F	F	F	F

We have already studied several of these functions. For example, $f_2(P,Q)$ is the 'or' function, expressed most simply by P v Q but also expressed by ~(~P & ~Q). $f_{15}(P,Q)$ is the 'and' function, expressed by (P & Q), and by ~(~P v ~Q). Notice that $f_5(P,Q)$ is the 'not both' function expressed by ~(P & Q). And $f_{12}(P,Q)$ is the 'neither-nor' function, expressible by (~P & ~Q) or by ~(P v Q). The 'exclusive or' function is represented here by $f_9(P,Q)$.

The number of different truth-functions at each level of complexity grows exponentially. It can be proven mathematically that there are 256 different 8-row truth-functions, 65,536 different 16-row truth-functions, 4,294,967,296 different 32-row truth-functions, 18,446,744,073,709,551,616 different 64-row truth-functions, and so on without limit.

Thus, there are truth-functions so large that their tables, if written out, would have billions and billions of rows of **T**'s and **F**'s. Indeed, there are truth-functions so large that their tables, if written out, would stretch 15 billion light years across the universe. Perhaps on a planet orbiting a star in some distant galaxy there exists an "advanced" civilization that thinks and speaks in terms of truth-functional operators expressed on tables containing billions of rows of truth-values.

●— EXPRESSIVE COMPLETENESS

Recall that one formula of TL expresses the same function as that expressed by another formula so long as the two formulas have matching final columns on their truth-tables. And now for a very remarkable fact: Any truth-function, no matter how complex, no matter how large its table, no matter how strange its operators, may be expressed using just the ampersand, tilde, and wedge, plus the other vocabulary elements of TL. In other words, the artificial language TL, with just three of its operators and its three rules of grammar, is sufficient to express an infinity of truth-functions. The set of operators consisting of the ampersand, tilde, and wedge thus may be said to be *expressively complete for truth-functionality*. When you reflect upon how syntactically simple TL is and upon how complex truth-functions can get, this seems a remarkable claim. The remainder of this appendix constitutes an argument for this claim.

Suppose we wish to represent an arbitrarily given truth-function using just the ampersand, tilde, and wedge, plus any constants and parenthetical devices that may be necessary. How should we proceed? In order to develop a general procedure, we must consider three cases.

Case 1 First, suppose we want to represent, using just the ampersand–tilde–wedge combination, a function containing only one **T** in its final column. Function f_{14} is an example of such a function:

P	Q	f_{14} (P,Q)
T	T	F
T	F	T
F	T	F
F	F	F

The particular configuration of **T**'s and **F**'s displayed on this table is what makes this function the function that it is. In order to express this function with a formula containing only ampersand, tilde, and wedge operators, we must write a formula that contains only ampersands, tildes, and wedges (plus any constants and parenthetical devices that may be necessary), and that has the same truth-table configuration of **T**'s and **F**'s. Since f_{14} takes the value **T** only when P is **T** and Q is **F**—at row 2—we can express f_{14} with just:

$$P \mathbin{\&} \mathord{\sim} Q$$

For this conjunction also takes the value **T** only when P is assigned **T** and Q is assigned **F**:

So, P & ~Q is just another way of expressing the function named 'f₁₄'. Both expressions, 'P & ~Q' and 'f₁₄ (P,Q)' name the truth-function whose configuration of **T**'s and **F**'s happens to be represented by the table for f_{14} (**P,Q**).

Call a row of a truth-table a "**T**-row" if the final column at that row contains a **T**. In general, if an arbitrarily given function has only one **T** in its final column, we can express that function by following this two-step procedure:

    a. Conjoin the components.

    b. Negate just those components that are assigned an **F** in the one **T**-row.

Case 2   For our second case, suppose we wish to use just ampersand, tilde, and wedge operators to represent a function whose final column contains more than one **T**. Function f_{10} is an example of such a function:

P	Q	f_{10} (P,Q)
T	T	F
T	F	T
F	T	F
F	F	T

This function takes the value **T** in either of two cases: (i) when, in row two, **P** is assigned **T** and **Q** is assigned **F**, and (ii) when, in row four, **P** and **Q** are both assigned **F**. Now, let us apply the two-step procedure from Case 1 above. When **P** is assigned **T** and **Q** is assigned **F**, as in case (i), then the function P & ~Q is true:

And when P and Q are both assigned **F**, as in case (ii), ~P & ~Q takes the value **T**:

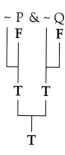

Putting these two cases together, function f_{10} takes the value **T** when, in row two, **P** is assigned **T** and **Q** is assigned **F** or when, in row four, **P** and **Q** are both assigned **F**. But then

$$(P \ \& \ {\sim}Q) \ v \ ({\sim}P \ \& \ {\sim}Q)$$

will express this function, for it too will take the value **T** when, in row two, P is assigned **T** and Q is assigned **F** or when, in row four, P and Q are both assigned **F**. In other words, f_{10} and (P & Q) v (~P & ~Q) will have identical final columns and so are two ways of expressing the same truth-function.

In general, if the final column of a function contains more than one **T**, we can express the function using only ampersand, tilde, and wedge operators if we:

a. Produce the conjunction for each **T**-row as described in Case 1.

b. Disjoin these conjunctions.

Case Three For our third case, suppose the final column of a function's table contains no **T**'s at all, that is, it shows all **F**'s. How do we express such a function in terms of ampersand, tilde, and wedge operators? We simply produce an ampersand–tilde–wedge function of the same number of components that is itself contradictory. The desired contradiction can easily be produced by the following two-step procedure:

a. Conjoin the components.

b. Conjoin to this the negation of any one component.

The resulting ampersand–tilde–wedge function will be contradictory and therefore will express the function in question.

Suppose we wish to construct a formula using the ampersand, tilde, and wedge, for $f_{11}(\mathbf{P},\mathbf{Q})$. The table for $f_{11}(\mathbf{P},\mathbf{Q})$ is:

P	Q	f_{11} (P,Q)
T	T	F
T	F	F
F	T	T
F	F	T

First, we produce a conjunction for each **T**-row:

$$\sim P \ \& \ Q \quad (\text{row } 3)$$
$$\sim P \ \& \sim Q \quad (\text{row } 4)$$

Next, we disjoin these:

$$(\sim P \ \& \ Q) \ v \ (\sim P \ \& \sim Q)$$

Notice that this function expresses f_{11}(**P,Q**), for the final column of $(\sim P \ \& \ Q) \ v$ $(\sim P \ \& \sim Q)$ is:

P	Q	$(\sim P \ \& \ Q) \ v \ (\sim P \sim Q)$
T	T	F
T	F	F
F	T	T
F	F	T

In summary, here is the procedure we used:

Step 1. Examine the final column of the function's truth-table. If it contains exactly one **T**, conjoin the components and negate just those components that are assigned an **F** in the **T**-row.

Step 2. If the final column contains more than one **T**, produce the conjunction for each **T**-row as described in step (1) and then disjoin these conjunctions.

Step 3. If the final column contains no **T**'s, simply conjoin each component and then conjoin to this the negation of any one component.

Any truth-function, no matter how complex, will have one, more than one, or no **T**'s in its final column. In either of these cases, we can construct a TL sentence that expresses the function. Since this procedure allows us to express any truth-function, of any size, using only the ampersand, tilde, and wedge operators, plus vocabulary elements of TL, it follows that the set of operators consisting of just the ampersand, tilde, and wedge is expressively complete for truth-functionality.

Perhaps on some faraway planet in some remote galaxy there are logic students who work with truth-tables containing billions of rows. Nevertheless, if these beings journeyed to Earth they could express their most complex truth-functions in terms of the language TL, with three of its operators and its three simple rules of grammar.

PROBLEM

Here's a challenging problem. Earlier in the chapter you learned how to translate conjunctions into disjunctions and disjunctions into conjunctions. Using the principles of that section, write an addendum showing that the set consisting of the tilde and ampersand alone is expressively complete for truth-functionality. Do the same for the tilde and wedge combination.

3

TRUTH-FUNCTIONAL
NATURAL DEDUCTION

1. RULES OF INFERENCE

You were introduced to truth-functional logic in Chapter 2, where you spent many a happy evening filling in truth-table rows with **T**'s and **F**'s. Truth-tables are valuable for a number of reasons. First, they provide us with decision procedures for truth-functional logic, procedures so mechanical that they are easily written into computer programs. Second, truth-tables can be used to provide a detailed introduction to truth-functional logic. However, truth-tables have one major drawback. As we evaluate bigger and bigger formulas, the size of the truth-table increases exponentially. A formula with two atomic components may require only a four-row table, but a formula containing four atomic components requires a sixteen-row table, six components calls for a 64-row table, eight components requires 256 rows, and so on. Consequently, when evaluating large formulas and complicated arguments, the truth-table method involves the construction of tables so large that the method becomes impractical.

Several alternative methods have been developed in recent years that accomplish essentially the same things truth-tables accomplish but that are much simpler to work with when solving complicated problems. Two commonly taught alternatives to the truth-tables are: (a) the method of natural deduction (covered in this chapter), and (b) the truth-tree method (covered in Chapter 4). Logic instructors typically use truth-tables to introduce the fundamentals of truth-functional logic and then, in order to deal with the more complicated problems, use just one of the two alternatives—either truth-trees or natural deduction—although some instructors cover both methods.

The natural deduction method, which is the subject of this chapter, gets its name from the fact that with this method we deduce a conclusion from a set of premises through a series of valid inferences corresponding to natural patterns of reasoning. The patterns of reasoning we will be referring to will be spelled out in the form of principles called *inference rules*.

●— INFERENCE RULES

An *inference rule* is a rule specifying that a particular conclusion may be inferred when certain premises are given. Such a rule consists of two parts. The *premise section* specifies the premises you must begin with, and the *conclusion section* specifies the conclusion that you may infer from the given premises. The application of a valid rule of inference always results in a valid inference. Recall from Chapter 1 that a person draws an inference when he or she asserts a conclusion on the basis of one or more premises. An inference is a *valid inference* just in case the argument formed out of the premises and conclusion of the inference constitutes a valid argument.

The following four valid inference rules reflect patterns of reasoning we frequently follow in everyday thought. In the examples below, arguments and inferences will be displayed by listing the premises one by one above a line, with the conclusion placed underneath the line.

The Disjunctive Syllogism Pattern   Consider the following valid argument:

1. Either we will run or we will swim.
2. We won't run.
3. So, we will swim.

In TL, using obvious abbreviations, this becomes:

1. R v S
2. ~R
3. S

This argument is an instance of the following valid argument form:

— The Disjunctive Syllogism Form →●

$$P \vee Q$$
$$\underline{\sim P}$$
$$Q$$

The validity of this form is easily proven on a truth-table:

PQ	P v Q	~P	Q
TT	T	F	T
TF	T	F	F
FT	T	T	T
FF	F	T	F

Notice that this truth-table contains no invalidating row.

The following inference rule reflects the Disjunctive Syllogism argument form:

The Disjunctive Syllogism Rule

From any sentences **P v Q** *and* **~P**, *one may infer the sentence* **Q**.

The Disjunctive Syllogism rule ("DS" for short) tells us that from a disjunction and the negation of the left disjunct, we may infer the right disjunct. We may abbreviate this rule by writing:

$$
\begin{array}{c}
\text{P v Q} \\
\underline{\text{~P}} \\
\text{Q}
\end{array}
$$

The formulae above the line represent the premise section of the rule and the formula below the line represents the conclusion section. Each of the following is an instance of the DS argument form:

1.	A v B	1.	(A & B) v (E & F)	1.	~(A ⊃ B) v ~(H ≡ S)
2.	~A	2.	~(A & B)	2.	~~(A ⊃ B)
3.	B	3.	(E & F)	3.	~(H ≡ S)

If you have trouble seeing that each of these is an instance of the DS form, then it might be helpful to review the discussion in Chapter 2 of an argument form and the distinction drawn there between a form and its instances.

We naturally reason in accord with this inference rule—whether we are aware of it or not—from a very early age. For instance, suppose a little child is told that today the family will either stay home and everyone will work in the yard or the family will go to the circus. If Mom or Dad then announces, "It's raining so we aren't going to stay home and work in the yard," nobody needs to tell the child what inference to draw. The child immediately infers that the family will be going to the circus.

The Modus Ponens Pattern This argument is clearly valid:

1. If it's raining, the roof is wet.
2. It's raining.
3. So, the roof is wet.

Using obvious abbreviations, this argument may be expressed in TL as:

1. R ⊃ W
2. R
3. W

This instantiates the following valid argument form:

— **The Modus Ponens Form** →•

$$P \supset Q$$
$$\underline{P}$$
$$Q$$

The validity of this form is established by the table below:

PQ	P⊃Q	P	Q
TT	T	T	T
TF	F	T	F
FT	T	F	T
FF	T	F	F

The following inference rule corresponds to the Modus Ponens argument form:

The Modus Ponens Rule

From any sentences **P** ⊃ **Q** *and* **P**, *one may infer the sentence* **Q**.

The Modus Ponens rule ("MP" for short) tells us that from a conditional and the antecedent of that conditional, we may infer the consequent of the conditional. To abbreviate this rule, we write:

$$P \supset Q$$
$$\underline{P}$$
$$Q$$

Each of the following inferences instantiates the Modus Ponens form:

1. A ⊃ B	1. (A & B) ⊃ (E v F)	1. (A ≡ E)⊃(SvH)
2. A	2. (A & B)	2. (A ≡ E)
3. B	3. (E v F)	3. (S v H)

We also naturally think in accord with this inference rule from an early age. For instance, suppose a little child is told that if the family stays home and works in the yard today, then everyone gets to go on a picnic tomorrow. After the yard is finished, nobody needs to tell the child what inference to draw. The child expects to go on a picnic.

The Modus Tollens Pattern The following argument is valid, although it may not appear to be valid at first glance:

1. If it's raining, then the roof is wet.
2. The roof is not wet.
3. So, it is not raining.

In TL, with obvious abbreviations, this is:

1. R⊃W
2. ~W
3. ~R

If the two premises are both true, then the conclusion must be true. This argument is an instance of the argument form known as *Modus Tollens*:

$$P \supset Q$$
$$\underline{\sim Q}$$
$$\sim P$$

This form is proven valid by the following table:

PQ	P⊃Q	~Q	~P
TT	T	F	F
TF	F	T	F
FT	T	F	T
FF	T	T	T

'Modus Tollens' is Latin for 'method of denial'. The form gets its name from the fact that the second premise is a denial of the consequent of the first premise. This valid argument form gives rise to the following valid inference rule:

The Modus Tollens Rule

From any sentences **P ⊃ Q** and **~Q**, one may infer the sentence **~P**.

The Modus Tollens rule ("MT" for short) tells us that from a conditional along with the negation of the consequent of the conditional, we may infer the negation of the antecedent. The abbreviation of the rule is:

$$P \supset Q$$
$$\underline{\sim Q}$$
$$\sim P$$

Here are some instances of the MT argument form:

1. A ⊃ B	1. (A&B) ⊃ (E&F)	1. ~(A&B) ⊃~(SvH)
2. ~B	2. ~(E&F)	2. ~~(SvH)
3. ~A	3. ~(A&B)	3. ~~(A&B)

We do often reason this way. Imagine there are reports that the King of Ruritania has stepped down and has fled the country. Furthermore, the government and state-run media are rumored to be in the hands of his opponents. Now, assume it is true that:

> If the King remains in power, he will appear on the 6 o'clock state television program tonight.

Assume next that you turn on your TV and the King is not on the 6 p.m. state television program. It definitely follows that the King no longer remains in power. This natural inference is an example of Modus Tollens.

The Hypothetical Syllogism Pattern In everyday reasoning, we often link several things together in a chain of 'if, then' links. For instance:

1. If it rains, then the roof gets wet.
2. If the roof gets wet, then the ceiling leaks.
3. So, if it rains, then the ceiling leaks.

This argument instantiates the natural and valid argument form named *Hypothetical Syllogism*:

$$P \supset Q$$
$$\underline{Q \supset R}$$
$$P \supset R$$

The validity of this form is easily shown:

PQR	P⊃Q	Q⊃R	P⊃R
TTT	T	T	T
TTF	T	F	F
TFT	F	T	T
TFF	F	T	F
FTT	T	T	T
FTF	T	F	T
FFT	T	T	T
FFF	T	T	T

Notice that the table does not contain an invalidating row.

The inference rule corresponding to this argument form is the following:

The Hypothetical Syllogism Rule

From sentences $P \supset Q$ and $Q \supset R$, one may infer the sentence $P \supset R$.

The Hypothetical Syllogism rule ("HS" for short) tells us that if we begin with two conditionals that are such that the consequent of the first one matches the antecedent of the second, we may infer a third conditional whose antecedent is the antecedent of the first conditional and whose consequent is the consequent of the second conditional. In abbreviated form, the rule is:

$$P \supset Q$$
$$\underline{Q \supset R}$$
$$P \supset R$$

Each of the following instantiates the HS form:

1.	$A \supset B$	1.	$(A\&B) \supset (EvF)$	1.	$\sim(AvB) \supset \sim(S\&H)$
2.	$\underline{B \supset C}$	2.	$\underline{(EvF) \supset (G\&H)}$	2.	$\underline{\sim(S\&H) \supset R}$
3.	$A \supset C$	3.	$(A\&B) \supset (G\&H)$	3.	$\sim(AvB) \supset R$

●— A NATURAL DEDUCTION PROOF

Let us now see how these rules may be used. Consider the following argument:

> Either Annette will swim or Bobby will swim. But Annette won't swim. If Bobby swims, then Darlene will swim. And if Darlene will swim then Ernest won't swim. If Henrietta will be the lifeguard, then Ernest will swim. Therefore, Henrietta will not be the lifeguard.

Using obvious abbreviations, this argument may be expressed within TL as:

1. $A \vee B$
2. $\sim A$
3. $B \supset D$
4. $D \supset \sim E$
5. $\underline{H \supset E}$
   $\sim H$

Is the argument valid? That is, if the five premises are true, must the conclusion be true? Well, one way to find out would be to run a truth-table test on the argument. However, the table for this argument would have 16 rows. There's another way to proceed, one that is closer to the way we naturally reason. Think through the following series of inferences.

Step 1: Look at the first two premises of the argument. What logically follows from just these two sentences? From the sentence A v B and the sentence ~A, the sentence B validly follows according to the Disjunctive Syllogism rule. That is, if we apply the Disjunctive Syllogism rule to premises 1 and 2, the rule allows us to derive the sentence B. The inference from A v B and ~A, to B is therefore a valid inference.

Step 2: We just deduced sentence B from the first two premises. Now, consider premises 3 and 4. What validly follows from these two? From the sentence B ⊃ D, and sentence D ⊃ ~E, B ⊃ ~E validly follows according to the Hypothetical Syllogism rule. That is, if we apply the Hypothetical Syllogism rule to premises 3 and 4, we are allowed to derive B ⊃ ~ E. We've now deduced B ⊃~E from the premises.

So far, the line of reasoning we've developed is the following:

1.   AvB
2.   ~A
3.   B⊃D
4.   D⊃~E
5.   H⊃E
_____
6.   B   (from lines 1, 2 according to DS)
7.   B⊃~E (from lines 3, 4 according to HS)

When we write down the rule by which an inference is made and the lines the rule was applied to when we made the inference, we are stating the *justification* for the inference. The justifications for the inferences made at lines (6) and (7) are written in parentheses to the right of those two lines.

Step 3: Next, consider lines 6 and 7. From the sentences B and B ⊃ ~E , what follows? Sentence ~E follows according to the Modus Ponens rule. That is, the application of the Modus Ponens rule to lines 6 and 7 results in the deduction of the sentence ~E. Therefore, the inference from sentence B and sentence B ⊃ ~E, to sentence ~E, is a valid inference.

We've now deduced ~E from the premises:

1.   A v B
2.   ~A
3.   B ⊃ D
4.   D ⊃ ~E
5.   H ⊃ E
_____
6.   B   (from 1,2 by DS)
7.   B ⊃ ~E   (from 3,4 by HS)
8.   ~E   (from 6,7 by MP)

Step 4: Finally, notice that lines 5 and 8 match the premise section of the Modus Tollens rule. If we apply Modus Tollens to lines 5 and 8, we may derive ~H. Thus, from lines 5 and 8, ~H validly follows according to Modus Tollens:

9.   ~H   (from 5,8 by MT)

The above series of valid inferences may be summarized thus:

1. A v B
2. ~A
3. B ⊃ D
4. D ⊃ ~E
5. H ⊃ E
———————————————
6. B   (from 1,2 by DS)
7. B ⊃ ~E   (from 3,4 by HS)
8. ~E   (from 6,7 by MP)
9. ~H   (from 5,8 by MT)

We began with the premises of the argument and, following only valid rules of inference, reasoned to the argument's conclusion. This *proves* that the argument is a valid argument. Let us see why. If the premises of the argument—lines 1 through 5—are true, then certainly lines 1 and 2 are true, since 1 and 2 are premises of the argument. If lines 1 and 2 are true, then line 6 must be true, since 6 follows from 1 and 2 according to a valid rule of inference. So, if 1 through 5 are true, then 6 must be true. If 1 through 6 are true, 7 must be true, since 7 follows from just lines 3 and 4 by a valid rule of inference. If 1 through 7 are true, then 8 must be true, since 8 follows from 6 and 7 by a valid rule of inference. Finally, if 1 through 8 are true, then 9, which is the argument's conclusion, must be true, since 9 follows from just 5 and 8 by a valid rule of inference. Therefore, if the premises are true, then the conclusion must be true, which is to say that the argument is a valid argument. In general:

> *If the conclusion of an argument is deduced from the premises, using only valid rules of inference, then the argument must be valid.*

Before going any further, we should remind ourselves of something. Why formulate valid inference rules? What is their value? If our inferences accord with valid inference rules, our arguments will be valid. Valid inference rules, then, are guides to correct reasoning. But why is correct reasoning important? It is by drawing correct inferences that we discover truths about the world. Since truth is important, it is important that our inferences be correct inferences.

●— THE SYSTEM TD

In this chapter, you will learn how to use a *natural deduction system* to prove truth-functional arguments valid and to prove sentences tautological. A natural deduction system consists of a formal language plus a set of deduction rules. The natural deduction system TD (for "truth-functional deduction"), which we are going to work with here, consists of two parts:

i. The language TL
ii. A set of deduction rules.

The set of deduction rules for TD consists of the inference rules and replacement rules presented in this chapter.

A *proof in TD* is a sequence of sentences of TL each of which is either a premise or an assumption or follows from one or more previous sentences according to a TD inference or replacement rule, and in which (i) every line (other than a premise) has a justification and (ii) any assumptions have been discharged. (The use of assumptions will be explained later in this chapter.) The conclusion of a proof is the last line of the proof. An argument is *valid in TD* if and only if there exists a proof in TD whose premises are the premises of the argument and whose conclusion is the conclusion of the argument. An argument is *invalid in TD* if and only if it is not valid in TD. Recall that an argument is *truth-functionally valid* just in case it can be proven valid using nothing more than the methods of truth-functional logic. Every argument that is valid in TD also counts as truth-functionally valid, since TD employs the methods of truth-functional logic.

●— A HELPFUL CONCEPT: THE OPPOSITE OF A FORMULA

It will be helpful at this point to introduce the idea of the *opposite* of a formula. If **P** is a sentence of TL, then the opposite of **P** is the sentence ~**P** and the sentence ~**P** is the opposite of the sentence **P**. Another way to put this is:

> *If a formula has a tilde as its main connective, the opposite of the formula is the formula with the tilde removed. If a formula does not have a tilde as its main connective, the opposite of the formula is the formula with a tilde applied to the whole.*

For example, A is the opposite of ~A, and ~A is the opposite of A. Also, ~(A&B) is the opposite of (A&B), and (A&B) is the opposite of ~(A&B). Here are additional examples:

FORMULA:	OPPOSITE:
(AvB)	~(AvB)
(~A&~B)	~(~A&~B)
~(A⊃B)	(A⊃B)
~~(H&G)	~(H&G)
~(AvB)	(AvB)
~(~Av~B)	(~Av~B)
[(H&B)v(SvG)]	~[(H&B)v(SvG)]

Disjunctive Syllogism and Modus Tollens may now be reformulated as follows:

The Disjunctive Syllogism Rule

> *From a disjunction and another sentence that is the opposite of one of the disjuncts, one may infer the other disjunct.*

In each of the examples below, the sentence below the line was derived from the sentences above the line in accord with this rule.

1.	~Av~B	1.	(A&B)v~(E≡F)	1.	~Hv~S	1.	AvB
2.	B	2.	(E≡F)	2.	H	2.	~B
3.	~A	3.	(A&B)	3.	~S	3.	A

The Modus Tollens Rule

From a conditional sentence and another sentence that is the opposite of the consequent, one may infer the opposite of the antecedent.

For example:

1.	~B⊃~E	1.	~(AvB)⊃~(H&S)	1.	P⊃Q
2.	E	2.	(H&S)	2.	~Q
3.	B	3.	(AvB)	3.	~P

As you will soon see, it is easier to work with Modus Tollens and Disjunctive Syllogism when they are stated in terms of the concept of the opposite of a formula.

●— SOME EXAMPLES OF PROOFS USING THE FIRST FOUR RULES

We are finally ready to begin constructing proofs. The following are examples of proofs employing the first four inference rules. In each of the following proofs, the sentences above the line are the premises and the sentence after the slash is the argument's conclusion. Each line derived from the premises is written under the line and the justifications are written to the right of each derived line.

(1)
1. P⊃Q
2. Q⊃R
3. S⊃P / S⊃R
4. P⊃R   HS 1,2
5. S⊃R   HS 3,4

(2)
1. ~A
2. AvB
3. B⊃S / S
4. B   DS 1,2
5. S   MP 3,4

(3)
1. PvQ
2. ~H⊃~P
3. H⊃S
4. ~S / Q
5. ~H   MT 3,4
6. ~P   MP 2,5
7. Q   DS 1,6

(4)
1. A⊃(B⊃E)
2. S⊃(E⊃J)
3. FvA
4. ~F
5. S / B⊃J
6. A DS 3,4
7. B⊃E   MP 1,6
8. E⊃J   MP 2,5
9. B⊃J   HS 7,8

It is very important that you understand the following. When these inference rules are applied, they must be applied to *whole* lines only. For example, if MP is to be applied to two lines, one of those lines must—as a whole—be an instance of the form **P⊃Q** and the other line—as a whole—must be the antecedent of the first line. Notice that in proof (4), B⊃E and E⊃J on lines 1 and 2 "fit" the Hypothetical Syllogism pattern. However, when HS is applied, it must be applied to whole lines. Therefore, if we are to apply HS to B⊃E and E⊃J, those formulas cannot be parts of bigger formulas; rather, each must constitute an entire line by itself. Consequently, in proof (4) we first used Modus Ponens to detach B⊃E from line 1 and bring it down to line 7, where it sits by itself. We then used Modus Ponens to bring E⊃J down onto its own line, line 8. Then, once B⊃E and E⊃J were situated on lines of their own, Hypothetical Syllogism was applied to those two lines to derive the conclusion, B⊃J.

● EXERCISE 3.1 ●

The proofs in this exercise require the use of only the first four inference rules.

Part I The following proofs are complete except that the justifications have not been filled in. Supply justifications by filling in the blanks with the appropriate rules and line numbers.

(1) 1. P ⊃ (Q & R)
     2. (Q & R) ⊃ S
     3. (P ⊃ S) ⊃ ~G
     4. G v H / H
     5. P ⊃ S   _____
     6. ~G   _____
     7. H   _____

(2) 1. ~P v ~ (Q v R)
     2. S ⊃ P
     3. H v S
     4. ~H
     5. ~(Q v R) ⊃ A / A
     6. S   _____
     7. P   _____
     8. ~(Q v R)   _____
     9. A   _____

(3) 1. P ⊃ (S & Q)

(1) 2. ~P ⊃ (B v R)
     3. ~(S & Q)
     4. (B v R) ⊃ (Q v H) / Q v H
     5. ~P   _____
     6. B v R   _____
     7. Q v H   _____

(4) 1. S v ~ (H & S)
     2. S ⊃ ~G
     3. P ⊃ G
     4. M ⊃ (H & S)
     5. P / ~M
     6. G   _____
     7. ~S   _____
     8. ~(H & S)   _____
     9. ~M   _____

(5) 1. (H & S) ⊃ ~(F v G)
     2. M v (H & S)

3. M ⊃ R

4. <u>~R</u> / ~(F v G)

5. ~M _____

6. H & S _____

7. ~(F v G) _____

(6) 1. (H & G) ⊃ (F ⊃ R)

2. P & Q

3. A v ~(F ⊃ R)

4. <u>~A</u> / ~(H & G)

5. ~(F ⊃ R) _____

6. ~(H & G) _____

(7) 1. ~H v ~Z

2. ~(S ≡ G) ⊃ (F v S)

3. ~~H

4. <u>~(F v S) ⊃ Z</u> / ~~(F v S)

5. ~Z _____

6. ~~(F v S) _____

(8) 1. A ⊃ (P v S)

2. B ⊃ ~P

3. S ⊃ (G ⊃ H)

4. H ⊃ R

5. A

6. <u>B</u> / G ⊃ R

7. P v S _____

8. ~P _____

9. S _____

10. G ⊃ H _____

11. G ⊃ R _____

(9) 1. (A & B) ⊃ ~S

2. A & B

3. S v [H ⊃ (F ⊃ M)]

4. ~M

5. <u>H</u> / ~F

6. ~S _____

7. H ⊃ (F ⊃ M) _____

8. F ⊃ M _____

9. ~F _____

(10) 1. ~A

2. <u>~A⊃[~A⊃(E⊃A)]</u> / ~E

3. ~A⊃(E⊃A) _____

4. E⊃A _____

5. ~E _____

(11) 1. A⊃B

2. B⊃R

3. <u>~(A⊃R)vG</u> / G

4. A⊃R _____

5. G _____

(12) 1. P

2. ~G

3. P⊃(A⊃~B)

4. <u>Gv(~B⊃E)</u> / A⊃E

5. A⊃~B _____

6. ~B⊃E _____

7. A⊃E _____

(13) 1. (A&B)v(E⊃S)

2. G⊃E

3. <u>~(A&B)</u> / G⊃S

4. E⊃S _____

5. G⊃S _____

(14) 1. AvE

2. ~A

3. <u>E⊃[~A⊃(E⊃S)]</u> / S

4. E   _____

5. ~A⊃(E⊃S) _____

6. E⊃S   _____

7. S   _____

(15) 1. (S&E)⊃~G

2. H⊃(S&E)

3. ~(H⊃~G)v~R

4. <u>~B⊃R</u> / B

5. H⊃~G _____

6. ~R   _____

7. B   _____

(16) 1. H⊃S

2. S⊃~R

3. ~B⊃F

4. <u>(H⊃~R)⊃(L⊃~B)</u> / L⊃F

5. H⊃~R _____

6. L⊃~B _____

7. L⊃F _____

(17) 1. H

2. H⊃(FvS)

3. <u>(FvS)⊃(H⊃B)</u> / B

4. H⊃(H⊃B) _____

5. H⊃B   _____

6. B   _____

(18) 1. Av(B⊃G)

2. Gv~A

3. ~G

4. <u>G⊃R</u> / B⊃R

5. ~A _____

6. B⊃G _____

7. B⊃R _____

(19) 1. (AvB)⊃(E&S)

2. (E&S)⊃G

3. H⊃(AvB)

4. <u>M⊃~(H⊃G)</u> / ~M

5. (AvB)⊃G _____

6. H⊃G   _____

7. ~M   _____

(20) 1. (A⊃B)⊃(E⊃A)

2. (E⊃B)⊃[(E⊃~H)⊃(J⊃A)]

3. ~H⊃B

4. <u>A⊃~H</u> / J⊃B

5. A⊃B   _____

6. E⊃A   _____

7. E⊃~H   _____

8. E⊃B   _____

9. (E⊃~H)⊃(J⊃A) _____

10. J⊃A   _____

11. J⊃B   _____

Part II Each of the following symbolized arguments is valid. Provide a proof of each.

(1) 1. (A & B) ⊃ S

2. H ⊃ R

3. <u>(A & B)</u> / S

(2) 1. R v G

2. H & (P ⊃ Q)

3. <u>~G</u> / R

(3) 1. ~(F ≡ S)
    2. R v M
    3. ~(F ≡ S) ⊃ D / D

(4) 1. ~(A & B)
    2. R ⊃ (A & B)
    3. H / ~R

(5) 1. F ⊃ ~ (H & B)
    2. A ⊃ (H & B)
    3. A / ~F

(6) 1. A v (B & C)
    2. A ⊃ F
    3. ~(B & C) / F

(7) 1. P v S
    2. ~P
    3. ~S v G / G

(8) 1. ~(A & B) v ~(E ≡ F)
    2. H ⊃ (A & B)
    3. E ≡ F / ~H

(9) 1. F ⊃ S
    2. S ⊃ G
    3. (F ⊃ G) ⊃ M / M

(10) 1. R ⊃ H
     2. H ⊃ S
     3. S ⊃ G
     4. (R ⊃ G) ⊃ F / F

(11) 1. ~F ⊃ ~S
     2. ~S ⊃ ~G
     3. H v ~(~F ⊃ ~G) / H

(12) 1. B ⊃ (A & G)
     2. R ⊃ B
     3. [R ⊃ (A & G)] ⊃ S/S

(13) 1. (H v B) ⊃ ~(S ≡ F)
     2. R ⊃ (H v B)
     3. ~(S ≡ F) ⊃ Q / R ⊃ Q

(14) 1. P ⊃ Q
     2. Q ⊃ ~R
     3. P
     4. A ⊃ ~B
     5. A
     6. R v (B v S)
     7. S ⊃ L / L

(15) 1. A ⊃ (B ⊃ E)
     2. A
     3. ~E / ~B

(16) 1. A ⊃ B
     2. B ⊃ E
     3. B ⊃ R
     4. ~R
     5. B v A / E

(17) 1. Q v A
     2. (Q v A) ⊃ S
     3. ~P ⊃ [P v (Q ⊃ ~S)]
     4. S ⊃ [~S v (P ⊃ ~S)]/A

(18) 1. A ⊃ B
     2. B ⊃ W
     3. ~W
     4. P ⊃ Q
     5. S ⊃ ~Q
     6. ~A ⊃ (P v Z)
     7. S / Z

Part III Symbolize the following English arguments and then use natural deduction to prove each valid.

1. If Ann goes swimming, then Bob will go swimming. Either Ann or Chris will go swimming. Bob won't go swimming. So, Chris will go swimming. (A, B, C)

2. If Ann goes swimming, then Bob will go swimming. If Bob goes swimming, then Chris will go swimming. But Chris won't swim. So, Ann won't swim. (A, B, C)

3. If either Ann or Bob is home, then it's the case that either Bob or Chris is home. Either it's the case that if Bob is home then Chris is home, or Ann is home. If it's the case that if Bob is home then Chris is home, then either Ann or Bob is home. Ann is not home. So, either Bob or Chris is home. (A, B, C)

4. If either Pat or Quinn works late, then Rita and Winona will work late. If it's the case that Herman is on vacation, then either Pat or Quinn will work late. So, if Herman is on vacation, then Rita and Winona will work late. (P, Q, R, W, H)

5. If Ralph tells Alice he's sorry, then Alice will be forgiving. If Norton has a talk with Ralph, then Ralph will tell Alice he's sorry. Alice will not be forgiving. So, Norton won't have a talk with Ralph. (R, A, N)

6. If Pat gets a ticket, then if Pat has an accident then Pat's insurance will be canceled. Pat's insurance won't be canceled. If Pat doesn't slow down, then Pat will get a ticket. Either Pat's insurance will be canceled or Pat won't slow down. So, Pat won't have an accident. (T, A, I, S)

7. Elroy won't go jogging. If Lulu jogs, then Chris will jog. Either Lulu will jog or Elroy will jog. Therefore, Chris will jog. (E, L, C)

8. If Donovan doesn't perform tonight, then either the Jefferson Airplane or Fever Tree will perform. Fever Tree will not perform. If the Jefferson Airplane performs then Suzie will get to hear her favorite song. Donovan won't perform tonight. So, Suzie will get to hear her favorite song tonight. (D, J, F, S)

9. If it's the case that if Ann swims then Bubba will swim, then Chris will swim. Either Dave won't swim or Ann will swim. If it's not the case that Dave swims, then if Ann swims then Bubba will swim. Ann won't swim. So, Chris will swim. (A, B, C, D)

10. Ann and Bob won't both sing. Either Chris or Rob will sing or Mary will sing. If Lucy doesn't sing, then it's not the case that either Chris or Rob will. Either Ann and Bob will both sing or Lucy won't sing. Consequently, Mary will sing. (A, B, C, R, M)

11. If Wimpy eats a hamburger, then Popeye will eat another can of spinach. Either Olive Oyl will eat a peach or Popeye won't eat another can of spinach. Olive Oyl won't eat a peach so Wimpy won't eat a hamburger. (W, P, O)

12. Either Gilligan won't show up for dinner or either Mr. Howell will be upset or Mrs. Howell won't show up for dinner. If Gilligan doesn't show up for dinner, then Mr. Howell will be upset. Mr. Howell won't be upset. So, Mrs. Howell won't show up for dinner. (G, H, M)

13. If Moe makes a face, then Curly will make a face. If it's the case that if Moe makes a face then Larry won't make a face, then it follows that if Curly makes a face then the director will not be happy. If Curly makes a face, then Larry won't make a face. So, if Moe makes a face then the director will not be happy. (M, C, L, D)

14. Either Hogan bribes Schultz or if Schultz tells Klink then General Burkhart will find out. If it's the case that if Schultz tells Klink then General Burkhart will find out, then if General Burkhart finds out, then Hogan will get in big trouble. Hogan won't bribe Schultz. And Hogan won't get in big trouble. So, Schultz won't tell Klink. (H, S, B, T)

15. If Gilligan is late, then the skipper will be worried about his little buddy. If the skipper is worried about his little buddy, then the skipper won't eat. If the skipper won't eat, then Mrs. Howell will become concerned. If it's the case that if Gilligan is late then Mrs. Howell will become concerned, then the professor will argue that Mrs. Howell's peace of mind is dependent on Gilligan's punctuality. So, the professor will argue that Mrs. Howell's peace of mind is dependent on Gilligan's punctuality. (G, S, E, M, P)

●— FOUR MORE RULES OF INFERENCE

There is no limit to the number of valid inference rules that can be formulated. However, as you will see later on, many of these rules are redundant. That is, once we have formulated a certain number of valid rules to start with, adding additional rules will merely allow inferences that the original group of rules already allow. For practical purposes, and as far as proofs of validity are concerned, a surprisingly small number of rules are all that will ever be needed. Here are four more valid rules of inference:

The Simplification Rule   The following rule, called the *Simplification rule* (or "Simp" for short) is obviously a valid inference rule:

> From a sentence $P \& Q$, one may infer P. Also, from a sentence $P \& Q$, one may infer Q.

This rule may be abbreviated in two parts as:

$$\frac{P\&Q}{P} \qquad \frac{P\&Q}{Q}$$

In each of the following examples, the sentence below the line has been derived from the sentence above the line in accord with the Simplification rule.

$$\frac{A \& B}{A} \qquad \frac{A \& B}{B} \qquad \frac{(AvB) \& (EvF)}{(EvF)} \qquad \frac{\sim A \& \sim B}{\sim A}$$

This rule corresponds to the two valid argument forms:

$$\frac{P \& Q}{P} \qquad \frac{P \& Q}{Q}$$

The Conjunction Rule The next rule, called the *Conjunction rule* (or "Conj" for short) is also obviously a valid rule of inference:

> From any sentence **P** and another sentence **Q**, one may infer the sentence **P & Q**.

In abbreviated form, this is:

$$\begin{array}{c} P \\ \underline{Q} \\ P\&Q \end{array}$$

In each of the following examples, the Conjunction rule applied to the sentences above the line permits the deduction of the sentence below the line.

1.	A	1.	(AvB)
2.	B	2.	(E&F)
3.	A&B	3.	(AvB)&(EvF)

1.	~A
2.	~B
3.	~A&~B

This rule corresponds to the valid argument form

$$\begin{array}{c} P \\ \underline{Q} \\ P\&Q \end{array}$$

The Addition Rule The next rule, the *Addition rule*, strikes many as invalid at first:

> From any sentence **P**, one may infer **P v Q**.

That is, from a sentence **P**, one may infer the disjunction of **P** with any sentence **Q**. For example, from B one may infer BvG, one may infer Bv(G&D), one may infer BvJ, and so on.

In each of the following, the Addition rule has been applied to the sentence above the line in order to derive the sentence below the line.

$$\frac{A}{AvH} \qquad \frac{A}{AvS} \qquad \frac{(A\&B)}{(A\&B)v(H\&S)} \qquad \frac{J{\supset}E}{(J{\supset}E)vW} \qquad \frac{B}{Bv{\sim}R}$$

Abbreviated, the Addition rule (or "Add" for short) becomes:

$$\frac{P}{PvQ}$$

If this rule doesn't seem valid to you, think of it this way. Imagine you are on a game show and you will win a 24-volume series on the history of symbolic logic if you correctly guess the truth-value of the disjunction behind the curtain. You are allowed only one piece of information: the left disjunct of the disjunction is a true sentence. Now, what's the truth-value of the disjunction as a whole? Clearly, the disjunction must be true, for if one disjunct is true, then no matter what the truth-value of the other disjunct, the whole disjunction must be true. Thus, if a sentence **P** is true, then **P** disjoined to any sentence **Q** produces a true sentence. In other words, if a sentence **P** is true then **PvQ** is true as well.

The Constructive Dilemma Rule   The next rule, *Constructive Dilemma*, is the most difficult to apply of our first eight rules. First, consider the following argument:

1. If it rains, then the roof will get wet.
2. If it snows, then the lawn will get white.
3. It will either rain or snow.
4. So, either the roof will be wet or the lawn will be white.

If we symbolize this English argument in TL we get:

1. R⊃W
2. S⊃L
3. <u>RvS</u>
4. WvL

This TL argument is an instance of the following argument form:

$$\begin{array}{l} P{\supset}Q \\ R{\supset}S \\ \underline{PvR} \\ QvS \end{array}$$

A truth-table test will reveal this to be a valid form. This gives rise to the Constructive Dilemma rule:

> From any sentence $P \supset Q$, another sentence $R \supset S$, and a third sentence $P \lor R$, one may infer the sentence $Q \lor S$.

Each of the following arguments is an instance of the Constructive Dilemma form:

A⊃B	H⊃(G&B)	R⊃~S
E⊃F	S⊃(A&M)	H⊃~G
AvE	HvS	RvH
BvF	(G&B)v(A&M)	~Sv~G

Notice that the Constructive Dilemma rule (or "CD" for short) requires the presence of three things:

1. a conditional
2. another conditional
3. a disjunction whose left disjunct is the antecedent of the one conditional and whose right disjunct is the antecedent of the other conditional.

From this, you are permitted to infer a disjunction whose left disjunct is the consequent of the one conditional and whose right disjunct is the consequent of the other conditional.

● PROOFS USING THE EIGHT RULES

We now have eight valid inference rules to work with. Here are some examples of proofs using various combinations of these rules.

5. 1. P⊃H
   2. (HvC)⊃(D&E)
   3. P /D
   4. H MP 1,3
   5. HvC   Add 4
   6. D&E   MP 2,5
   7. D Simp 6

6. 1. F⊃L
   2. (M⊃N)&S

   3. N⊃T
   4. FvM / LvT
   5. M⊃N   Simp 2
   6. M⊃T   HS 3,5
   7. LvT   CD 4,1,6

7. 1. ~M&N
   2. P⊃M
   3. Q&R
   4. (~P&Q)⊃A /AvT

5. ~M   Simp 1
6. ~P   MT 2,5
7. Q   Simp 3
8. ~P&Q   Conj 6,7
9. A   MP 4,8
10. AvT   Add 9

8. 1. T⊃G
2. A
3. ~A / G
4. AvT   Add 2
5. T   DS 3,4
6. G   MP 1,5

Some Advice on Learning to Construct Proofs   When first learning to work proofs, it is natural to wonder if there are mechanical rules we can follow that will tell us how to begin and just what steps to take. After all, there are mechanical rules—decision procedures—that tell us precisely how to construct a truth-table. Decision procedures for the construction of proofs can be formulated, but they are so complex that they are extremely difficult to work with. Consequently, there is no practical set of step-by-step instructions telling you precisely and mechanically how to construct any and every valid proof. Constructing a proof is therefore something of an art or a skill. It requires ingenuity and creativity. Since skills typically improve with practice, you can expect your abilities to improve as you complete more and more proofs.

The first step in learning to build proofs is this: Develop an ability to spot the inference rule patterns when you look at the premises of an argument. Once you recognize a pattern, you will be able to infer the next line in the proof. And once you have inferred or brought down something, a new pattern will usually form, enabling you to bring down another step, which will probably form a new pattern and make possible another step. Eventually, this should take you to the conclusion. This process involves the mental skill of *pattern recognition*. This is a skill that usually develops only with practice.

The second step in learning to build proofs is this: Learn to apply the rules accurately. The rules must be applied very precisely or else incorrect sequences of formulas will result. If you apply a rule incorrectly at, say, step 3 and bring down the wrong thing, this can throw off every subsequent step in your proof.

The third step in the process of learning to build proofs is this: Learn to strategize. Someone who is good at strategizing can look at an argument for a few minutes and plan out in advance the steps that will likely lead to the conclusion. This is a skill that only develops after extended practice and once one fully understands the natural deduction process. Strategy is discussed in the next section.

Ultimately, the only way to become skilled at constructing proofs is to practice them on your own. Most people find logic proofs a bit overwhelming at first. However, if you refuse to get discouraged, and if you put some work into it, you will find that your ability to complete a proof increases with each one that you "solve." You will also learn a lot by simply watching as proofs are worked on the board in class.

●— PROOF STRATEGIES

When constructing proofs for the first time, it is natural to ask: Where do I start? How do I know which steps to take? Since there are no practical decision procedures that tell you how to construct a proof, you will have to proceed by a combination of trial and error, intuition, pattern recognition, general problem-solving skills, and a strategy we will call "working backward from the conclusion." Although they are not decision procedures, the following *strategies* will help you construct successful proofs.

The Trial and Error Strategy Suppose you begin a proof and are not sure what to start with or what direction to go. Look for any of the inference rule patterns among the premises. When you find one, make the required move, whether or not it seems to take you to the conclusion. Often, if you make enough moves, even though you have no overall plan or direction in mind, you will eventually arrive at the conclusion. Call this the *trial and error strategy*. For example, consider the following argument:

1. $(A \lor E) \supset {\sim}G$
2. $G \lor (R \& H)$
3. $(A \lor E)$ /H

Once you have become familiar with the inference rule patterns, you should be able to spot the MP pattern on lines 1 and 3. This gives you:

1. $(A \lor E) \supset {\sim}G$
2. $G \lor (R \& H)$
3. $A \lor E$ /H
4. ${\sim}G$ MP 1,3

Next, look for another pattern. Notice that lines 2 and 4 instantiate the premise section of DS. Thus:

1. $(A \lor E) \supset {\sim}G$
2. $G \lor (R \& H)$
3. $A \lor E$ /H
4. ${\sim}G$ MP 1,3
5. $R \& H$ DS 2,4

Finally, notice that line 5 fits the premise section of Simp, which allows you to infer the conclusion:

6. H Simp 5

If you are proceeding by the trial and error strategy, then the one most important piece of advice is this:

When in doubt: Do It.

If you spot a pattern and see a "move" you could make, but you don't see where it would take you, don't hesitate . . . just *do it*! Once you infer another line, another pattern will probably open up and you will see something else you can do, which will probably lead to another inference, which will probably lead to another, until you eventually reach the conclusion. We could also name the trial and error strategy the "Jerry Rubin" strategy. In 1970, Rubin wrote a book titled *Do It!* Although Rubin's book wasn't exactly a work of symbolic logic (it was advertised as the *Communist Manifesto* of the Yippie movement), and although Rubin probably didn't intend it, the title he chose does sum up a proof strategy for truth-functional natural deduction.

The Decomposition Strategy Another proof strategy may be called the *decomposition strategy*. Break larger sentences down into smaller sentences by deriving parts of the larger sentences. Use these smaller parts to derive additional lines, until you finally derive the conclusion. For example, consider the following symbolized argument:

1. (PvQ)⊃[(A&B)v(EvF)]
2. ~(EvF)
3. P / B

Let us begin by deriving something that we can use to break up line 1:

4. PvQ Add 3

Using this, plus MP, we can now bring down the consequent of 1:

5. (A&B)v(EvF) MP 1,4

Notice that line 2 is the opposite of the right disjunct of line 5. Consequently, we can use DS to break up 5 as follows:

6. A&B DS 2,5

The last move is obvious:

7. B Simp 6

Notice that in each step after line 4 we broke a larger sentence down by deriving one of its parts.

The Strategy of Working Backward from the Conclusion Let's suppose we are trying to prove the following symbolized argument valid:

1. H ⊃ (B ⊃ A)
2. ~S v B
3. ~H ⊃ E
4. S
5. ~E /A

Ultimately, we want to reach—at the bottom of the proof—the conclusion 'A'. Let's begin with the conclusion and then trace a series of steps backward to the first step in the proof. First, where in the premises is the conclusion formula? Notice that 'A' is at the end of line 1. How might we derive it? Well, on line 1 the sentence 'A' is the consequent of a conditional. What rule would let us derive the consequent of a conditional? Modus Ponens. What would MP require? First, the conditional must be all by itself on a line. Second, another line must consist of just the antecedent of the conditional. So, we will need to first bring the conditional B⊃A down onto a line by itself, and then we will need a line consisting of 'B' by itself. From those two lines, 'A' will follow. The inference would look like this:

$$\begin{array}{l} B \supset A \\ \underline{B} \\ A \end{array}$$

How might we "detach" or bring down $B \supset A$? Notice that on line (1), $B \supset A$ is itself the consequent of a conditional. If we could get the antecedent of that conditional on a line by itself, we could then use MP to bring down $B \supset A$. The inference would look like this:

$$\begin{array}{l} H \supset (B \supset A) \\ \underline{H} \\ B \supset A \end{array}$$

But how might we get H on a line by itself? First, find H in the premises. In line 3, ~H is the antecedent. Modus Tollens would let us infer H from line 3, if we also had on another line the opposite of the consequent of line 3. The inference would look like this:

$$\begin{array}{l} {\sim}H \supset E \\ \underline{{\sim}E} \\ H \end{array}$$

So we need '~E' on a line by itself. How will we get that? If you look closely at the proof, you will see ~E on line 5. Next, we will need to derive B in order to apply MP to $B \supset A$. How will we get B? Notice that B is the right disjunct of the disjunction on line 2. If we can derive or find, on a line by itself, the opposite of the left disjunct of that disjunction, we can then infer B, which we may then use to bring down our conclusion, A. The opposite of the left disjunct of line 2 is S. So, we will need S in order to break B out of line 2. Looking around, we find S on line 4. Putting this together in the proper order, we get:

1. $H \supset (B \supset A)$
2. $\sim S \vee B$
3. $\sim H \supset E$
4. S

5.	~E/A	
6.	H	MT 3,5
7.	B ⊃ A	MP 1,6
8.	B	DS 2,4
9.	A	MP 7,8

Approach a proof the way you would a game of strategy such as Chess or Monopoly. Constantly think about the various possibilities open to you at each step. Look ahead to where you want to be and think of ways to get there from where you are. In a game of strategy, a skilled player develops the ability to look several steps ahead and plans moves in terms of that goal.

Some Additional Suggestions Concerning Strategy Here are some additional suggestions:

1. If you have a conditional, try to find or derive the antecedent. Once you have the antecedent, apply MP and derive the consequent. Or, try to find or derive the opposite of the consequent. Once you have that, derive the opposite of the antecedent through an application of MT.

2. If you have a disjunction, try to find or derive the opposite of one disjunct. By applying DS to the disjunction and the opposite of one disjunct you may derive the other disjunct.

3. If your conclusion is a conjunction, try using the Conjunction rule.

4. Simplify conjunctions and use the parts thus derived to bring down other formulas.

5. If you have a disjunction and can't apply DS, try to spot a CD pattern. This would require two conditionals that have as antecedents the two disjuncts of the disjunction. CD would then allow you to infer the disjunction of the two consequents of the two conditionals.

6. If you have more than one conditional, watch for HS.

7. If you have two conditionals along with a disjunction, watch for CD.

8. If a letter or formula appears just once in the premises and doesn't link with anything else in the proof, it might be a "useless" component. Not every element of the premises must be used in the derivation of the conclusion.

9. If the conclusion contains an element not found in the premises, you will have to use Add to derive the conclusion.

10. A premise may be used more than once. Also, it is not always necessary to use every premise.

11. Try to isolate an atomic component on a line by itself. In many cases, if you can derive a single letter, or a single negated letter, you can use it to break down other lines in the proof.

● LOGICAL FALLACIES

Each of the inference rules we've used represents a valid argument form. There are also invalid argument forms, and since some of these appear deceptively similar to certain valid forms, invalid forms are often easily confused with valid forms. For instance, the following invalid argument form looks deceptively like the Modus Ponens form:

$$P \supset Q$$
$$\underline{Q}$$
$$P$$

The reader can easily verify the invalidity of this form by performing a simple truth-table test. This logical error or invalid form is named the *fallacy of affirming the consequent*. Be careful not to confuse Modus Ponens with the fallacy of affirming the consequent.

For example, consider this argument:

If Pete is a member of the mafia, then Pete is Italian.
Pete is Italian.
Therefore, Pete is a member of the mafia.

Symbolized in TL, this is an instance of the fallacy of affirming the consequent:

$$P \supset I$$
$$\underline{I}$$
$$P$$

In times past, individual Italian-Americans were discriminated against on the basis of this type of fallacious reasoning.

For another example, the following invalid argument form, called the *fallacy of denying the antecedent*, looks similar to Modus Tollens:

$$P \supset Q$$
$$\underline{\sim P}$$
$$\sim Q$$

Once again, the invalidity of this form may easily be proven with a simple truth-table. Be careful not to confuse this form with the Modus Tollens form.

For instance, consider the argument:

If Chris swims, then Pat will swim.
But Chris won't swim.
So, Pat won't swim.

Symbolized in TL, this is an instance of the fallacy of denying the antecedent:

$$C \supset P$$
$$\underline{\sim C}$$
$$\sim P$$

When arguing within a natural language, one must constantly be on the look-out for invalid or fallacious patterns of reasoning, because what appears at first glance to be a valid argument form sometimes turns out to be an invalid pattern of reasoning.

EXERCISE 3.2

Part I Using any of the first eight rules, supply justifications.

(1)
1. (A v B) ⊃ (G v S)
2. (G v S) ⊃ H
3. (H v S) ⊃ M
4. (A v E) ⊃ Q
5. A & R / M & Q
6. A _____
7. A v B _____
8. G v S _____
9. H _____
10. H v S _____
11. M _____
12. A v E _____
13. Q _____
14. M & Q _____

(2)
1. H ⊃ S
2. B
3. R
4. (B & R) ⊃ ~S
5. ~H ⊃ Z / Z v Q
6. B & R _____
7. ~S _____

8. ~H _____
9. Z _____
10. Z v Q _____

(3)
1. (H v P) ⊃ (S v W)
2. S ⊃ O
3. W ⊃ K
4. H / O v K
5. H v P _____
6. S v W _____
7. O v K _____

(4)
1. A v B
2. C
3. [(A v B) & C] ⊃ Q
4. S ⊃ ~Q
5. (~S v H) ⊃ X / X v E
6. (A v B) & C_____
7. Q _____
8. ~S _____
9. ~S v H _____
10. X _____
11. X v E _____

(5) 1. $(S \lor P) \supset Z$

2. $\sim Z \lor H$

3. $H \supset A$

4. $S \& G$

5. $A \supset M$

6. $O \supset N / M \lor N$

7. S _____

8. $S \lor P$ _____

9. Z _____

10. H _____

11. A _____

12. $A \lor O$ _____

13. $M \lor N$ _____

(6) 1. $(P \supset Q) \& (R \supset S)$

2. $A \lor (P \lor R)$

3. $\sim A \& [(Q \lor S) \supset \sim R] / P$

4. $\sim A$ _____

5. $P \lor R$ _____

6. $P \supset Q$ _____

7. $R \supset S$ _____

8. $Q \lor S$ _____

9. $(Q \lor S) \supset \sim R$ _____

10. $\sim R$ _____

11. P _____

(7) 1. $P \& R$

2. $(P \lor \sim S) \supset Q$

3. $(R \& Q) \supset H / H \& P$

4. P _____

5. $P \lor \sim S$ _____

6. Q _____

7. R _____

8. $R \& Q$ _____

9. H _____

10. $H \& P$ _____

(8) 1. $P \supset Q$

2. $P \lor R$

3. $R \supset (R \supset S)$

4. $\sim Q / S \lor Q$

5. $\sim P$ _____

6. R _____

7. $R \supset S$ _____

8. S _____

9. $S \lor Q$ _____

Part II Each of the following arguments is valid. Using the first eight inference rules, supply proofs.

(1) 1. $A \supset B$

2. $B \supset R$

3. $\sim R / \sim A \& \sim B$

(2) 1. $(A \lor B) \supset G$

2. A

3. $G \supset S / S$

(3) 1. $(H \& S) \supset \sim (F \equiv S)$

2. $B \supset (F \equiv S)$

3. H

4. $S / \sim B$

(4) 1. $A \& (R \supset S)$

2. $S \supset Q$

3. (R ⊃ Q) ⊃ O / O

(5) 1. B ⊃ (S v R)

2. S ⊃ P

3. R ⊃ G

4. H & B / P v G

(6) 1. H ⊃ ~(S v G)

2. ~(R v W) ⊃ (S v G)

3. F & H

4. R ⊃ (P & A)

5. W ⊃ M

6. ~M / A

(7) 1. H v ~S

2. [(H v ~S) v G] ⊃ M

3. ~M v R / R v X

(8) 1. ~F ⊃ ~S

2. H & S

3. F ⊃ B

4. ~B v G / G

(9) 1. F v B

2. [H & (F v B)] ⊃ ~S

3. Q & H / ~S

(10) 1. A ⊃ (P & S)

2. B ⊃ F

3. A

4. [(P & S) v F] ⊃ G / G

(11) 1. (H v S) ⊃ D

2. (R v M) ⊃ Q

3. A ⊃ ~(D & Q)

4. H & R / ~A

(12) 1. A v (B ⊃ C)

2. ~A ⊃ (~H ⊃ P)

3. ~H v B

4. ~A & Q / P v C

(13) 1. (A ⊃ B) & (A ⊃ C)

2. A

3. (B v C) ⊃ Z / Z

(14) 1. (S ⊃ I) & (S ⊃ P)

2. ~(I v P) v G

3. H & S / G

(15) 1. (A ⊃ B) & (S ⊃ Q)

2. P ⊃ R

3. (A v P) & (S v E)/BvR

(16) 1. (A ⊃ P) & X

2. F v (P ⊃ F)

3. ~F / ~A v Z

(17) 1. S ⊃ [P & (Q v S)]

2. [P v (A & B)]⊃(~N & ~O)

3. S & Z / ~N & P

(18) 1. (P ⊃ Q) & (R ⊃ S)

2. P v R

3. (Q ⊃ R) & (S ⊃ A)

4. ~(R v A) v E / E

(19) 1. (P ⊃ Q) & (~P ⊃ S)

2. Q ⊃ B

3. [(P ⊃ Q) & (Q ⊃ B)] ⊃
    [(P & B) v (~P & ~B)]

4. (P & B) ⊃ A

5. (~P & ~B) ⊃ H / A v H

(20) 1. (A & S) ⊃ Z

2. ~P ⊃ A

3. ~S ⊃ P

4. ~P & B / Z v Q

(21)　1.　A ⊃ B

　　　2.　~B & Q

　　　3.　~S v ~G

　　　4.　(~A & ~B) ⊃ [(~S ⊃ A) &
　　　　　(~G ⊃ X)] / X

(22)　1.　P v Q

　　　2.　~Q

　　　3.　(~Q v F) ⊃ (P ⊃ M)

　　　4.　(M v H) ⊃ (S ⊃ Q) / ~S

(23)　1.　(A v B) ⊃ [(E ⊃ G) &
　　　　　(P ⊃ Q)]

　　　2.　(A v E) ⊃ (E v P)

　　　3.　A & B / G v Q

(24)　1.　~A & B

　　　2.　~A & (P v ~Q)

　　　3.　(P ⊃ S) & (~Q ⊃ ~H)

　　　4.　(S ⊃ A) & (~H ⊃ ~Z) / ~Z

2. RULES OF REPLACEMENT

●— INTRODUCTION

We have been using rules that apply to *whole* lines of proofs. The rules that will be introduced in this section—*replacement rules*—may be applied to whole lines or to *parts* of lines.

　　Suppose we have placed a symbolized argument consisting of two premises and a conclusion on top of a truth-table:

P Q R	premise	premise	conclusion

Suppose further that the argument is valid. Then the table contains no invalidating row. That is, no row shows true premises and a false conclusion.

　　Now, suppose we replace a premise with a sentence that is logically equivalent to the one we removed. Will the reconstituted argument remain valid? Certainly. Since the replacement sentence is equivalent to the sentence that was removed, it follows that the replacement sentence's final column matches the final column of the sentence that was removed. Therefore, the columns under the three main connectives on the table must be unchanged. Consequently, if the argument was valid before the replacement, it will be valid after it as well.

　　Suppose we remove just a *subformula* of a premise and replace the subformula with a logically equivalent expression. A subformula is a wff that constitutes a part of a bigger formula. In the following examples, the subformulas are marked with brackets:

(P v ~Q) v (~P & Q)

(P v ~Q) ⊃ [P & (Q v R)]

However, in the following examples, the brackets do not mark off subformulas:

If a subformula is removed from a premise of a valid argument and replaced with an equivalent formula, will the argument remain valid? Clearly it will remain valid. Substituting one subformula for another, when the two are logically equivalent, won't affect the final columns of the argument's truth-table. If the argument was valid before the replacement, it will be valid afterward.

The general principle just demonstrated may be summarized thus:

> *If a subformula of a premise—or an entire premise—is removed from a valid argument and replaced by an equivalent expression, the reconstituted argument will remain valid.*

This gives rise to a new type of rule called a *replacement rule*.

●— THE FIRST FOUR REPLACEMENT RULES

The Commutative Rule   Any sentence PvQ is equivalent to the sentence QvP. In a disjunction, the order of the two disjuncts doesn't affect the truth-value of the disjunction. And certainly any sentence P&Q will be equivalent to the sentence Q&P. The order of the conjuncts doesn't affect the truth-value of the conjunction. Our first replacement rule, the *Commutative rule*, reflects these two truths. The Commutative rule is the combination of the following two rules:

— **The Commutative Rule of Disjunction** —●

$$(P \lor Q) \; // \; (Q \lor P)$$

— **The Commutative Rule of Conjunction** —●

$$(P \& Q) \; // \; (Q \& P)$$

The Commutative Rule of Disjunction should be understood as:

> *Anywhere in a natural deduction, a sentence **PvQ**—whether it is an entire line or a subformula of a line—may replace or be replaced by the sentence **QvP**.*

The replacement is made by rewriting the line and making the replacement as the line is being rewritten. The double slash mark in the formulation of the rule is to be interpreted as 'may replace or be replaced by', and it should be understood that the operation may be performed on a whole line or on a subformula of a line, at any point in a proof. This should become clear after some examples are examined.

Similarly, the Commutative Rule of Conjunction is to be understood as:

> *Anywhere in a natural deduction, a sentence **P&Q**—whether it is an entire line or a subformula of a line—may replace or be replaced by the sentence **Q&P**.*

And along with this goes the understanding that the operation may be performed on a whole line or on a subformula of a line at any point in a proof.

As you recall, when you apply one of the eight inference rules, the rule must be applied to an entire line. The replacement rules are different. There are two ways to apply a replacement rule: to an entire line, or to a part of a line, that is, to a subformula of a line. If we apply a replacement rule to an entire line, we simply rewrite the whole line, writing the replacement in place of the original line. For example, in the following proof, as we go from line 2 to line 3, the Commutative rule ("Comm" for short) is applied, and the entire line consisting of BvA is replaced with AvB:

1. (AvB)⊃E
2. BvA / E
3. AvB   Comm 2
4. E   MP 1,3

After AvB was derived at line 3, we applied MP to lines 1 and 3 to derive E.

If we apply a replacement rule to a part of a line—to a subformula—we rewrite the whole line and make the replacement. In the following proof, as we go from line 1 to line 3, line 1 is rewritten in its entirety. As it is rewritten, the subformula AvB inside line 1 is replaced by the subformula BvA:

1. (AvB)⊃E
2. BvA/ E
3. (BvA)⊃E Comm 1
4. E   MP 2,3

Notice that as we wrote in line 3, we rewrote all of line 1 and replaced AvB with BvA. The inference was justified by the Commutative rule. Thus, line 3 is exactly like line 1 except that the subformula AvB was replaced with the formula BvA.

In each of the following, the Commutative rule is applied to a whole line:

A v B	A & B	(E≡F)vG	~(A&B)&~E
B v A	B & A	Gv(E≡F)	~E&~(A&B)

However, in each of the following, the Commutative rule is applied to a subformula within a line:

$$(H\&R)\supset G \qquad (SvG)\&F$$
$$(R\&H)\supset G \qquad (GvS)\&F$$

The Associative Rule Any sentence **(PvQ)vR** is equivalent to the sentence **Pv(QvR)**, and any sentence **(P&Q)&R** is equivalent to the sentence **P&(Q&R)**. In a triple disjunction or triple conjunction, the placement of the parentheses does not affect the truth-value of the whole. This can easily be confirmed on a truth-table. Thus:

— **The Associative Rule for Disjunction** ⟶•

$$(PvQ)vR \; // \; Pv(QvR)$$

— **The Associative Rule for Conjunction** ⟶•

$$(P\&Q)\&R \; // \; P\&(Q\&R)$$

We'll lump these two rules together and call their combination the *Associative rule*. Thus, the Associative rule ("Assoc" for short) may be summarized as:

> *Anywhere in a natural deduction, a sentence* **(PvQ)v R**—*may replace or be replaced by the sentence* **Pv(Q v R)**, *and a sentence* **(P&Q)&R** —*may replace or be replaced by the sentence* **P&(Q & R)**.

The following inferences are permitted by the Associative rule:

(AvB)vE	A&(B&E)	[(A&B)v(E&S)]v(G&H)
Av(BvE)	(A&B)&E	(A&B)v[(E&S)v(G&H)]

Here is a proof employing this new rule:

1. (AvB)v E
2. ~A&~B/ E
3. Av(BvE) Assoc 1
4. ~A   Simp 2
5. BvE   DS 3,4
6. ~B   Simp 2
7. E   DS 5,6

Notice that at line 3 the Associative rule allowed us to shift the brackets in order to set things up for the application of DS.

If a pair of parentheses has a tilde attached, then the Associative rule does not apply. Thus, in the following case, you may not apply the Associative rule:

$$\sim(PvQ)vR$$

In sum, according to the Commutative rule, if the main operator is a wedge—or if it is an ampersand—then we can reverse the order of the components without affecting the truth-value of the formula. According to the Associative rule, if the operators are all ampersands or all wedges, then we may shift the brackets without affecting the truth-value of the formula.

There are no associative or commutative rules for the horseshoe. A truth-table will confirm that $(P{\supset}Q){\supset}R$ is not equivalent to $P{\supset}(Q{\supset}R)$ and that $P{\supset}Q$ is not equivalent to $Q{\supset}P$.

DeMorgan's Rule   In Chapter 2 you saw that any sentence $\sim(P\&Q)$ is equivalent to the sentence $\sim Pv\sim Q$. (To say "Pat and Quinn are not both home" is just to say "Either Pat is not home or Quinn is not home.") You also saw that any sentence $\sim(PvQ)$ is equivalent to the sentence $(\sim P\&\sim Q)$. (To say "It's not the case that either Pat or Quinn is home" is to say "Pat is not home and Quinn is not home.") These generalizations are reflected in the following four replacement rules:

$$\sim(P\&Q) \quad // \quad \sim Pv\sim Q$$
$$\sim(PvQ) \quad // \quad \sim P\&\sim Q$$
$$(P\&Q) \quad // \quad \sim(\sim Pv\sim Q)$$
$$(PvQ) \quad // \quad \sim(\sim P\&\sim Q)$$

These four rules display a common pattern. In general, you can convert an ampersand to a wedge or a wedge to an ampersand if you apply the following three-step algorithm:

> To replace a sentence containing a wedge with one containing an ampersand, or vice versa, make your replacement after completing the following three steps:
>
> 1. Change the ampersand to a wedge or the wedge to an ampersand.
> 2. Take the opposite of each side of the ampersand or wedge.
> 3. Take the opposite of the formula as a whole.

Augustus DeMorgan, a nineteenth-century mathematician, was the first to formulate the basic principle of this algorithm. It is therefore known today as *DeMorgan's rule*.

For example, let us apply DeMorgan's rule (or "DM" for short) to $\sim(\sim AvB)$. Here are the three steps:

> 1. Change the ampersand to a wedge or the wedge to an ampersand:
>
> $$\sim(\sim A\&B)$$

2. Take the opposite of each side of the ampersand or wedge:

~(A&~B)

3. Take the opposite of the formula as a whole:

A&~B

So, ~(~AvB) is equivalent to A&~B. This can easily be confirmed on a truth-table. If we apply DM to ~P&Q, the three steps are:

1. ~PvQ
2. Pv~Q
3. ~(Pv~Q)

Thus, ~P&Q is equivalent to ~(Pv~Q). This too can easily be confirmed on a truth-table.

Here are additional applications of the DeMorgan rule:

~(Pv~Q)	~(~P&Q)	(~PvQ)	(P&~Q)
(~P&Q)	(Pv~Q)	~(P&~Q)	~(~PvQ)

Notice that DeMorgan's rule applies *only* to conjunctions and disjunctions—it does not apply to horseshoes and triple bars.

Applications Let us now put some of the first three replacement rules to work. Consider the following argument:

1. (A&B)v~(EvF)
2. (~Av~B)
3. ~F⊃(AvE) /Ev(AvS)

This argument is valid. However, using only the eight inference rules, it cannot be proven valid in TD. This is where the replacement rules are indispensable. Notice that line 2 displays a DM pattern. By applying DM to line 2, we can convert that line into the opposite of the left disjunct of line 1:

4. ~(A&B) DM 2

This allows us to break line 1 down by DS:

5. ~(EvF) DS 1,4

Since line 5 instantiates one of the DM forms, we apply DM to get:

6. ~E&~F DM 5

This is fortunate, for now line 6 can be broken down by Simp:

> 7. ~E Simp 6
> 8. ~F Simp 6

Next, notice that lines 3 and 8 instantiate the premise section of MP. This gives us:

> 9. AvE MP 3,8

Now we can use Comm to turn 9 into:

> 10. EvA Comm 9

Using Add, we get S into place:

> 11. (EvA)vS Add 10

Finally, we use Association to shift the parentheses and derive our conclusion:

> 12. Ev(AvS) Assoc 11

Distribution   Do you remember the distributive principle of multiplication? One way to put that principle is:

$$a \times (b + c) = (a \times b) + (a \times c)$$

For example,

$$3 \times (5+6) = (3 \times 5) + (3 \times 6)$$

We have a similar principle in logic. The principle, which is called the *Distribution rule*, or "Dist" for short, has two parts:

> a. [P v (Q & R)] // [(P v Q) & (P v R)]
> b. [P & (Q v R)] // [(P & Q) v (P & R)]

This rule reflects the fact that any sentence **Pv(Q&R)** is equivalent to the sentence **(PvQ)&(PvR)** and any sentence **[P&(QvR)]** is equivalent to the sentence **[(P&Q)v(P&R)]**. (This can be confirmed on an eight-row truth-table). Here's a proof employing Distribution at its very first step:

> 1. Av(B&C)
> 2. (AvB)⊃E / E
> 3. (AvB)&(AvC) Dist 1
> 4. AvB   Simp 3
> 5. E   MP 2,4

In this proof, Dist was used to replace Av(B&C) with (AvB)&(AvC).

In the next proof, a line of the form **P&(QvR)** replaces a line of the form **(P&Q)v(P&R)**:

1. (A&B)v(A&C)
2. A⊃E /E
3. A&(BvC) Dist 1
4. A   Simp 3
5. E   MP 2,4

Notice that when we apply the Distribution rule to a sentence, the main oper-ator switches from the ampersand to the wedge or from the wedge to the ampersand.

EXERCISE 3.3

Part I Using any of the eight inference rules plus any of the first four replacement rules, supply justifications for the following:

(1)
1. Av(B&C)
2. (AvB)⊃S
3. ~Sv(QvR) /RvQ
4. (AvB)&(AvC) _____
5. AvB   _____
6. S   _____
7. QvR   _____
8. RvQ   _____

(2)
1. (AvB)vC
2. ~B&R
3. A⊃F
4. C⊃S /~(~F&~S)
5. (BvA)vC_____
6. Bv(AvC)_____
7. ~B   _____
8. AvC   _____
9. FvS   _____
10. ~(~F&~S)_____

(3)
1. ~(AvB)
2. ~B⊃Z
3. H⊃~Z
4. ~H⊃(Q&R)/~(~Qv~R)
5. ~A&~B_____
6. ~B   _____
7. Z   _____
8. ~H   _____
9. Q&R   _____
10. ~(~Qv~R)_____

(4)
1. H&(SvG)
2. (~Hv~S)
3. G⊃P /P
4. (H&S)v(H&G) _____
5. ~(H&S)   _____
6. H&G   _____
7. G   _____
8. P   _____

(5)

1.	A&B	
2.	A⊃S	
3.	~~S⊃Q	
4.	(QvP)⊃H	
5.	~HvT/T	
6.	A	_____
7.	S	_____

8.	~~S	_____
9.	Q	_____
10.	QvP	_____
11.	H	_____
12.	~~H	_____
13.	T	_____

Part II Use the inference rules and the first four replacement rules to prove the following arguments valid:

(1)
1. (AvB) ⊃~E
2. Bv(GvA)
3. ~G/~E

(2)
1. (H⊃S)v~R
2. (AvB)vQ
3. ~B&~S
4. (AvQ)⊃R/~H

(3)
1. [(AvB)vC]⊃S
2. Bv(AvC)
3. S⊃Q/Q

(4)
1. (AvB)⊃G
2. R&H
3. ~(~A&~B)/G

(5)
1. ~(AvB)
2. ~B⊃E
3. E⊃S/S

(6)
1. Av(B&C)
2. (AvC)⊃~S
3. H⊃S/ ~H

(7)
1. (H&S)v(H&P)
2. H⊃(GvM)/(GvM)

(8)
1. A&(BvC)
2. (A&B)⊃H
3. (A&C)⊃O/HvO

(9)
1. ~(~Av~B)
2. B⊃~~H
3. H⊃S/ S

(10)
1. (R⊃S)v~G
2. ~~G
3. (R⊃S)⊃~~P
4. H⊃~P/~H

(11)
1. R⊃~(HvB)
2. Hv(A&B)
3. ~~S⊃R/~S

(12)
1. (G≡H)v~F
2. ~(Av~F)
3. (G≡H)⊃P/P

(13)
1. ~(R&S)
2. ~R⊃Q
3. ~S⊃Z
4. ~Q
5. Z⊃(A≡B)/A≡B

(14) 1. (P⊃Q)&(R⊃S)

2. (QvS)⊃~H

3. <u>Pv(R&B)</u> /~H

(15) 1. ~R⊃~S

2. ~~S

3. ~~R⊃H

4. <u>(H⊃A)&(Z⊃O)</u>/AvO

(16) 1. Av(B&C)

2. ~B

3. <u>~AvE</u>/E

(17) 1. S⊃~P

2. <u>(P&S)v(P&F)</u>/F

(18) 1. ~(~H&E)

2. <u>A&(B&E)</u>/HvS

(19) 1. ~A

2. <u>(A&B)v(H&S)</u>/S

(20) 1. K

2. ~(A&B)&~(E&F)

3. <u>(H&K)⊃[(A&B)v(E&F)]</u>/
~H

(21) 1. H⊃S

2. K⊃H

3. <u>~(AvS)</u>/~(KvA)

(22) 1. ~(AvB)

2. (~BvC)⊃(E&G)

3. <u>S⊃~G</u>/~SvX

(23) 1. A⊃(BvC)

2. ~(B&E)

3. <u>~(Cv~E)</u>/~A

(24) 1. QvR

2. [(PvQ)vR]⊃~S

3. <u>H⊃S</u>/~H

(25) 1. H⊃(E&P)

2. Av(B&~C)

3. A⊃~E

4. <u>~C⊃~P</u>/~H

(26) 1. (AvB)⊃~(HvE)

2. Av(B&H)

3. <u>(A&~E)⊃(P&Q)</u>/QvX

(27) 1. ~(A&B)

2. ~A⊃(P&Q)

3. <u>~B⊃(Q&S)</u>/Q

(28) 1. RvS

2. ~(A&R)

3. <u>~(A&S)</u>/~A

(29) 1. (AvB)⊃~E

2. (H&A)v(B&E)/A

(30) 1. A⊃(B&H)

2. <u>(HvS)⊃Z</u>/A⊃Z

●— RULES OF REPLACEMENT: SIX MORE RULES

The Double Negation Rule If Pat is *not* a *non*-Hindu, then Pat is Hindu. If someone's tractor "*ain't* got *no* traction," then that tractor has traction. In English, if one negation operator applies directly to another, the two cancel each other out and the result is as if neither of the two negation operators existed. The rule in logic concerning double negation corresponds to ordinary English usage:

— **Double Negation** ——•

$$\sim\sim P \;//\; P$$

For example, by Double Negation (or "DNG" for short), we can replace

$$\sim\sim AvB$$

with

$$AvB$$

and we can replace

$$\sim\sim P \supset Q$$

with

$$P \supset Q$$

Here are two short proofs requiring Double Negation:

1.	$\sim\sim P \supset S$			1.	$\sim\sim P \supset S$	
2.	$\underline{P}$ /S			2.	$\underline{P}$ /S	
3.	$\sim\sim P$	DNG, 2		3.	$P \supset S$	DNG 1
4.	S	MP 1,3		4.	S	MP 2,3

It is important that you understand the following: Two negatives cancel each other out only in case the first applies directly to the second, with no intervening parenthetical device. Thus, in each of the following cases Double Negation is applied to the sentence above the line in order to derive the sentence below the line:

$$\frac{\sim\sim A \& \sim\sim B}{A \& B} \qquad\qquad \frac{\sim\sim(AvB)}{(AvB)} \qquad\qquad \frac{Jv\sim\sim G}{JvG}$$

However, Double Negation does not remove the tildes in the following cases:

$$\sim(\sim A \& B) \qquad\qquad Av\sim(\sim B \supset G)$$

Notice that in the two cases immediately above, the first tilde does not apply to the second tilde. Rather, the first tilde applies to the parenthesis that comes between the two tildes.

Transposition Any sentence P⊃Q is equivalent to the sentence ~Q⊃~P. This gives us our next replacement rule, the *Transposition rule*:

$$P⊃Q \ // \ {\sim}Q⊃{\sim}P$$

Here's one way to remember this rule: When you apply Transposition to a formula P⊃Q, P and Q trade places and, in the process, each becomes the opposite of what it was prior to the switch.

Here are some examples of inferences allowed by the Transposition rule:

A⊃B	~S⊃~G	(A&B)⊃~(HvE)	~A⊃B
~B⊃~A	G⊃S	(HvE) ⊃~(A&B)	~B⊃A

Here is a proof employing Transposition ("Trans" for short):

1. E⊃(A⊃B)
2. (~B⊃~A)⊃S
3. (E⊃S)⊃H/ H
4. (A⊃B)⊃S Trans 2
5. E⊃S HS 1,4
6. H MP 3,5

Implication A sentence P⊃Q is equivalent to the sentence ~PvQ. This is the basis of our next replacement rule, the *Implication rule*:

$$P⊃Q \ // \ {\sim}PvQ$$

In each example below, the sentence below the line replaces the sentence above the line in accord with the Implication rule ("Imp" for short).

A⊃B	EvS	~H⊃~G	~Fv~P
~AvB	~E⊃S	Hv~G	F⊃~P

So, when the operator is either a horseshoe or a wedge, you may take the opposite of the left side, change the horseshoe to a wedge or the wedge to a horseshoe, provided that you leave the right side as it is.

Here is a proof requiring Implication:

1. A⊃(G⊃E)
2. (Ev~G)⊃S/ A⊃S
3. A⊃(~GvE) Imp 1
4. A⊃(Ev~G) Comm 3
5. A⊃S HS 2,4

This proof could have been worked this way as well:

1. A⊃(G⊃E)
2. (Ev~G)⊃S/ A⊃S
3. (~GvE)⊃S Comm 2
4. (G⊃E)⊃S Imp 3
5. A⊃S HS 1,4

Exportation Any sentence (P&Q)⊃R is equivalent to the sentence P⊃(Q⊃R). Corresponding to this, we shall have the *Exportation rule*:

$$(P\&Q) \supset R \; // \; P \supset (Q \supset R)$$

In each of the following examples, the Exportation rule has been applied to the sentence above the line in order to derive the sentence below the line:

(A&B)⊃E	[(HvS)&G)]⊃M
A⊃(B⊃E)	(HvS)⊃(G⊃M)

Here is a proof involving Exportation ("Exp" for short):

1. A⊃(B⊃S) / B⊃(A⊃S)
2. (A&B)⊃S Exp 1
3. (B&A)⊃S Comm 2
4. B⊃(A⊃S) Exp 3

The Tautology Rule Since any sentence **P** is equivalent to the sentence PvP, and since any sentence **P** is also equivalent to the sentence P&P, the following rule, the *Tautology rule* ("Taut" for short), is certainly valid:

$$P \; // \; PvP$$
$$P \; // \; P\&P$$

Here is a proof employing both versions of this rule:

1. P⊃(S⊃G)
2. PvP
3. SvS / GvG
4. P   Taut 2
5. S   Taut 3
6. S⊃G   MP 1,4
7. G   MP 5,6
8. GvG   Taut 7

This proof contains a more interesting application of the Tautology rule:

1. ~(A&E)
2. ~(A&B)
3. EvB/~A
4. ~Av~E   DM 1
5. ~Av~B   DM 2
6. ~Ev~A   Comm 4
7. ~Bv~A   Comm 5
8. E⊃~A   Imp 6
9. B⊃~A   Imp 7
10. ~Av~A   CD 3,8,9
11. ~A   Taut 10

The Equivalence Rule Any sentence **P≡Q** is equivalent to the sentence **(P⊃Q)&(Q⊃P)**. A sentence **P≡Q** is also equivalent to the sentence **(P&Q) v(~P&~Q)**. Our final replacement rule, called the *Equivalence rule*, reflects this:

$$P≡Q \ // \ (P⊃Q)\&(Q⊃P)$$

$$P≡Q \ // \ (P\&Q)v(~P\&~Q)$$

The following inferences are among the inferences permitted by the Equivalence rule ("Equiv" for short):

A≡B
(A⊃B)&(B⊃A)

A≡B
(A&B)v(~A&~B)

The following proof uses this rule.

1. A≡B
2. ~B/~A
3. (A⊃B)&(B⊃A) Equiv 1
4. A⊃B   Simp 3
5. ~A   MT 2,4

Are Replacement Rules Worth the Bother? Do we really need the replacement rules? They are sometimes difficult to work with. However, without the replacement rules, our proof system is limited, for there are many valid arguments that can't be proven valid unless we use the replacement rules. For example, using only the rules of inference, we could not construct a proof of the following simple and obviously valid argument:

$$1. \quad (AvB) \supset C$$
$$\underline{2. \quad (BvA)/C}$$

Notice that MP does not apply to this argument. In order to prove this argument valid, we must employ the Commutative rule. Two alternative proofs are possible:

I. 1. (AvB)⊃C
    2. (BvA)/C
    3. AvB Comm 2
    4. C   MP 1,3

II. 1. (AvB)⊃C
    2. (BvA)/C
    3. (BvA)⊃C Comm 1
    4. C   MP 2,3

●— SOME INTERESTING PATTERNS OF REASONING

The proofs below display some interesting patterns of reasoning that sometimes show up within longer proofs. An understanding of these proofs may help you plan strategy in cases of more complex proofs.

(1) 1. Q/ P⊃Q
     2. Qv~P   Add 1
     3. ~PvQ   Comm 2
     4. P⊃Q   Imp 3

(2) 1. PvQ
     2. ~PvQ/ Q
     3. Qv~P   Comm 2
     4. ~Q⊃~P   Trans 3
     5. ~P⊃Q   Imp 1
     6. ~Q⊃Q   HS 4,5
     7. QvQ   Imp 6
     8. Q   Taut 7

(3) 1. ~P / P⊃Q
     2. ~PvQ   Add 1
     3. P⊃Q   Imp2

(4) 1. P&~P/ Q
     2. P   Simp 1
     3. ~P Simp 1

     4. PvQ   Add 2
     5. Q   DS 3,4

(5) 1. P⊃Q / P⊃(QvS)
     2. ~PvQ   Imp 1
     3. (~PvQ)v S   Add 2
     4. ~Pv(QvS)   Assoc 3
     5. P⊃(QvS)   Imp 4

(6) 1. P⊃Q
     2. P⊃R/P⊃(Q&R)
     3. ~PvQ   Imp 1
     4. ~PvR   Imp 2
     5. (~PvQ)&(~PvR) Conj 3,4
     6. ~Pv(Q&R) Dist 5
     7. P⊃(Q&R)   Imp 6

(7) 1. P⊃Q
     2. S⊃Q/(PvS)⊃Q
     3. ~PvQ   Imp 1
     4. ~SvQ   Imp 2

5. (~PvQ)&(~SvQ)   Conj 3,4   7. P⊃R   Imp 6

6. Qv(~P&~S)   Dist 5   (9) 1. P⊃Q

7. (~P&~S)vQ   Comm 6   2. PvQ / Q

8. ~(~P&~S)⊃Q   Imp 7   3. QvP   Comm 2

9. (PvS)⊃Q   DM 8   4. ~Q⊃P   Imp 3

(8) 1. (PvQ)⊃R/P⊃R   5. ~Q⊃Q HS 1,4

    2. ~(PvQ)vR   Imp 1   6. QvQ   Imp 5

    3. (~P&~Q)vR   DM 2   7. Q   Taut 6

    4. (Rv~Q)&(Rv~P)   Dist 3   (10) 1. ~P⊃P /P

    5. Rv~P   Simp 4   2. PvP   Imp 1

    6. ~PvR   Comm 5   3. P   Taut 2

• EXERCISE 3.4 •

Part I Using any of the inference and replacement rules, fill in the justifications for the following.

(1) 1. P⊃Q   4. R⊃S/ S& (~P&~Q)

    2. (~Q⊃~P)⊃S   5. (P⊃Q)&(Q⊃P) _____

    3. (SvH)⊃[(A&B)⊃C]   6. P⊃Q _____

    4. A/~C⊃~B   7. R   _____

    5. ~Q⊃~P_____   8. S   _____

    6. S _____   9. (P&Q)v(~P&~Q) _____

    7. SvH _____   10. ~(P&Q)_____

    8. (A&B)⊃C _____   11. ~P&~Q _____

    9. A⊃(B⊃C) _____ 12. S&(~P&~Q) _____

  10. B⊃C _____ (3) 1. A⊃B

  11. ~C⊃~B_____   2. E⊃B

(2) 1. P≡Q   3. ~(~A&~E)

    2. (P⊃Q)⊃R   4. B⊃G/G

    3. (~Pv~Q)   5. AvE _____

6. BvB _____

7. B _____

8. G _____

(4) 1. ~S⊃S

2. ~SvG

3. G⊃~(PvQ)

4. ~P⊃Z/Z

5. SvS _____

6. S _____

7. G _____

8. ~(PvQ) _____

9. ~P&~Q _____

10. ~P _____

11. Z _____

(5) 1. A⊃B

2. (~AvB)⊃P

3. H⊃~P

4. ~H⊃Q

5. (QvQ)⊃(SvS)/S

6. ~AvB _____

7. P _____

8. ~H _____

9. Q _____

10. QvQ _____

11. SvS _____

12. S _____

Part II Use any of the replacement and inference rules to prove the following:

(1) 1. H⊃~S

2. (~Hv~S)⊃F

3. F⊃B/B

(2) 1. ~A⊃B

2. A⊃E

3. B⊃S

4. (EvS)⊃X/X

(3) 1. P⊃Q

2. (~Q⊃~P)⊃S

3. F⊃~S/~F

(4) 1. H⊃S

2. ~P⊃~S/H⊃P

(5) 1. ~(A&~B)⊃C

2. A⊃B

3. Hv~C/H

(6) 1. A⊃B

2. A⊃~B

3. ~A⊃G/G

(7) 1. (PvP)⊃S

2. P&F/S

(8) 1. A⊃(B⊃C)/B⊃(A⊃C)

(9) 1. H⊃(A⊃E)

2. (Ev~A)⊃S/H⊃S

(10) 1. (A&E)⊃S

2. (E⊃S)⊃F/A⊃F

(11) 1. ~PvQ

2. ~QvP/P≡Q

(12) 1. B≡G

2. (B&G)⊃Z

3. (~B&~G)⊃Z/Z

(13) 1. P≡Q
     2. ~(P&Q)/~P&~Q

(14) 1. A⊃(B&C)
     2. B⊃(C⊃S)/A⊃S

(15) 1. (A⊃B)&(E⊃B)
     2. ~(~A&~E)/B

(16) 1. ~[(P⊃Q)&(Q⊃P)]
     2. (A&S)⊃(P≡Q)
     3. A/~S

(17) 1. (A⊃B)⊃[(TvQ)&(H≡P)]
     2. (TvQ)⊃[(H≡P)⊃X]/(A⊃B)⊃X

(18) 1. A⊃(B&E)/(A⊃B)&(A⊃E)

(19) 1. H⊃~H
     2. ~(H&S)⊃G/~(Hv~G)

(20) 1. A⊃B
     2. A⊃~B/~A

(21) 1. Av~Q
     2. ~AvG/Q⊃G

(22) 1. ~A⊃S
     2. ~E⊃S
     3. ~(A&E)/SvH

(23) 1. (P⊃Q)⊃R
     2. R⊃~R/P

(24) 1. Pv(Q&S)
     2. P⊃S/S

(25) 1. P⊃(S&H)
     2. R⊃(~S&~H)
     3. PvR/H⊃S

(26) 1. A⊃B/A⊃(BvE)

(27) 1. A⊃~(B⊃G)
     2. AvG/ A≡~G

(28) 1. (A⊃B)⊃G
     2. (Q⊃B)⊃~G/~B

(29) 1. (AvB)⊃G
     2. (Q⊃G)⊃S
     3. B/S

(30) 1. (P⊃Q)⊃(Q⊃P)
     2. (P≡Q)⊃~(A&~B)
     3. Q&A/A&B

(31) 1. A≡B
     2. AvB
     3. A⊃(B⊃E)/E

3. CONDITIONAL and INDIRECT PROOF

●─ CONDITIONAL PROOF

Suppose you are working on a proof of an argument consisting of premises **P**₁ . . . .
Pₙ and a conclusion **R** ⊃ **S**. Call the argument 'argument 1':

Argument 1:

1.	P_1	
2.	P_2	
•	•	
•	•	
•	•	
n.	P_n	/R ⊃ S

Now, suppose you are having difficulty deducing the conclusion by natural deduction and decide to try a different argument, argument 2, which has as premises the same formulas, $P_1....P_n$, plus the formula R, the formula that appears as the antecedent of argument 1's conclusion. Suppose the conclusion of argument 2 is the formula S, which appears as the consequent of argument 1's conclusion:

$$
\begin{array}{lll}
1. & P_1 \\
2. & P_2 \\
& \cdot & \cdot \\
\text{Argument 2:} & \cdot & \cdot \\
& \cdot & \cdot \\
n. & P_n \\
n+1 & R & /S \\
\hline
\end{array}
$$

Next, suppose you succeed in proving argument 2 valid. Would such a proof of argument 2 tell us anything about argument 1? Yes. Consider the following. First, if argument 2 is valid, then there is no possibility that its premises are true and its conclusion is false. That is to say, there is no possibility of the following distribution of truth-values:

$$
\begin{array}{lll}
P_1 & = & T \\
P_2 & = & T \\
& \cdot & \cdot \\
\text{Argument 2:} & \cdot & \cdot \\
P_n & = & T \\
R & = & T \\
S & = & F \\
\end{array}
$$

Turning to argument 1, argument 1 is invalid if and only if there is a possibility that its premises are true while its conclusion is false. That is, argument 1 is invalid if the following distribution of truth-values is a possibility:

$$
\begin{array}{lll}
P_1 & = & T \\
P_2 & = & T \\
& \cdot & \cdot \\
& \cdot & \cdot \\
\text{Argument 1:} & P_n & = & T \\
R & = & T \\
S & = & F \\
\end{array}
$$

However, we learned from our successful proof of argument 2 that this particular distribution of truth-values is impossible. It follows that it is impossible for the premises of argument 1 to be true and its conclusion false. Therefore, if argument 2 is valid, then argument 1 must also be valid.

The Rule of Conditional Proof incorporates this reasoning into an inference rule, which may be stated as follows:

RULE OF CONDITIONAL PROOF

If, at any point in a proof, you wish to derive a conditional **P ⊃ Q**, then you may:

1. Indent and assume the antecedent of the conditional you seek to prove. (Write as justification "AP" for "assumed premise.")

2. Using this assumption plus, if needed, any available lines occurring earlier, derive the consequent of the conditional.

3. Draw a line around the indented steps just taken. The line drawn around the steps is called a *discharge line* and the indented lines enclosed within the discharge line constitute a *conditional proof sequence*. The completion of this step is called "discharging" the assumption.

4. End the indentation and infer the conditional whose antecedent is the assumption you began the conditional proof sequence with and whose consequent is the latest line in that sequence. (Write as justification "CP" and the line numbers of the indented lines within the sequence.)

5. Once the steps within a conditional proof sequence have been discharged and the conditional has been derived, those indented lines may no longer be used in justification of further lines in the proof. That is, once an assumption is discharged and the conditional derived, the lines in the assumption's conditional proof sequence are no longer "available" to be used in justification of further lines in the proof.

Step 6 deserves emphasis. It is important to understand that once the discharge line has been drawn and the conditional derived, the indented sentences enclosed within the discharge line cannot be used in the derivation of future steps in the proof. This is because in a conditional proof, we have no reason to suppose the indented steps follow validly from the premises. If we have no reason to suppose that the indented lines follow from the premises, then we have no reason to suppose that lines that follow from the indented lines follow from the premises either. However, for reasons given above, the conditional derived as a result of a conditional proof sequence does validly follow from the premises, so this conditional may be used in the derivation of future steps if needed.

The Conditional Proof rule stated above may be abbreviated as follows:

> If, from the indented assumption **P**, plus any previous lines that are available, the sentence **Q** is derived, you may end the indentation and assert **P ⊃ Q**.

Here is an application of this rule. In the following proof, we need to derive H⊃G. Employing the conditional proof strategy, we first indent and assume H. We then derive G. From this, we infer H⊃G according to the Conditional Proof rule ("CP" for short):

1. (HvS) ⊃ (B&~E)
2. GvE / H ⊃ G
3. | H   AP
4. | HvS   Add 3
5. | B&~E MP 1, 4
6. | ~E   Simp 5
7. | G   DS 2, 6
8. H ⊃ G   CP 3–7

Notice that we began by assuming the antecedent of the conditional we were seeking to prove. We then applied Addition and added S in order to derive the antecedent of line 1. This allowed us to "break up" line 1 and derive B&~E. We then applied Simp and derived ~E, which allowed us to infer the consequent of the conditional we were trying to prove. At step 8, we discharged our assumption and inferred the conditional whose antecedent was our assumption and whose consequent was the last line reached within the conditional proof sequence.

Make sure you understand that in the above proof, lines 3 through 7 do not follow from the premises; they merely follow from the assumption, which itself does not follow from the premises. However, for reasons given earlier, line 8 does follow from the premises—that's what the conditional proof sequence establishes.

There are a number of things you can do with the Conditional Proof rule. For instance, the assumption for a conditional proof sequence does not have to be the first line of a proof. In the following proof, the conditional proof sequence begins in the middle of the proof:

1. Av(B&C) / ~B ⊃ A
2. (AvB) & (AvC)   Dist 1
3. (AvB)   Simp 2
4. (AvC)   Simp 2
5. | ~B   AP
6. | A   DS 3, 5
7. ~B ⊃ A   CP 5–6

Furthermore, the conditional that is established at the end of a conditional proof sequence does not have to be the last line of the proof. In the following proof, the conclusion is not a conditional and the conditional proof sequence in the argument's proof is not used to derive the conclusion of the argument. Rather, the conditional proof is used to derive a formula in the middle of the proof, and this formula is used to derive the conclusion of the argument:

1. (A⊃S) ⊃ (H&J)
2. A⊃ (QvS)
3. Q⊃ S / J

4.	A	AP
5.	QvS	MP 2, 4
6.	~Q⊃S	Imp 5
7.	~S⊃Q	Trans 6
8.	~S⊃S	HS 3,7
9.	SvS	Imp 8
10.	S	Taut 9
11. (A⊃S)		CP 4–10
12. H&J		MP 1,11
13. J		Simp 12

Notice that in the above argument, the conditional proof method is used to derive A⊃S. Once A⊃S is derived, it—A⊃S—is used to bring down H&J. Finally, from H&J, the conclusion follows by Simplification.

Here's another example of this. The conditional established at the end of the following conditional proof sequence is not the last line of the proof.

1. A⊃ (B&C)		
2. (BvH) ⊃ R		
3. ~(A&~R) ⊃ S/ S		
4.	A	AP
5.	B&C	MP 1,4
6.	B	Simp 5
7.	BvH	Add 6
8.	R	MP 2,7
9. A⊃R		CP 4–8
10. ~AvR		Imp 9
11. ~(A&~R)		DM 10
12. S		MP 3,11

Remember that once an assumed premise is discharged and the conditional proof sequence has been concluded, the lines within that sequence may not be appealed to if further steps in the proof are inferred. So, in the above proof, after step 9, the lines within the Conditional Proof sequence may not be referred to as we derive further steps. Thus, the following proof is an incorrect proof:

1. A⊃ (B&G)		
2. (BvH) ⊃ F		
3. (F ⊃ R) / R		
4.	A	AP
5.	B&G	MP 1, 4
6.	B	Simp 5
7.	BvH	Add 6
8.	F	MP 2, 7
9. A⊃F		CP 4–8
10. F		MP 4, 9 (Incorrect! Lines 4–8 cannot be used here.)
11. R		MP 3, 10

The proof is incorrect because step 10 violates the Conditional Proof rule.

Conditional Proof may be used twice in a proof, as the following example illustrates:

```
1. (AvB) ⊃ E
2. (EvS) ⊃ A /A ≡ E
3.          │ A          AP
4.          │ AvB        Add 3
5.          │ E          MP 1, 4
6. A⊃E        CP 3–5
7.          │ E          AP
8.          │ EvS        Add 7
9.          │ A          MP 2, 8
10. E ⊃ A     CP 7–9
11. (A⊃E) & (E ⊃ A)   Conj 6, 10
12. A ≡ E  Equiv 11
```

It is extremely important that you understand the following point: You cannot end a proof on an indented line of a conditional proof sequence. That is, the assumption of a conditional proof sequence must be discharged before a proof is ended. So, the last line of a proof can never be an indented conditional proof sequence line. If you could legally end a proof on an indented line, without discharging the assumption, then you could derive any arbitrary conclusion from any premise. For example, the following would be a correct proof if you were allowed to end a proof without discharging your assumption:

1.   Ann is 18. (A)

2.   Therefore, Wyoming is a state. (W)

```
1. A / W
2.      │ W          AP
3.      │ WvW        Add 2
4.      │ W          Taut 3
```

This argument is obviously invalid. We were able to construct a "proof" of it only because we ended on an indented line.

———————————— • EXERCISE 3.5 • ————————————

Use the conditional proof method to prove each of the following:

(1) 1. A⊃(B&C)/A⊃C   (3) 1. P⊃(Q⊃R)

(2) 1. P/Q⊃Q   2. (Q⊃R)⊃(Q⊃S)/P⊃(Q⊃S)

(4) 1. (A&B)⊃C

 2. A⊃B/A⊃C

(5) 1. ~QvZ

 2. Z⊃A/Q⊃(Z&A)

(6) 1. (PvQ)⊃S

 2. A⊃P/A⊃(SvT)

(7) 1. A⊃(B⊃E)

 2. A⊃(H⊃E)

 3. ~E/A⊃~(BvH)

(8) 1. (PvQ)⊃(A&B)

 2. (BvE)⊃(O&S)/P⊃O

(9) 1. H⊃(S&L)

 2. ~LvQ

 3. X⊃~Q

 4. ~X⊃B/H⊃B

(10) 1. A⊃(BvC)

 2. E⊃S

 3. B⊃C/A⊃B

(11) 1. A⊃P

 2. A⊃(P⊃B)

 3. P⊃(B⊃T)/A⊃T

(12) 1. A⊃(B&C)

 2. B⊃(A&G)/A≡B

(13) 1. P⊃(Q⊃R)

 2. S⊃(R⊃B)

 3. (Q⊃B)⊃(Hv~S)/(P&S)⊃H

(14) 1. P⊃Q

 2. (P&R)⊃E

 3. (Q&E)⊃S/P⊃(R⊃S)

●─ INDIRECT PROOF

Mathematicians in ancient times proved that √2 is an irrational number. The type of proof they employed is known as a *reductio ad absurdum* proof. Essentially, such a proof works this way. Suppose you want to prove some claim **P**. Begin your proof by assuming the opposite of the claim you wish to prove. So, begin your proof with ~**P**. (In English, the opposite of a sentence is what results when you attach a negation operator to the sentence. For example, the opposite of 'It is snowing' is 'It is not the case that it is snowing'). Next, demonstrate that this assumption implies a contradiction. Since only a contradictory statement can imply a contradiction, this shows that the assumption, ~**P**, is itself necessarily false. This is said to "reduce" the assumption to an "absurdity." It then follows that the opposite of the assumption, namely, **P**—the claim you originally sought to prove—must itself be necessarily true. Reductio ad absurdum proofs are also called *indirect proofs* because in such a proof, the conclusion is derived in an indirect way, by first deriving a contradiction from the denial or opposite of the conclusion and then inferring the conclusion from that.

The proof of the irrationality of √2 may be put as follows:

1. Our goal is to prove that √2 is irrational. Let us begin by assuming the opposite of the conclusion we wish to prove:

Suppose √2 is rational. (Assumption)

2. Since every rational number may be expressed as a ratio of two mutually prime numbers, it follows that

$$\frac{a}{b} = \sqrt{2}$$

where a and b are mutually prime. (Two numbers are mutually prime just in case they have no common factor other than 1.)

3. If we perform the same arithmetical operation on both sides of an equality, the result remains an equality. Thus, multiplying each side by b, we get:

$$a = \sqrt{2}b$$

4. Squaring each side, we get:

$$a^2 = 2b^2$$

5. Any number that is equal to twice some number must itself be an even number. Consequently, since a^2 has been shown to be twice some number, it follows that:

a² is even

6. If 2 is a factor of a^2, then 2 is a factor of a as well. That is, if a^2 is twice some number, then a is also twice some number. Therefore, if a^2 is even, then a is also even. Thus:

a is even

7. Since a and b are mutually prime, they have no common factor. So if one is even then the other is odd. We have already established that a is even. Therefore:

b is odd

8. Since a is an even number, we know a is twice some number. Call this number k. Therefore,

$$a = 2k$$

9. Substituting 2k for a in step 4, we get:

$$(2k)^2 = 2b^2$$

10. If we square 2k, we get:

$$4k^2 = 2b^2$$

11. Dividing each side by 2, we get:

$$2k^2 = b^2$$

12. It follows that

b² is even

since b² is twice some number.

13. Since any factor of b² is a factor of b, if 2 is a factor of b² then 2 is a factor of b. That is, if b² is even, b is also even. Thus:

b is even

14. Combining 7 and 13, we have proven that on the basis of the assumption, it follows that b is odd and b is even. It follows, that is, that:

b is odd and it is not the case that b is odd.

15. The consequence reached at step 14 is, of course, contradictory. This proves that our initial assumption, namely, that √2 is *rational*, implies a contradiction. But only that which is itself contradictory can imply a contradiction. So, our initial assumption must itself be contradictory. Therefore, the opposite of our assumption must be a necessary truth. Consequently:

It is necessarily true that √2 is irrational.

The indirect proof method is embodied in the following inference rule:

RULE OF INDIRECT PROOF

Anywhere in a proof, you may:

1. indent and assume the opposite of the conclusion you seek to prove, writing as justification 'AP'.

2. using the assumption plus any previous steps that are available, derive a contradiction. The contradiction can be either:

    i. one line consisting of a sentence **P** and a second line consisting of ~**P**.

    or

    ii. a single line of the form **P&~P**.

If you accomplish the two steps above, you may then:

3. end the indentation, draw a line around the indented steps, and assert the opposite of the assumption. (This counts as "discharging" the

assumption.) Write as justification 'IP' for 'indirect proof' and cite the indented lines.

4. The indented lines produced are called an *indirect proof sequence*. Once an assumption has been discharged and the opposite of the assumption has been derived, the lines in the indirect proof sequence may no longer be used for the derivation of future lines in the proof.

The Indirect Proof rule stated above may be abbreviated by the following:

> *Anywhere in a proof, if you indent, assume a sentence **P**, and derive a contradiction, then you may end the indentation and assert the opposite of **P**.*

Here is an application of the Rule of Indirect Proof:

```
1. HvS
2. ~S ⊃ ~ H/S
3.  ┌ ~S         AP
4.  │ ~H         MP 2, 3
5.  │ H          DS 1, 3
6.  └ H&~H       Conj. 4, 5
7. S IP 3–6
```

Did you notice that we began at line 3 by assuming the opposite of our conclusion? From that, we used MP to derive ~H. This allowed us to derive H by DS. The Conjunction rule then allowed us to infer H&~H, which is an explicit contradiction. Since our assumption implied a contradiction, we discharged that assumption and inferred its opposite, which was the conclusion we originally sought to prove.

In the following indirect proof, the conclusion we seek to prove is ~Av~B. Since the method requires that we assume the opposite of our conclusion, our assumed premise will be ~(~Av~B):

```
 1. A ⊃ (Ev~B)
 2. B ⊃ (~EvG)
 3. (A&B) ⊃ ~G/ ~Av~B
 4.  ┌ ~(~Av~B)    AP
 5.  │  A&B        DM 4
 6. *│  ~G         MP 3,5
 7.  │  A          Simp 5
 8.  │  Ev~B       MP 1,7
 9.  │  B          Simp 5
10.  │  E          DS 8,9
11.  │  ~EvG       MP 2,9
12. *└  G          DS 10,11
13. ~Av~B          IP 4–11
```

The contradiction appears in the form of the two starred lines.

The Indirect Proof method may be used exactly as the Conditional Proof method was used. That is, the assumption for an Indirect Proof sequence does not have to be the first line of a proof and the conclusion established at the end of an Indirect Proof sequence does not have to be the last line of the proof. Also, once an assumed premise is discharged and the indirect proof sequence has been concluded, the lines within that sequence may not be appealed to if further steps in the proof are inferred.

―――――――――――――― • **EXERCISE 3.6** • ――――――――――――――

Use the Indirect Proof method to prove the following:

(1) 1. Pv(Q&E)
     2. P⊃E/E

(2) 1. P⊃(A&B)
     2. (AvE)⊃R
     3. (EvP)/R

(3) 1. (A⊃A)⊃B
     2. (BvG)⊃C/C

(4) 1. A⊃(B⊃E)
     2. B⊃(E⊃~B)/~Av~B

(5) 1. (HvS)⊃(A&B)
     2. (BvF)⊃K
     3. HvF/K

(6) 1. A⊃B
     2. Q⊃~B
     3. ~K⊃(A&Q)/KvS

(7) 1. (HvB)⊃(A&E)
     2. (EvS)⊃(G&~H)/~H

(8) 1. P⊃(Q⊃R)
     2. P⊃Q

        3. ~Kv(RvP)/~KvR

(9) 1. (~PvQ)⊃(A&B)
     2. (AvZ)⊃(B⊃P)/P

(10) 1. (P&Q)vE
     2. ~EvQ/P⊃Q

(11) 1. (P&Q)⊃(EvF)
     2. (P⊃E)⊃(Q⊃F)
     3. Q/F

(12) 1. (PvQ)⊃(E&H)
     2. (HvA)⊃(Bv~E)
     3. (BvK)⊃~(P&H)/~P

(13) 1. P⊃[(A⊃A)⊃(KvH)]
     2. H⊃~(Xv~X)/P⊃K

(14) 1. A⊃(S⊃G)
     2. A⊃S
     3. Q⊃(AvG)/Gv~Q

(15) 1. P⊃(~Q⊃S)
     2. (P⊃Q)⊃S/S

NESTED CONDITIONAL and INDIRECT PROOFS

It is permissible to construct a conditional proof or an indirect proof inside another conditional or indirect proof. That is, you may construct a conditional proof within a conditional proof, an indirect proof within an indirect proof, a conditional proof within an indirect proof, and an indirect proof within a conditional proof. Call such proofs *nested* proofs. Since this type of proof can get complicated, it might be helpful to have some rules to guide us. First, in order to make things more precise, let us say that the *scope* of an assumption consists of the assumption itself along with all of its indented sentences. One conditional or indirect proof sequence lies inside the scope of another if its assumption is within the scope of the other proof sequence. When constructing nested proofs, the following rules must be followed:

1. You may not discharge an assumption unless all other assumptions within the assumption's scope have been discharged.

2. Every assumption must be discharged before a proof is completed.

Here is an example of a nested conditional proof:

1. (A⊃B) ⊃ (EvF)
2. (EvF) ⊃ (S ⊃H) / (A⊃B) ⊃ (S ⊃H)
3.   A⊃B AP
4.   EvF   MP 1,3
5.   S   AP
6.   S ⊃ H   MP 2,4
7.   H   MP 5,6
8.   S ⊃ H CP 5-7
9. (A ⊃ B) ⊃ (S ⊃ H) CP 3-8

Notice that the conclusion of this argument is a conditional whose antecedent is (A⊃B) and whose consequent is (S⊃H). Notice further that the consequent (S⊃H) is itself a conditional. At step 3 we assumed the antecedent of the conclusion. At step 5 we assumed the antecedent of the consequent of the conclusion, that is, we assumed the antecedent of (S⊃H). We then completed the inner conditional proof sequence and closed it off. Finally, we completed the outer conditional proof sequence and closed it off.

Concerning the nested proof above, the first of the two rules given above shows that you cannot end the first conditional proof sequence—the sequence whose assumption is A⊃B—and discharge its assumption until the conditional proof within has been completed and its assumption has been discharged. The second rule simply reminds you that you cannot end a proof on an indented line; all assumptions must be discharged before you derive the final conclusion.

Consider the following nested proof.

1. G ⊃ (~AvB)
2. ~Ev~B / E ⊃ (~Gv~A)
3.   E   AP
4.   ~B   DS 2,3
5.   G   AP
6.   ~AvB   MP 1,5
7.   ~A   DS 4,6
8.   G ⊃ ~A CP 5-7
9. E ⊃ (G ⊃~A) CP 3-8
10. E ⊃ (~Gv~A) Imp 9

When we are deriving lines inside the inner conditional proof sequence, we are allowed to use lines within the outer conditional proof sequence, for its assumption has not yet been discharged. Thus at step 6 in the proof immediately above, we can appeal to steps 1 through 5 if we need to. However, once the inner proof sequence has been closed off and its assumption has been discharged, we cannot derive lines from the lines within its "box"—lines 5 through 7—as we continue on in the outer conditional proof sequence. Thus, after step 8, we cannot derive lines from lines 5–7 as we finish the rest of the proof. Similarly, after step 9, we cannot use the lines within the outer conditional proof—lines 3 through 8—to derive further lines in the proof.

Generally, you will need to nest two conditional proofs if the conclusion you seek to prove is a conditional whose consequent is itself a conditional. In such a case, the consequent of the conditional you are seeking to prove will itself contain an antecedent and consequent, which is to say that the conditional you are trying to derive is a conditional of the form $P \supset (Q \supset R)$. If you want to use a nested proof to derive a formula of this form, then:

    i. indent and assume the antecedent, P, of the whole conditional;

    ii. indent again and assume the antecedent, Q, of the conditional constituting the consequent of the whole conditional.

The overall structure of the nested proof in this type of case will look like this:

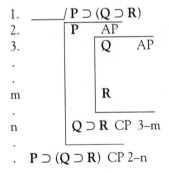

Some arguments require a triple-nested proof. Here is an example of the overall structure of such a proof:

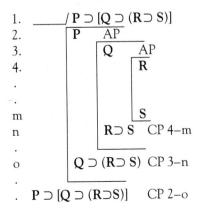

You may also nest an indirect proof sequence within a conditional proof sequence or a conditional proof sequence within an indirect proof sequence. The rules allow both types of proofs. Here is an indirect proof that is nested within a conditional proof:

```
 1. L ⊃ [~M ⊃ (N&O)]
 2. ~N & P / L ⊃ (M&P)
 3.     │ L         AP
 4.     │ ~M ⊃ (N&O)  MP 1,3
 5.     │   │ ~M      AP
 6.     │   │ N&O     MP 4,5
 7.     │   │ N       Simp 6
 8.     │   │ ~N      Simp 2
 9.     │ M           IP 5–8
10.     │ P           Simp 2
11.     │ M&P         Conj 9,10
12. L ⊃ (M&P)         CP 3–11
```

─────────────────── • EXERCISE 3.7 • ───────────────────

Nested proofs are either required or recommended on the following problems:

(1) 1. P⊃(Q⊃R)

     2. R⊃(Q⊃S) / P⊃(Q⊃S)

(2) 1. A⊃(B⊃C)

     2. (CvH)⊃K / A⊃(B⊃K)

(3) 1. A⊃(B&E) / (P⊃A)⊃(P⊃E)

(4) 1. A⊃(B⊃P)

     2. A⊃(Q⊃~X)

     3. X⊃(BvQ) / A⊃(X⊃P)

(5) 1. H⊃(S&T)

     2. B⊃(A&G) / (T⊃B)⊃(H⊃G)

(6) 1. A⊃B

     2. (A&B)⊃Q

     3. (H&Q)⊃S / A⊃(H⊃S)

(7) 1. A⊃[(PvQ)⊃(M&N)]

     2. (NvS)⊃G / A⊃(Q⊃G)

4. PROOFS OF TAUTOLOGIES

In Chapter 2, we used truth-tables to prove sentences tautological. This section presents two ways to prove, using natural deduction, that a sentence is a tautology.

● USING INDIRECT PROOF TO PROVE A TAUTOLOGY

Suppose we assume some sentence **P** and proceed to construct, using no premises whatsoever, an indirect proof sequence. If we derive a contradiction, this proves that our assumption is contradictory, since only that which is contradictory implies a contradiction. It follows from this that the opposite of our assumption must be tautological, for in truth-functional logic, the opposite of a contradiction is a tautology.

So, to prove that a sentence **P** is tautological with this method, construct an indirect proof sequence using no premises, with the opposite of **P** as the assumption. If you reach a contradiction, you've proven that **P** is tautological. For example:

```
1. _____ / P ⊃ P
2.        ┌ ~( P ⊃ P)  AP
3.        │ ~(~P v P)  Imp 2
4.        └   P& ~P    DM 3
5. P ⊃ P     IP 2–4
```

Notice that the proof contains no premises.

Here are two more examples:

```
1. _____ / ~(P&~P)
2.        ┌ P&~P      AP
3.        │ P         Simp 2
4.        └ ~P        Simp 2
5. ~(P&~P)   IP 2–4
```

```
1. _____ / Pv~P
2.        ┌ ~( Pv~P)  AP
3.        └ ~P&P      DM 2
4. Pv~P      IP 2–3
```

USING CONDITIONAL PROOF TO PROVE A TAUTOLOGY

A material conditional is tautological just in case its truth-table contains all **T**'s in the final column. Suppose we assume the antecedent of a particular conditional and proceed to construct, using no premises whatsoever, a conditional proof sequence that ends with the consequent of the conditional. This proves that if the antecedent of the conditional is true, then the consequent of the conditional must be true as well. It follows that it is not possible for the antecedent to be true and the consequent false. Therefore, the conditional must be a tautology, since the only way a conditional can be false is if its antecedent is true when its consequent is false.

So, to prove a conditional sentence tautological with this method, construct a conditional proof sequence using no premises, with the antecedent of the conditional as the assumption. If you succeed in deriving the consequent, you will have proven that the conditional is tautological. Here is an extremely short example:

```
1. _____ / P ⊃ P
2.        ┌ P AP
          └
3. P ⊃ P   CP 2
```

Here is another:

1. _____/ P ⊃ (PvQ)
2. | P   AP
3. | PvQ   Add 2
4. P ⊃ (PvQ)   CP 2-3

Let us prove that

$$[P⊃(Q⊃R)] ⊃ [(P⊃Q)⊃(P⊃R)]$$

is a tautology. Before we begin, notice that this is a conditional whose antecedent is itself a conditional and whose consequent is itself a conditional as well. Indeed, the consequent of the antecedent is itself a conditional, and the antecedent and consequent of the consequent are each also conditionals. When beginning a proof such as this, it is recommended that you proceed in two steps: (1) set up the overall structure of the proof; and (2) fill that structure in. In order to set up the overall structure, we must arrange the assumed premises in the proper order. Let us first assume the antecedent of the formula as a whole:

1. | P ⊃ (Q ⊃ R)   AP

Next we assume within this sequence the next antecedent, that is, the antecedent of the whole formula's consequent:

1. | P ⊃ (Q ⊃ R)   AP
2. | | (P ⊃ Q)   AP

Then, within this nested sequence, we assume the last antecedent:

1. | P ⊃ (Q ⊃ R)   AP
2. | | (P ⊃ Q)   AP
3. | | | P   AP

We now have the proof's overall structure worked out. The completed proof follows:

1.	P ⊃ (Q ⊃ R) AP	
2.	(P ⊃ Q) AP	
3.	P	AP
4.	Q ⊃ R	MP 1,3
5.	Q	MP 2,3
6.	R	MP 4,5
7.	(P ⊃ R)	CP 3–6
8.	(P ⊃ Q) ⊃ (P ⊃ R)	CP 2–7

9. [P ⊃ (Q ⊃ R)] ⊃ [(P ⊃ Q) ⊃ (P ⊃ R)]   CP 1–8

A proof whose last line is **P** and which contains no premises is called a *premise-free proof* of **P**. If a premise-free proof of **P** can be constructed in TD, then **P** is called a *theorem of TD*. In such a case, **P** is also said to be 'truth-functionally true' because using only the methods of truth-functional logic, **P** can be proven true. In this case, **P** is also called 'logically true', since **P** can be proven true using only the methods of logic—without investigating the physical world.

A natural deduction proof demonstrates that the conclusion of the proof must be true if the premises are true. If a proof begins with no premises, and the assumption is discharged, then this demonstrates that the conclusion must be true absolutely and not just if certain premises are true.

EXERCISE 3.8

Construct premise-free proofs for the following tautologies.

1. _____/ ~[(A ⊃ ~A) & (~A ⊃ A)]

2. _____/ [(A v B) & ~A] ⊃ B

3. _____/ [(P ⊃ Q) & (Q ⊃ R)] ⊃ (P ⊃ R)

4. _____/ ~(B & G) ⊃ (~B v ~G)

5. _____/ ~A v (B v A)

6. _____/ (A ⊃ B) ⊃ (~A v B)

7. _____/ (A ≡ B) ⊃ [(A ⊃ B) & (B ⊃ A)]

8. _____/ [(P & Q) ⊃ R] ⊃ [P ⊃ (Q ⊃ R)]

9. _____/ ~(A & ~A) v (B & ~B)

10. _____/ ~(P & Q) v ~(~P & ~Q)

11. _____/ ~P ⊃ (P ⊃ Q)

12. _____/ ~{P & ~[(P ⊃ Q) ⊃ Q]}

13. _____/ ~P ⊃ ~(Q & P)

14. _____/ (P ⊃ Q) v (Q ⊃ P)

15. _____/ (P ⊃ ~P) v (~P ⊃ P)

16. _____/ A ⊃ (Q ⊃ A)

17. _____/ A ⊃ [(A ⊃ B) ⊃ B]

18. _____/ [~(A & ~B) & ~B] ⊃ ~A

19. _____/ [(P ⊃ Q) ⊃ P] ⊃ P

20. _____/ (~A ⊃ B) v (A ⊃ E)

21. _____/ (A ⊃ B) ⊃ [(A & E) ⊃ (B & E)]

22. _____/ A ≡ [A & (S v ~S)]

23. _____/ (A ⊃ B) v (~B ⊃ A)

24. _____/ [(A ⊃ B) & (A ⊃ R)] ⊃ [A ⊃ (B & R)]

25. _____/ (A ⊃ B) ⊃ [(A ⊃ ~B) ⊃ ~A]

26. _____/ A ⊃ [(B & ~B) ⊃ K]

5. THE LAW OF NONCONTRADICTION

We noted in Chapter 1 that within logic, anything counts as logically possible except that which is contradictory. Contradictions are logically impossible. Recall
that an explicit contradiction may be defined as any statement of the form **P&~P**.
We used a truth-table in Chapter 2 to prove that any statement of the form **P&~P**
is necessarily false, that is, can't possibly be true. The principle that contradictions
are impossible is called the *Law of Noncontradiction* (or "LNC" for short). The LNC
was formulated in ancient times by the world's first mathematicians. Some logic students have a skeptical attitude toward the LNC. Now that you understand the
nature of a formal proof, you have the background to understand one important
consideration in support of the LNC: If a reasoning process begins with a contradictory premise, no matter what the contradiction is, then by applying the obviously
valid rules of Simplification, Addition, and Disjunctive Syllogism, any arbitrarily
chosen conclusion validly follows. For example:

> 1. <u>A&~A/B</u>
> 2. A   Simp 1
> 3. AvB Add 2
> 4. ~A   Simp 1
> 5. B   DS 3,4

This proof is correct no matter what sentence A and B abbreviate.[1]

So, any systematic reasoning that proceeds from a contradiction will be a piece of reasoning in which anything and everything validly follows. In such a case, no conclusion drawn—no matter how absurd—can possibly be the wrong conclusion to draw. However, any set of logical principles must specify some distinction between drawing the correct conclusion and drawing an incorrect conclusion. After all, that's the whole point of logical theory. After reflecting on these considerations, perhaps you will agree that if we do not presuppose the LNC, no useful system of reasoning can even be formulated.

——————• SOME INTERESTING PROBLEMS •——————

1. Show that MT is superfluous in our natural deduction system. That is, show how to derive a formula ~P from formulas P⊃Q and ~Q without using MT.

2. Show that DS is superfluous. (Hint: Use MT or any of the other rules in your argument.)

3. Show that HS is superfluous.

4. Which other rules may be proven superfluous?

5. Consider the following arguments carefully. For each argument, write out in English the strategy you would employ if you were to construct a natural deduction proof of the argument.

> 1. 1. A⊃(B&C)
>   2. (CvE)⊃H /A⊃H
>
> 2. 1. Av(H&R)
>   2. (A⊃L)&(L⊃R)/ R
>
> 3. 1. (~A&E)vB
>   2. (E&D)⊃A
>   3. A⊃~B / D⊃[~(~A&E)⊃~E]

---

[1] For instance, if we let A abbreviate '2 + 2 = 4' and B abbreviate 'the earth is only 5 minutes old,' the proof shows that from the contradiction it follows that the earth is only five minutes old.

TRUTH-FUNCTIONAL
TRUTH-TREES

1. PRELIMINARIES

The beginnings of truth-functional logic were presented in Chapter 2. The truth-tables covered in that chapter are valuable for a number of reasons. First, they provide us with decision procedures for truth-functional logic, procedures so mechanical that they are easily written into computer programs. Second, truth-tables can be used to provide a detailed and effective introduction to truth-functional logic. However, the tables have one major drawback. As we evaluate bigger and bigger formulas, the size of the truth-table increases exponentially. A formula with two atomic components requires only a four-row table, but a formula containing four atomic components requires a 16 row-table, six components entails a 64-row table, eight components calls for 256 rows, and so on. Consequently, when evaluating large formulas and complicated arguments, the truth-table method involves the construction of tables so large that the method becomes impractical.

Fortunately, several alternative methods developed in recent years accomplish essentially the same things truth-tables accomplish but are much simpler to work with when solving complicated problems. Two such commonly taught alternatives are: (a) the truth-tree method (covered in this chapter) and (b) the method of natural deduction (covered in Chapter 3). Logic instructors typically use truth-tables to introduce the fundamentals of truth-functional logic and then, in order to deal with more complicated problems, cover just one of the two alternatives, either truth-trees or natural deduction. Some instructors cover both alternative methods.

This chapter's goal is to set out the truth-tree method for truth-functional logic. Before we take up our main topic, it will be helpful to first lay some conceptual groundwork. Truth-trees will make a lot more sense to you if you first think your way through the following two logical processes:

1. The construction of a decision tree

2. Reductio ad absurdum argumentation

●─ A DECISION TREE

If you have a complicated decision to make, it can sometimes help to organize the possibilities in the form of a *decision tree*. Let us begin by making an assumption. Suppose you have decided to travel to another country. What must be the case if you go through with your decision? First, you would have to save some extra money. Second, you would have to get time off from work. Our initial assumption and these two consequences may be listed as follows:

    1.   take a trip   (assumption)
    2.   save extra money
    3.   get time off

However, suppose that in order to save the extra money, you will first have to pay off a loan you took out. But in order to do that, you will have to work extra hours. You will also need to get a passport. So, given our assumption, we must add to our list:

    4.   pay off loan
    5.   work extra hours
    6.   get passport

Suppose further that you have decided to travel to either Europe or Asia. We must add:

    7.   Europe or Asia

Should you go by boat or plane? And should you travel alone or take your friend Elmo, who complains about everything and is the world's pickiest eater? Thus:

    8.   boat or plane
    9.   alone or with Elmo

Your choices may be organized on what is called a decision tree:

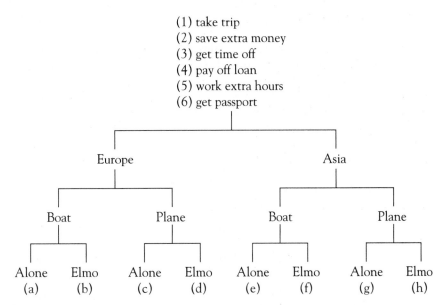

Notice that you have eight different trips to choose from, each represented by a different path from the first line—containing the assumption—to the bottom of the list. On the diagram above, the eight paths—representing the eight possible trips—run from the assumption to the eight endpoints labeled (a) through (h). The diagram is an example of a decision tree. Each of the eight paths on the decision tree constitutes a distinct "branch" of the tree. The six sentences at the top, prior to the split in the tree, constitute the "trunk" of the tree.

Suppose you talk to your boss and find out that if you are to get time off from work, it will be only in the winter. In other words, in order for (3) to happen, you must travel in the winter. We must add the information that the trip will be a winter trip to the bottom of each of the eight branches on the tree:

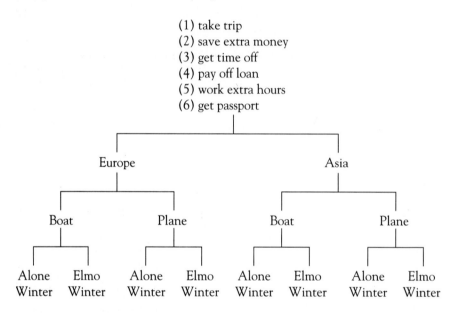

(1) take trip
(2) save extra money
(3) get time off
(4) pay off loan
(5) work extra hours
(6) get passport

Notice that when we draw a logical consequence from step 3—a step on the tree's trunk—we enter that consequence at the bottom of each branch.

• SOME QUESTIONS •

1. How many possible trips do you have to choose from if you add to the tree:

   a. Travel first class or economy class.

   b. Elmo would travel to Asia but not to Europe.

   c. The trip must begin on a Monday.

   d. If you go to Asia, you must choose between the XYZ tour and the ABC tour.

●— REDUCTIO AD ABSURDUM ARGUMENTATION

In ancient times, mathematicians proved that $\sqrt{2}$ is an irrational number. The type of argument they employed is known as a *reductio ad absurdum* argument. Essentially, such an argument works this way. Suppose you want to prove some claim **P**. Begin your argument by assuming the opposite of the claim you wish to prove. In English, the "opposite" of a claim is what results when you attach a negation operator to the claim. For instance, the opposite of 'It is raining' is 'It is not the case that it is raining'. If **P** is a sentence of TL, the opposite of **P** is the sentence **~P** and the opposite of **~P** is **P**. Begin by assuming **~P**. Then, demonstrate that this assumption implies a contradiction. Since only a contradictory statement can imply a contradiction, this shows that the assumption, **~P**, is itself contradictory or necessarily false. This is said to "reduce" the assumption to an "absurdity." It then follows that the opposite of the assumption, namely, **P**—the claim you originally sought to prove—must itself be necessarily true. Reductio ad absurdum arguments are also called *indirect proofs* or *indirect arguments* because in such an argument, the conclusion is derived in an indirect way, by first deriving a contradiction from the denial of the conclusion and then inferring the conclusion from that.

The following indirect proof establishes the irrationality of $\sqrt{2}$.

1. Our goal is to prove that $\sqrt{2}$ is irrational. We begin by assuming the opposite of the conclusion we wish to prove:

   Suppose $\sqrt{2}$ is rational. (assumption)

2. Since every rational number may be expressed as a ratio of two mutually prime numbers, it follows that

   $\underline{a} = \sqrt{2}$

   b

   where a and b are mutually prime. (Two numbers are mutually prime just in case they have no common factor other than 1.)

3. If we perform the same arithmetical operation on both sides of an equality, the result remains an equality. Thus, multiplying each side by b, we get:

   $a = \sqrt{2}b$

4. Squaring each side, we get:

   $a^2 = 2b^2$

5. Any number that is equal to twice some number must itself be an even number. Consequently, since a^2 has been shown to be twice some number, it follows that:

   a^2 is even

6. If 2 is a factor of a^2, then 2 is a factor of a as well. That is, if a^2 is twice some number, then a is also twice some number. Therefore, if a^2 is even, then a is also even. Thus:

   a is even

7. Since a and b are mutually prime, they have no common factor. Thus, if one is even, then the other is odd. We've already established that a is even. Therefore:

b is odd

8. Since a is an even number, we know a is twice some number. Call this number k. Therefore,

a=2k

9. Substituting 2k for a in step 4, we get:

$(2k)^2 = 2b^2$

10. If we square 2k, we get:

$4k^2 = 2b^2$

11. Dividing each side by 2, we get:

$2k^2 = b^2$

12. It follows that

b^2 is even

since b^2 is twice some number.

13. Since any factor of b^2 is a factor of b, if 2 is a factor of b^2 then 2 is a factor of b. That is, if b^2 is even, b is also even. Thus:

b is even

14. Combining 7 and 13, we have proven that on the basis of the assumption, it follows that b is odd and b is even. It follows, that is, that:

b is odd and it is not the case that b is odd.

15. The consequence reached at step 14 is, of course, contradictory. This proves that our initial assumption, namely, that √2 is rational, implies a contradiction. But only that which is itself contradictory can imply a contradiction. So, our initial assumption must itself be contradictory. Therefore, the opposite of our assumption must be a necessary truth. Consequently:

It is necessarily true that √2 is irrational.

The next step in preparation for the truth-tree method is to apply the reductio ad absurdum method to a problem in truth-functional logic.

● APPLYING THE REDUCTIO METHOD
   IN TRUTH-FUNCTIONAL LOGIC

Are (~A & ~B) and ~(R v B) consistent? We could simply construct a truth-table and find out the answer. However, that would require drawing up an eight-row truth-table. The reductio ad absurdum method offers a simpler way to solve the problem. Remember that two sentences are consistent just in case there's a possibility that both sentences are true.[1] So two sentences are consistent if and only if there exists a *truth-value assignment*—an assignment of truth-values to the components of the sentences—that makes both sentences true. The question, then, may be put this way: Is there any truth-value assignment that makes both sentences true?

We will begin with an assumption: Let us assume that the sentences are indeed consistent. If the two sentences are consistent, then there's a truth-value assignment on which both are true. Therefore, let us record our opening assumption by listing the two sentences and assigning **T** to the *main connective* of each:

$$1. \quad \sim\!A \,\&\, \sim\!B$$
$$\mathbf{T}$$
$$2. \quad \sim\!(R \lor B)$$
$$\mathbf{T}$$

These two lines constitute our initial assumption. Next, let us write down what must be the case if this initial assumption is correct. Let's focus on the first sentence, ~A & ~B. What must be the case if this formula is assigned **T**? If a conjunction is assigned **T**, then both of its conjuncts must also be assigned **T**, since a conjunction is true if and only if both conjuncts are true. We shall record this consequence of our initial assumption by breaking sentence (1) down into its two components and assigning **T** to each component:

$$3. \quad \sim\!A$$
$$\mathbf{T}$$
$$4. \quad \sim\!B$$
$$\mathbf{T}$$

To keep track of what we've done so far, we can place a checkmark after line 1 to record the fact that we've written down what must be the case if line 1 is correct. The check signifies that we've drawn the consequences we need to draw from this line and are finished with it:

$$1. \quad \sim\!A \,\&\, \sim\!B \checkmark$$
$$\mathbf{T}$$

Let us next focus on sentence 2, ~(R v B). What must be the case if this is also to be assigned **T**? If a negated formula is assigned **T**, the component that is negated must itself be assigned **F**. So, if the assignment on line 2 is correct, the component R v B must itself be assigned **F**. We therefore check off line 2 and add the following consequence to our line of reasoning:

$$5. \quad R \lor B$$
$$\mathbf{F}$$

---

[1] We'll often be attributing truth and falsity to things. Recall the simplifying convention stipulated in Chapter 2: If a sentence *expresses* a truth (or falsity), we will simplify things by saying the sentence itself is true (or false). However, remember we mean by this that the sentence expresses the truth.

Turning now to line 3, what must be the case if ~A is assigned **T**? Clearly, if ~A is assigned **T**, then A must itself be assigned **F**. We therefore check off line 3 and add the following to our list:

6. A
    F

Consider next line 4, which assigns **T** to ~B. If this is correct, then we must assign **F** to B. Consequently, we check off line 4 and add the following line to our list:

7. B
    F

Next, we turn to line 5. What must be the case if we assign **F** to R v B? A disjunction is false if and only if both disjuncts are false. Consequently, if the assignment at line 5 is correct, then we must assign **F** to R and **F** to B:

8. R
    F
9. B
    F

Putting this all together, we have:

1. ~A & ~B ✔
    T
2. ~(R v B) ✔
    T
3. ~A ✔
    T
4. ~B ✔
    T
5. R v B ✔
    F
6. A
    F
7. B
    F
8. R
    F
9. B
    F

Recall from Chapter 2 that an atomic sentence of TL is a sentence of TL containing no connectives. Thus lines 6 through 9 each contain one atomic sentence. We have reached a point where every sentence on the list is either checked off or is an atomic sentence. Nothing can be broken down any further, and our list is at an end.

What does this list of truth-value assignments indicate? Recall our goal: We wanted to know if there is any truth-value assignment that makes the two sentences (~A & ~B) and ~(R v B) both true. We began by assuming that both sentences are true. The lines following the assumption simply recorded what must be the case if that initial assumption is correct. In other words, the lines on our list tell us what truth-values must be assigned if both sentences are true.

Consequently, let us write down the truth-values assigned to the atomic sentences on the list and then try to form from this a truth-value assignment making both (~A & ~B) and ~(R v B) true. According to the list we've put together, if we assign **F** to A, **F** to B, and **F** to R, then the initial assumption will be correct: both sentences will be true. This shows that there is a possibility both sentences are true, which establishes that the two sentences are consistent.

We can restate this result in terms of truth-tables. Recall that each row of a truth-table specifies a truth-value assignment for the sentence or sentences on top of the table. A truth-value assignment corresponds to a row of a truth-table. So, since our list tells us that there exists a truth-value assignment making the two sentences true, we know that there exists a row on a truth-table making both sentences true. If two sentences are placed on a truth-table, and there is a row on which both compute true, then the two sentences are consistent.

For another example, suppose we want to know whether (A&B) and ~(AvB) are consistent or inconsistent. Let us begin by assuming that the two are consistent. This assumption may be recorded by assigning **T** to the main connective of each sentence:

1. A & B
      T
2. ~(A v B)
      T

This constitutes our initial assumption. Next, let us begin writing down what truth-value assignments must be made if this initial assumption is correct. Turning first to line 1, if a conjunction is true, both of its conjuncts must be true. So, we must assign **T** to each conjunct of 1:

3. A
      T
4. B
      T

Of course, we also check line 1 to record that we are through with it. We've written down what must be true if line 1 is correct and no longer need consider that line again in our list.

If line 2 is correct, what must be the case regarding its components? Clearly, if ~(AvB) is assigned **T**, then AvB itself must be assigned **F**. Let us check off line 2 and add the following to our list:

$$5. \quad A \vee B$$
$$F$$

Next, if line 5 is correct, what follows? A disjunction is only false if both of its disjuncts are false. Consequently, we check line 5 and assign **F** to each disjunct:

$$6. \quad A$$
$$F$$
$$7. \quad B$$
$$F$$

Every sentence on our list is now either checked off or is an atomic sentence. Nothing remains to be broken down into smaller parts. Let us now see if we can form a truth-value assignment making the initial assumption correct. Remember that the lines on our list were written down to record what must be the case if our initial assumption is correct. The list tells us that if we are to assign truth-values so as to make our initial assumption correct, we must assign **T** to A, **F** to A, **T** to B, and **F** to B. However, to do this would be contradictory, for this would be to assign sentence A a **T** and an **F** simultaneously. No truth-value assignment assigns both **T** and **F** to a component. We may conclude from this that no assignment of truth-values makes the initial assumption correct. That is, no assignment of truth-values makes sentences (A&B) and ~(AvB) true together. It follows that these two sentences are inconsistent.

Each list above began with an initial assumption and then traced the logical consequences of that assumption in the form of lines that had to be true if the initial assumption was correct. In each case, we followed the reductio ad absurdum strategy of making an assumption and then examining the logical consequences of that assumption.

2. TRUTH-TREE RULES

This method is getting difficult to work with. Besides being tedious, writing so many little T's and F's can become confusing. Fortunately, there's an easier way to keep track of what's true and what's false as we reason from an assumption. Let's switch to a logical device called a *truth-tree*.

Rules for the Ampersand Suppose we begin by making an assumption. Let us assume that a sentence P&Q is true. Instead of placing a **T** under the main connective as before, let us simply place the sentence on the left side of a line as follows:

$$P\&Q\,\big|$$

If a sentence is placed on the left side of a line, let this represent the assumption that the sentence as a whole is true. Next, what must be the case regarding the components **P** and **Q** if this assumption is correct? That is, what is the logical consequence of our initial assumption? If a conjunction is assigned **T**, then both of its conjuncts must also be assigned **T**, since a conjunction is true if and only if both conjuncts are true. Therefore, **P** and **Q** must each be true as well. Accordingly, we should extend the line and place **P** and **Q** both on the true side below **P&Q**:

$$\begin{array}{c} ✔ \text{ P\&Q} \\ \text{P} \\ \text{Q} \end{array}\bigg|$$

The check beside **P&Q** records the fact that we have taken account of the information represented by that sentence's placement and drawn an inference from that.

The general rule in this case, which will be named *Ampersand Left*, may be put as follows:

Ampersand Left

If a sentence P&Q is situated on the true side of a line, you may carry down P and Q and place them individually on the true side of the line under the sentence P&Q.

More formally:

$$\begin{array}{c} ✔ \text{ P\&Q} \\ \text{P} \\ \text{Q} \end{array}\bigg|$$

Each of the following involves an application of Ampersand Left:

✔ A&B	✔(AvB)&(GvE)	✔ ~(A⊃B)&C
A	AvB	~(A⊃B)
B	GvE	C

Notice that we may apply this rule only to a sentence positioned on the left side of a line and only when the main connective is an ampersand.

Assume next that a sentence **P&Q** is false. Instead of writing an **F** under the main connective, as we did in the previous section, let us place the sentence on the right side of the line:

$$P\&Q$$

If a sentence is placed on the right side of a line, let this represent the assumption that the sentence as a whole is false. So the left side of the line represents truth and the right side of the line represents falsity. Now, what follows from our assumption in this case? To begin with, if our assumption is correct, it does not follow that **P** and **Q** are both false. A conjunction is false just in case one, or the other, or both conjuncts are false. Thus, it follows only that one, or the other, or both of **P** and **Q** are false. Let us represent this inclusive disjunction of alternatives by splitting the line into two branches and placing the **P** and **Q** as follows:

You may be wondering whether we also need a third branch in this case upon which **P** and **Q** are both assigned **F**. The answer is that no such branch is needed, for this possibility is already covered by the two branches constituting the split. Remember that the split represents the inclusive disjunction that either **P** is false, or **Q** is false, or both are false.

The general rule in this case may be called *Ampersand Right*.

Ampersand Right

> If a sentence **P&Q** is situated on the false side of a line, you may split the line and place **P** on the false side of one branch and **Q** on the false side of the other branch.

More formally:

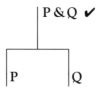

Each of the following involves an application of this new rule:

In every case, Ampersand Right is to be applied only to sentences positioned on the right side of the line and only to sentences whose main connective is the ampersand.

Notice that when we use Ampersand Left we "stack" the sentences and when we use Ampersand Right we "split" them.

Rules for the Wedge Suppose we assume that a sentence PvQ is false:

$$\Big| \text{ PvQ}$$

What must be the case if our assumption is correct? A disjunction is false if and only if both disjuncts are false. Therefore, if our assumption in this case is correct, then **P** and **Q** must both be false:

$$\Big| \begin{array}{l} \text{PvQ} \quad ✔ \\ \text{P} \\ \text{Q} \end{array}$$

The general rule, represented by the diagram immediately above, may be called *Wedge Right*.

Wedge Right

> *If a sentence* **P v Q** *is situated on the false side of a line, you may carry down* **P** *and* **Q** *and place them individually on the false side of the line under the sentence* **P v Q**.

Here are some applications:

$$\Big| \begin{array}{l} \text{AvB} ✔ \\ \text{A} \\ \text{B} \end{array} \qquad \Big| \begin{array}{l} \text{(E\&H)v\~(D⊃S)} \quad ✔ \\ \text{E\&H} \\ \text{\~(D⊃S)} \end{array}$$

Assume next that a sentence PvQ is true:

$$\text{PvQ} \Big|$$

What follows? It certainly does not follow that both **P** and **Q** are true. A disjunction is true just in case one or the other or both disjuncts are true. Consequently, if our initial assumption is correct, it follows merely that either **P** is true, or **Q** is true, or both are true. This consequence is represented by splitting the line as follows:

The split in this case represents the inclusive disjunction that either **P** or **Q** or both are true. Call the rule represented by this diagram *Wedge Left*.

Wedge Left

*If a sentence **P**v**Q** is situated on the true side of a line, you may split the line and place **P** on the true side of one branch and **Q** on the true side of the other branch.*

For example:

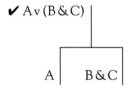

Rules for the Horseshoe Suppose we assume that a sentence **P⊃Q** is false. What follows? Well, if you look at the truth-table for **P⊃Q**, you will see that only one truth-value assignment results in an **F** in the final column, namely, that in which **P** is assigned true while **Q** is assigned false. This gives rise to the rule we will call *Horseshoe Right*:

Horseshoe Right

*If a sentence **P ⊃ Q** is situated on the false side of a line, you may carry down **P** and place it on the true side and you may carry down **Q** and place it on the false side.*

More formally:

$$P \quad \Big| \quad \begin{matrix} P{\supset}Q \ ✔ \\ Q \end{matrix}$$

For example:

$$A \quad \Big| \quad \begin{matrix} A{\supset}B \ ✔ \\ B \end{matrix} \qquad\qquad E \quad \Big| \quad \begin{matrix} E{\supset}(G\&S) \ ✔ \\ G\&S \end{matrix}$$

Suppose we assume next that a sentence **P⊃Q** is true:

$$P{\supset}Q \ \big|$$

What follows from this assumption? There are three possibilities:

1. If **P** itself is false, then no matter what **Q** is—whether **Q** is true or false—**P⊃Q** is automatically true. This can be verified by examining the truth-table for the horseshoe. So, if **P⊃Q** is true, it might be that **P** is false.
2. If **Q** itself is true, then no matter what **P** is, **P⊃Q** is true. This too can be verified by an examination of the horseshoe table. So, if **P⊃Q** is true, it might be that **Q** is true.
3. If **P** is false and **Q** is true, then **P⊃Q** is true. So, if **P⊃Q** is true, it might be that **P** is false and **Q** is true.

If we split the tree in the following way, the split will represent all three possibilities:

The split records the fact that if **P⊃Q** is true, then either **P** is false or **Q** is true, or both **P** is false and **Q** is true. Call the rule represented by this diagram *Horseshoe Left*.

Horseshoe Left

If a sentence **P ⊃ Q** *is situated on the true side of a line, you may split the line and place* **P** *on the false side of the left branch and* **Q** *on the true side of the other branch.*

Here are some applications of this rule:

Rules for the Triple Bar Suppose a sentence P≡Q is assumed true:

$$P{\equiv}Q \mid$$

Recall that a triple bar receives a **T** if and only if both sides have the same truth-value. Thus, if **P≡Q** is true, then either **P** and **Q** are both true or **P** and **Q** are both false:

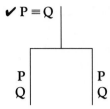

Call this rule *Triple Bar Left*. Here is an application:

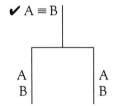

Consider next the case where a sentence **P≡Q** is assumed false:

$$| \ \mathbf{P{\equiv}Q}$$

If **P≡Q** is false, then **P** and **Q** must not both have the same truth-value. So, either **P** is true and **Q** is false, or **P** is false and **Q** is true. From this we get the rule *Triple Bar Right*.

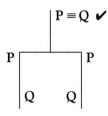

For example:

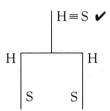

Rules for the Tilde The last rule to present is the simplest rule of all. If we assume a sentence **~P** is true, we write:

$$\text{~P} \Big|$$

If **~ P** is true, then **P** must itself be false. So, the obvious consequence of our assumption is:

$$\text{✔~P} \Big| \text{P}$$

Call this rule *Tilde Left*.

On the other hand, suppose we start by assuming that a sentence **~P** is false:

$$\Big| \text{~P}$$

It follows that **P** must itself be true:

$$\text{P} \Big| \text{~P ✔}$$

Call this rule *Tilde Right*.

3. TRUTH-TREE TESTS

● SOME TERMINOLOGY

Before putting these rules to work, some terminology will be necessary. The line diagrams we've been constructing here are called truth-trees and the ten rules developed above are *truth-tree rules*. A truth-tree has the shape of an upside-down tree. The top of the truth-tree is the tree's root. A *branch* of a truth-tree is a sequence of sentences that begins with a sentence at the top of the tree and continues to a tip of the tree. A *tip* of a truth-tree is formed by the last sentence at the bottom of a tree. Thus, a tree has as many branches as it has tips.

When a line running from the root to a tip turns into two branches, we say that the branch *splits*. The part of the tree above any splitting is the tree's *trunk*. Since the left side of any branch represents truth, we will call that side the "truth side." The right side, representing falsity, will be called the "false side." We will test sentences for the various logical properties by placing them on a tree's trunk. A sentence placed on the truth side will represent the *assumption* that the sentence is true, and a sentence placed on the false side will represent the assumption that the sentence is false.

Once we have drawn a consequence from a sentence, we will *dispatch* the sentence by placing a check beside it. Dispatched sentences are called "dead." Sentences not dispatched are called "live." Let us say that a branch with the same sentence appearing "live" on both of its sides is *closed*. In other words, if a branch of a tree assigns both **T** and **F** to a sentence, the branch is a closed branch. Closed branches will be marked with an "X" placed at their tip. If a branch is not closed, it is *open*. A *completed* branch is a branch consisting of only atomic sentences and checked-off sentences. In other words, on a completed branch, everything that can be broken down and checked off has been broken down and checked off. If every branch of a tree is closed, we shall say the tree is a *closed tree*. If at least one branch is completed and open, the tree shall be called an *open tree*. Finally, if every branch is either closed or is completed and open, we shall say the tree is a *completed* tree.

Truth-Tree Decision Procedures All of the material we have worked through so far has been mere preparation for the procedures that will be developed in this section. Using these procedures, we will be able to test sentences and arguments for the standard logical properties introduced in Chapter 2.

•— TESTING INDIVIDUAL SENTENCES

Tautology Testing What is the logical status of (A&B)v~(A&B)? In other words, is (A&B)v~(A&B) tautological, contradictory, or contingent? Recall from Chapter 2 that a truth-functional sentence is tautological if and only if it cannot possibly be false, it is contradictory if and only if it cannot possibly be true, and it is contingent just in case it is possibly true and possible false. Let us begin by assuming (A&B)v~(A&B) is false:

$$| \quad (A\&B)v\sim(A\&B)$$

Next, we apply the tree rules to the main connectives of each sentence. Since the main connective is the wedge, the first rule to apply is *Wedge Right*:

(A&B)v~(A&B) ✔

A&B
~(A&B)

Next, let us apply *Tilde Right* to ~(A&B):

(A&B)v~(A&B) ✔

A&B
~(A&B) ✔

(A&B)

In the next move, *Ampersand Left* is applied to the conjunction that sits on the true side:

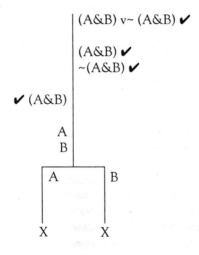

Finally, when we apply *Ampersand Right* to the last undispatched sentence, the tree closes:

We have now traced all the consequences of our initial assumption. The tree can't be extended any further because all compound sentences have been broken down to their simplest components. The tree is therefore completed. Since each branch has ended in a contradictory assignment of truth-values, the completed tree has closed. This tells us that our assumption implies a contradiction. Since only that which is contradictory can imply a contradiction, this proves our assumption is contradictory. Note carefully: This proves our assumption is contradictory, but it does not prove the sentence at the top of the tree is contradictory. The assumption was that (A&B)v~(A&B) is false. Since the assumption is contradictory, it follows that it is impossible that (A&B)v~(A&B) is false. It follows that (A&B)v~(A&B) is a tautology.

The general principle in the case of a tautology may be put as follows:

The Tautology Test

> *If a sentence is placed on the false side of a tree and the resulting completed tree is closed, the sentence is a tautology.*

More formally:

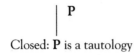

Closed: **P** is a tautology

If we had placed (A&B)v~(A&B) on the true side of a tree, the completed tree would have had at least one open branch. An open branch tells us it is possible our assumption is correct. Thus, if we had assumed our sentence true—by placing it on the true side—our completed tree would have told us that the sentence is possibly true. In the following tree, we begin by assuming (A&B)v~(A&B) is true. We then apply Wedge Left and split the tree. The tree is completed with applications of Ampersand Left, Tilde Left, and Ampersand Right.

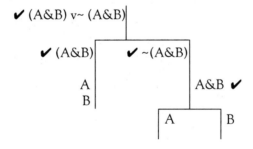

If a sentence is a tautology, it is said to be a *theorem* of truth-functional logic. Such a sentence is also said to be *truth-functionally true* or *logically true* because it can be proven true using just the methods of truth-functional logic, without performing scientific experiments or otherwise observing the physical world. So, if a sentence generates—according to the truth-functional rules of this chapter—a closed tree when placed on the false side of a tree, the sentence is a theorem of truth-functional logic.

Testing for Contradictions   Let's test (EvF)&~(EvF) to determine its logical status. We will start by assuming the sentence is true, which we may record by writing:

(EvF)&~(EvF) |

The resulting tree, when complete, closes:

✔ (EvF)&~(EvF)
EvF
✔ ~(EvF)
EvF
X

Our tree has assigned **T** to EvF and **F** to EvF. Thus, the assumption that (EvF)&~(EvF) is true has shown itself to be contradictory. Therefore, (EvF)&~(EvF) cannot possibly be true, which is to say that (EvF)&~(EvF) is contradictory. The general principle in the case of a contradiction is:

The Contradiction Test

> *If a sentence generates a closed tree when it is placed on the* true *side of a tree, then the sentence is contradictory.*

More formally:

Closed: **P** is a contradiction

If we had placed this sentence on the false side of a tree, the tree would have had at least one open branch. Since an open branch represents the possibility that the assumption is correct, this tree would have told us only that (EvF)&~(EvF) is possibly false:

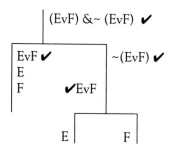

When to Close a Branch   When you build a tree, you may proceed in either of two ways: (1) you may stop and close off a branch as soon as the same sentence appears live on both sides, whether or not everything has been broken down into atomic sentences; or (2) you may extend each branch until everything has been broken down into atomic sentences. Thus, compare the following two trees:

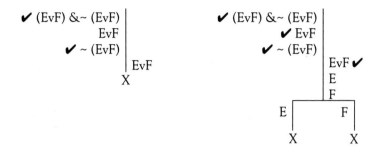

Notice that we get the same result in both cases. Whether we end a branch as soon as a contradiction arises, or whether we continue until every sentence has been broken down into atomic sentences, the result is the same: The branches close and the sentence is proven to be contradictory.

Testing for Contingency Recall that a sentence is contingent just in case it is possible for it to be true and possible for it to be false. In order to prove a sentence contingent, we must prove two things: (1) it is possible for the sentence to be true; (2) it is possible for the sentence to be false. Consequently, it takes *two* trees to prove that a sentence is contingent. Let's test (A&B)v(AvB).

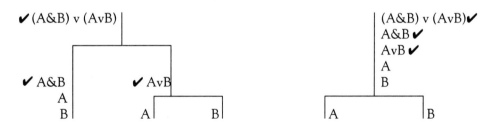

The tree on the left proves that the sentence is possibly true; the tree on the right proves the sentence possibly false. Putting these two conclusions together, we have proven that (A&B)v(AvB) is both possibly true and possibly false, which is to say that it is contingent. The general principle in the case of contingency is:

The Contingency Test

> *If a sentence generates a completed open tree when placed on the true side and also generates a completed open tree when placed on the false side, then the sentence is contingent.*

More formally:

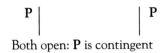

Both open: **P** is contingent

A REMINDER: THE RATIONALE BEHIND THE TREE TESTS

Now that we have completed a few trees, let us recall the rationale behind the tree method. The truth-tree method reflects the reductio ad absurdum strategy. We begin every tree with an assumption. This assumption is reflected in the initial positions that sentences receive at the top of the tree. When we place a sentence on a tree, we simply record an assumption concerning its truth-value. Now, as we apply the proper rules and extend the tree, we are simply deducing the logical consequences of our assumption. A closed branch means that we have reached a contradiction. If every branch of the tree closes, this indicates that our assumption implies a contradiction. It follows from this that the opposite of our assumption must be true. In other words, if a tree closes, it says, in effect, "Your assumption is impossible." And if a tree turns out to be an open tree, it is saying, in effect, "Your assumption is possible."

For example, if the assumption is that a sentence **P** is false, and the tree closes, this tells us that **P** can't possibly be false. It follows that **P** is necessarily true. If the assumption is that a sentence **P** is true, and the tree closes, this tells us that it's impossible for that sentence to be true. It follows that the sentence is contradictory.

Sometimes it is tempting to simply treat the tree method as a game in which the goal is to get the same sentence "live" on both sides of the line. This can be a helpful approach. However, it is important that you understand the rationale behind the tree method and behind each of the tree rules. It is sometimes easier to build the trees if, instead of relying mechanically on the rules themselves, you are able to think each move through logically.

EXERCISE 4.1

Use truth-trees to determine which of the following are tautological, which are contradictory, and which are contingent.

1. ~P ⊃ (P ⊃ Q)

2. Q ⊃ (P ⊃ Q)

3. ~(P ≡ P)

4. P ⊃ ~P

5. ~(P & ~P)

6. P v (P & ~P)

7. P ⊃ (P v Q)

8. (P & ~P) ⊃ Q

9. ~(P v ~P)

10. ~[P ⊃ (Q ⊃ P)]

11. $P \equiv \sim P$

12. $P \equiv (P \lor P)$

13. $(P \& Q) \supset Q$

14. $\sim(P \& Q) \lor Q$

15. $P \& (Q \& \sim P)$

16. $(P \supset P) \supset P$

17. $(P \lor Q) \supset P$

18. $(P \& Q) \supset (P \lor Q)$

19. $(P \supset Q) \supset Q$

20. $Q \supset (P \lor \sim P)$

21. $(P \supset Q) \supset (\sim P \lor Q)$

22. $(P \supset Q) \supset (\sim Q \supset \sim P)$

23. $(P \& Q) \supset (\sim P \supset \sim Q)$

24. $P \supset [Q \supset (P \supset Q)]$

25. $(P \supset Q) \equiv (\sim P \lor Q)$

26. $(P \supset Q) \equiv (\sim Q \supset \sim P)$

27. $(P \supset Q) \equiv (Q \supset P)$

28. $[(P \supset Q) \& P] \supset Q$

29. $[(P \lor Q) \& \sim P] \supset Q$

30. $(P \& Q) \supset \sim(\sim P \lor \sim Q)$

31. $\sim(P \& Q) \supset (\sim P \lor \sim Q)$

32. $\sim(P \lor Q) \supset (\sim P \& \sim Q)$

33. $\sim[P \supset (P \lor \sim P)]$

34. $[(P \supset Q) \& \sim Q] \supset \sim P$

35. $(P \equiv Q) \supset (P \supset Q)$

36. $(P \lor \sim P) \supset (Q \& \sim Q)$

37. $[(P \supset Q) \& (Q \supset R)] \supset (P \supset R)$

38. $(\sim P \& \sim Q) \supset \sim(P \& Q)$

39. $[(P \supset Q) \& Q] \supset P$

40. $(P \& \sim P) \lor (Q \& \sim Q)$

41. $(P \supset Q) \lor (Q \supset P)$

42. $(P \& Q) \& (\sim P \lor \sim Q)$

●— PROPER SPLITTING

In the next section, you will be working with trees containing several sentences. In order to prepare for this, consider the following problem. Suppose you have a tree with three sentences placed as follows:

$$
\begin{array}{c|c}
\text{PvQ} & \\
\text{AvB} & \text{Q\&P} \\
\end{array}
$$

Let's first dispatch PvQ:

$$
\begin{array}{c|c}
\checkmark \text{PvQ} & \\
\text{AvB} & \text{Q\&P} \\
\hline
\text{P} \quad\quad\quad \text{Q} &
\end{array}
$$

Now, how do we dispatch Q&P? When applying a rule to a sentence, we must write down the consequence on every branch underneath the spot on the tree occupied by the sentence being dispatched. Therefore, when we dispatch Q & P, we must split the tree on every tip underneath (Q&P):

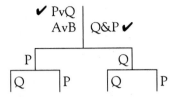

And when we dispatch AvB, we write down its consequences on each tip below where it sits:

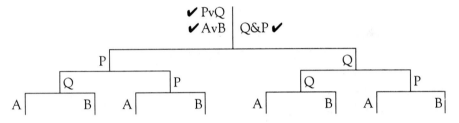

If you do not see why we must write down the consequences on every branch underneath the sentence we are applying a rule to, it might be helpful to go back to the section on decision trees and think through the decision tree for the hypothetical trip once more. A consequence of an item on the trunk of that tree was placed at the tip of each branch. Reviewing that decision tree again should also help make sense of the truth-tree method in general.

● PROVING AN ARGUMENT VALID

Suppose an English argument has been symbolized as follows:

1. (Pv~Q)v~R
2. ~(PvS)
3. ~(~Rv~H)
4. ~Q

The sentences above the line are the premises and the sentence below the line is the conclusion. Is this argument valid? Let us begin by making an assumption. Let us assume the argument is invalid. That is, assume it is possible that the premises are true and the conclusion is false. In order to record this assumption, we place the premises on the true side and the conclusion on the false side:

(Pv~Q)v~R
~(PvS)
~(~Rv~H)   ~Q

If this tree closes, this will indicate that our assumption is impossible. In that case, it would follow that it is impossible for the premises to be true and the conclusion false, which would prove that the argument is valid. However, if the tree is open, this will indicate that it is possible our assumption is correct. In that case, it would follow that it's possible for the premises to be true and the conclusion false, which would prove that the argument is invalid. The tree follows:

```
                    ✔(Pv~Q)v~R
                    ✔~(PvS)
                        ✔~(~Rv~H) │  ~Q ✔
                                     PvS ✔
                                     ~Rv~H ✔
                            Q
                                        P
                                        S
                            R          ~R ✔
                            H          ~H ✔

      ✔ Pv~Q │                                  ✔ ~R │
    P │            ✔ ~Q │                            │ R
    X              │ Q                            X
                   X
```

Since every branch of the completed tree closes, the tree is closed. The argument is therefore valid. The truth-tree test for deductive validity may be summed up as follows:

The Validity Test

> *Place the premises on the true side of a tree and the conclusion on the false side. If the completed tree closes, the argument is valid. If the completed tree is open, the argument is invalid.*

More formally:

$$\text{Premises} \mid \text{Conclusion}$$

Closed: The argument is valid
Open: The argument is invalid

●— FOUR COMMON PATTERNS OF REASONING

Let us now consider four patterns of reasoning we frequently follow in everyday thought.

The Disjunctive Syllogism Pattern   Consider the following valid argument:

Argument A
1.   Either we'll run or we'll swim.
2.   We won't run.
3.   So, we'll swim.

In TL, using obvious abbreviations, this becomes:

1.   RvS
2.   ~R
3.   S

This argument is an instance of the following valid argument form:

— **Disjunctive Syllogism Form** ——•

$$P \vee Q$$
$$\underline{\sim P}$$
$$Q$$

The validity of this form is easily proven on a truth-table:

PQ	PvQ	~P	Q
TT	T	F	T
TF	T	F	F
FT	T	T	T
FF	F	T	F

Notice that this truth-table contains no invalidating row. The following truth-tree proves the validity of argument A:

The Modus Ponens Pattern   This argument is clearly valid:

Argument B
1.   If it's raining, the roof is wet.
2.   It's raining.
3.   So, the roof is wet.

Using obvious abbreviations, this argument may be expressed in TL as:

1. R⊃W
2. R
3. W

This instantiates the following valid argument form:

— **Modus Ponens Form** ⟶•

$$P \supset Q$$
$$\underline{P}$$
$$Q$$

The validity of this form is established by the table below:

PQ	P⊃Q	P	Q
TT	T	T	T
TF	F	T	F
FT	T	F	T
FF	T	F	F

The following tree proves the validity of argument B:

The Modus Tollens Pattern   The following argument is valid, although it may not appear to be valid at first glance:

   Argument C
   1.   If it's raining, then the roof is wet.
   2.   The roof is not wet.
   3.   So, it is not raining.

In TL, with obvious abbreviations, this is:

1. R⊃W
2. ~W
3. ~R

If the two premises are both true, then the conclusion must be true. This argument is an instance of the argument form known as Modus Tollens:

$$P \supset Q$$
$$\underline{\sim Q}$$
$$\sim P$$

Modus Tollens is proven valid by the following table:

PQ	P⊃Q	~Q	~P
TT	T	F	F
TF	F	T	F
FT	T	F	T
FF	T	T	T

The following tree proves argument C valid:

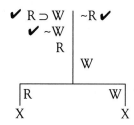

The Hypothetical Syllogism Pattern In everyday reasoning, we often link several things together in a chain of 'if, then' links. For instance:

Argument D
1. If it rains, the roof gets wet.
2. <u>If the roof gets wet, then the ceiling leaks.</u>
3. So, if it rains, then the ceiling leaks.

In TL, this is

1. R⊃W
2. <u>W⊃L</u>
3. R⊃L

This argument instantiates the valid argument form called *Hypothetical Syllogism*:

$$P \supset Q$$
$$\underline{Q \supset R}$$
$$P \supset R$$

The validity of this form is easily proven:

PQR	P⊃Q	Q⊃R	P⊃R
TTT	T	T	T
TTF	T	F	F
TFT	F	T	T
TFF	F	T	F
FTT	T	T	T
FTF	T	F	T
FFT	T	T	T
FFF	T	T	T

Notice that the table does not contain an invalidating row. The validity of argument D is easily proven on a truth-tree:

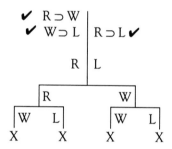

— STRATEGY

One question that arises at approximately this point is this: Which sentences should I dispatch first? The order in which you proceed won't affect the outcome of the tree you are constructing. You may begin wherever you wish and dispatch sentences in whatever order you wish. However, your tree will be less complex, and therefore simpler to build, if you will follow two suggestions:

1. Apply rules to sentences that do not require a split before you apply rules to those that do call for a split. Stack before you split.

2. Close off a branch as soon as a contradiction appears. There's no need to continue developing a branch once it has closed.

For example, consider the following argument:

1. (P&Q)v(RvS)
2. ~(R⊃S)
3. PvQ

In order to test this argument for validity, we may build the following tree:

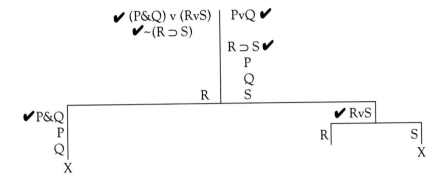

Since the completed tree is open, the argument is invalid.

If you will trace the course of the reasoning from the top of the tree to the bottom, you will notice that we first broke down only those lines that did not require a split in the tree. We then broke down the line requiring the split. If we had split the tree in the beginning, we would have had to break down all the other lines onto all three branches of our tree as follows:

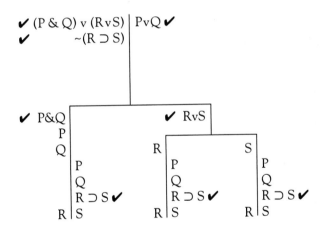

Compare the above two trees. Both end up proving the same thing. No matter what order you follow in breaking things down on a tree, the result will ultimately be the same. That is, if a tree closes when you break things down in one order, it will close if you break the same sentences down in any other order; and if it is open when you follow one order, it will be open if you follow any other order. The only difference between breaking things down in one order rather than another will be in the complexity of the resulting tree. For example, if you compare the previous two trees, you will probably agree that although both reach the same result, the first is simpler and easier to construct and the second is more complex and harder to construct.

EXERCISE 4.2

A. Use truth-trees to determine which of the following are valid arguments and which are not. In each case, the premises appear above the line and the conclusion appears below the line.

1. P ⊃ Q
    Q ⊃ P

2. P ⊃ Q
    ~Q
    ~P

3. ~(P & Q)
    ~P & ~Q

4. Q
    P v ~P

5. ~(P v Q)
    ~P v ~Q

6. P & ~P
    Q

7. P v Q
    ~P
    Q

8. ~(P & Q)
    P
    ~Q

9. ~(P & Q)
    ~P & Q

10. P ⊃ Q
    ~P
    ~Q

11. P ⊃ Q
    Q
    P

12. ~(P v Q)
    ~P & ~ Q

13. P ⊃ Q
    ~Q ⊃ ~P

14. ~P v ~Q
    ~(P v Q)

15. P ⊃ Q
    Q ⊃ R
    P ⊃ R

16. P ⊃ Q
    ~(Q v R)
    ~P

17. P ⊃ (Q v R)
    Q ⊃ S
    P ⊃ S

18. P ⊃ Q
    P ⊃ R
    P ⊃ (Q & R)

19. ~(A&B)vC
    ~A&(BvC)

20. ~[A&(B&C)]
    ~A&(~B&~C)

21. (P & Q) ⊃ R
    P ⊃ (Q ⊃ R)

22. P ≡ Q
    (P ⊃ Q) & (Q ⊃ P)

23. P ⊃ Q
    R ⊃ S
    ~P v ~R
    ~Q v ~S

24. A ⊃ (Ev~S)
    A&H
    Sv~E
    E

25. P ⊃ Q
    P ⊃ R
    Q ⊃ R

26. $\underline{P \vee (Q \& R)}$
    $(P \vee Q) \& (P \vee R)$

27. $P \supset Q$
    $\underline{Q \supset P}$
    $P \equiv Q$

28. $\underline{\sim P \& \sim Q}$
    $\sim(P \& Q)$

29. $\underline{\sim (P \& Q)}$
    $\sim P$

30. $A \vee B$
    $P \vee Q$
    $\underline{(A \vee P) \supset (B \& Q)}$
    $\sim B$

31. $P \supset R$
    $\underline{Q \supset R}$
    $(P \vee Q) \supset R$

32. $P \supset Q$
    $\underline{Q \supset R}$
    $R \supset P$

33. $\underline{\sim(P \& Q)}$
    $\sim P \vee \sim Q$

34. $A \supset (B \supset E)$
    $(B \supset E) \supset (A \supset B)$
    $\underline{A \& S}$
    B

35. $P \supset Q$
    $\underline{R \supset Q}$
    $P \supset R$

36. $A \vee B$
    $A \vee C$
    $\underline{\sim A \vee \sim E}$
    $B \vee C$

37. $\underline{P \& (Q \vee R)}$
    $(P \vee Q) \& (P \vee R)$

38. $\sim A \vee (\sim B \vee C)$
    $\underline{\sim B \vee (\sim C \vee H)}$
    $\sim A \vee H$

39. $P \supset Q$
    $R \supset S$
    $\underline{P \vee R}$
    $Q \vee S$

40. $P \supset Q$
    $R \supset S$
    $\underline{Q \vee S}$
    $P \vee R$

41. $(P \supset Q) \& (R \supset H)$
    $\underline{P \vee R}$
    $Q \vee H$

42. $P \supset (Q \vee R)$
    $\underline{(Q \vee R) \supset P}$
    $P \supset P$

43. $P \supset Q$
    $R \supset S$
    $\underline{\sim P \vee \sim R}$
    $\sim Q \vee \sim S$

44. $P \vee Q$
    $\underline{Q \vee R}$
    $P \vee R$

45. $P \supset Q$
    $\sim R \supset \sim S$
    $\underline{\sim Q \supset \sim H}$
    $\sim H \supset \sim S$

46. $(P \supset \sim Q)$
    $(\sim P \equiv Q)$
    $\underline{\sim P}$
    $\sim Q$

47. $(P \vee Q) \supset (Q \vee R)$
    $(Q \supset R) \vee P$
    $(Q \supset R) \supset (P \vee Q)$
    $\underline{\sim P}$
    $\sim Q \vee \sim R$

48. $\sim(A \& B) \vee C$
    $\sim H \vee G$
    $\sim B \vee H$
    $\underline{\sim(B \& G)}$
    $(A \& C) \vee (\sim A \& \sim C)$

B. Symbolize each of the following and then use truth-trees to test each for validity.

1. If Spock calls the ship, then Kirk will call the ship. If Kirk calls the ship, then Spock will call the ship. So, Kirk will call the ship if and only if Spock calls the ship.

2. If Ann and Bob both order burgers, then it's not the case that Elmer and Lulu will both order fish and chips. So, Ann and Bob won't both order burgers.

3. Either Edna brought the salad or Andy and Chris both brought salads. So, either Edna or Andy each brought a salad and either Edna or Chris each brought salad.

4. If Groucho sings, then if Chico sings then Harpo will play his harp. So, if Groucho and Chico sing, then Harpo will play his harp.

5. Mr. Ed speaks very well. Therefore, either Mr. Ed speaks very well or Muhammad Ali is the President of the United States.

6. Jane is an engineer. Melissa is a doctor. So, Jane is an engineer and Melissa is a doctor.

7. If the federal deficit increases, then Senator Smith's political future looks dim. So, either the federal deficit won't increase or Senator Smith's political future looks dim.

8. If Ilene swims, then Jolene will swim. So, if Jolene doesn't swim today, then Ilene won't swim today.

9. If Winnie wins a week's worth of *The Weekly*, then Monty will win a month's worth of *The Monthly*. If Daisy wins a day's worth of *The Daily*, then Yolanda will win a year's worth of *The Yearly*. Since Winnie will win her week's worth of *The Weekly* or Daisy will win her day's worth of *The Daily*, it follows that either Monty will win his month's worth of *The Monthly* or Yolanda will win her year's worth of *The Yearly*.

10. If the raindrops are 1 to 2 millimeters in size, the rainbow will be very bright violet, green, and red but have little blue in it. If the raindrops are 15 millimeters in size, the red of the primary bow will be weak. So, if the raindrops are 1 to 2 millimeters in size, then the red of the primary bow will not be weak.

11. If Elliott buys a pet turtle, then Lorraine will buy a pet hamster. Lorraine will buy a pet hamster. So, Elliott will buy a pet turtle.

12. If Joan puts roses in her garden, then Clyde will put daffodils in his garden. Clyde won't put daffodils in his garden. So Joan won't put roses in her garden.

13. Either Katie will do an experiment with her chemistry set or she will look at a drop of pond water with her microscope. She will look at a drop of pond water with her microscope. So, she won't do an experiment with her chemistry set.

14. Either Nathan will look at some galaxies with his telescope or Katie will try to synthesize an exotic chemical in her chemistry lab. If Katie tries to synthesize an exotic chemical in her chemistry lab then Nathan will not look at some galaxies with his telescope. Nathan will therefore look at some galaxies with his telescope.

●— TESTING FOR CONSISTENCY and INCONSISTENCY

Are A⊃B and ~(~AvB) consistent or inconsistent? Recall that two statements are inconsistent just in case it's not possible that both are true. Let's begin by assuming it is possible that both are true. We record this assumption by placing both sentences on the true side of a tree:

$$A⊃B$$
$$~(~AvB)$$

Since two statements are consistent just in case it's possible both are true, we are assuming that the two sentences are consistent. However, the tree closes:

The closed tree tells us that our assumption is impossible. Thus, it is impossible that the two sentences are both true. The two sentences are therefore inconsistent.

Suppose two sentences are placed on the true side of a tree and the completed tree is open. For example, suppose we begin a tree with:

$$A\&B$$
$$~(~Av~B)$$

The completed tree does not close:

✔ A&B
✔ ~(~Av~B)

> ~Av~B ✔
> ~A ✔
> ~B ✔

A
B
A
B

This proves that it is possible both sentences are true. It follows that the two sentences are consistent.

The general principle regarding consistency and inconsistency is:

The Consistency Test

> If two sentences are placed on the true side of a tree and the tree closes, then the two sentences are inconsistent. If two sentences are placed on the true side of a tree and the completed tree is open, the two sentences are consistent.

That is:

P
Q

Open: **P** and **Q** are consistent
Closed: **P** and **Q** are inconsistent

●— TESTING FOR IMPLICATION

Does A&B imply AvB? Recall that a sentence **P** implies a sentence **Q** if and only if it's not possible that **P** is true and **Q** false. Let's assume A&B does not imply AvB. Thus, let us suppose it is possible that A&B is true while AvB is false:

A&B | AvB

This initial assumption generates the following tree:

✔ A&B | AvB ✔
A

B | A

| B
X

The completed tree closes. So our assumption cannot possibly be correct. That is, it's not possible that A&B is true while AvB is false. It follows that A&B does imply AvB. The general principle in this case is the following:

The Implication Test

> If **P** is placed on the true side of a tree and **Q** is placed on the false side of a tree, and the tree closes, then **P** implies **Q**. If the completed tree is open in such a case, **P** does not imply **Q**.

That is:

$$\textbf{P} \mid \textbf{Q}$$

Closed: **P** implies **Q**
Open: **P** does not imply **Q**

TESTING FOR EQUIVALENCE

Is AvB equivalent to ~(~A&~B)? Recall that two sentences are equivalent just in case their truth-values match in every possible circumstance. So, if two sentences are equivalent, it is not possible that one is true and the other false; the two sentences are either both true or both false. It follows that in order to prove two sentences equivalent, we must prove two things: (1) it is not possible for the first to be true and the second false; (2) it is not possible for the second to be true and the first to be false. Since these are the only two ways two sentences could differ in truth-value, if we can prove (1) and (2), it will follow that the truth-values of the two sentences match in every circumstance, which will prove the two are equivalent.

Thus, we must construct two trees for the two sentences under consideration:

$$\text{AvB} \mid \text{~(~A\&~B)} \qquad\qquad \text{~(~A\&~B)} \mid \text{AvB}$$

The completed trees both close:

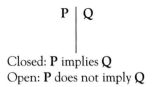

These two trees prove that it is not possible that one of the two sentences is true while the other is false. (AvB) and ~(~A&~B) are thus proven equivalent. However, if even one of these two trees had been an open tree, this would have proven that the two sentences are not equivalent.

The general principle for equivalence may be abbreviated as follows:

The Equivalence Test

Both closed: P and Q are equivalent
At least one tree open: P and Q are not equivalent

─────────────── • EXERCISES • ───────────────

Exercise 4.3

A.   In each case below, use a truth-tree to determine whether the sentence to the left of the comma implies the sentence to the right of the comma.

1.   P ⊃ Q, ~P v Q

2.   P v Q, P & Q

3.   P & Q, P v Q

4.   P ≡ Q, Q ⊃ P

5.   ~P , (P ⊃ Q)

6.   P ⊃ Q, Q ⊃ P

7.   P & Q, P

8.   P, P v Q

9.   Q, (P ⊃ Q)

10. P & Q, ~(~P v ~Q)

11. P ≡ Q, P ⊃ Q

12. P, P ⊃ P

13. ~(P v Q), P ⊃ Q

14. ~(~P & ~Q), P v Q

15. ~(P & Q), ~P & ~Q

16. ~(P ⊃ Q), ~(~P v Q)

17. ~(P v Q), ~P v ~Q

18. P & Q, P ⊃ Q

19. [(P ⊃ Q) & Q], P

20. P ⊃ Q, ~Q ⊃ ~P

21. [(P ⊃ Q) & ~P], ~Q

22. [(P ⊃ Q) & ~Q)], ~P

23. [(P ⊃ Q) & P], Q

24. ~(P & Q), ~P v ~Q

25. [(P v Q) & ~P], Q

26. (P ⊃ Q) & R, P ⊃ (Q & R)

27. P v (Q & R), (P v Q) & (P v R)

B.   In each case above, use a truth-tree to determine whether the two sentences are consistent or inconsistent.

Exercise 4.4

A.   Use truth-trees to determine, in the case of each pair of sentences below, whether the sentences are equivalent or not equivalent. Note that in each pair, a comma separates the members of the pair.

1.   P v Q , ~P & ~Q

2.   ~(P & Q) , ~P v ~Q

3.   P ⊃ Q , Q ⊃ P

4.   P & Q , Q & P

5.   P ≡ P , P ⊃ P

6.   ~(P v Q) , ~P & ~Q

7.   P v Q , ~(~P & ~Q)

8.   ~(P & Q) , ~P & ~Q

9.   ~(~P v Q) , ~(P & ~Q)

10. ~(P ⊃ Q) , ~P ⊃ ~Q

11. ~P ≡ ~Q , ~P ⊃ ~Q

12. P ≡ Q , [(P ⊃ Q) & (Q ⊃ P)]

13. P ⊃ (Q ⊃ R) , (P ⊃ Q) ⊃ R

14. (~P v ~Q) v R , ~(P & Q) v R

15. (P v Q) v R , P v (Q v R)

16. P & (Q v R) , (P & Q) v (P & R)

17. P ≡ Q , [(P & Q) v (~P & ~Q)]

18. (P ⊃ Q) v R , (~Q ⊃ ~P) v R

19. P ≡ (Q ≡ R) , (P ≡ Q) ≡ R

20. P ≡ (Q & R) , (Q & R) ≡ P

B.   For additional practice, use truth-trees to determine, in the case of each pair of sentences above, whether the sentences are consistent or inconsistent.

●— FURTHER MATTERS

Constructing a Truth-Value Assignment from a Tree   When we place a sentence on the left side of a tree, we are assuming that it is assigned **T**; if we place it on the right side, we are assuming that it is assigned **F**. Now, as we extend the tree, we are putting together a *truth-value assignment* that fits with our assumption. That is, a branch of a tree simply assigns **T**'s and **F**'s to the components of the sentence based upon the initial assumption. A completed open branch represents a truth-value assignment that yields the initial assumption. So, if our assumption assigns a sentence the truth-value **T**, then an open branch represents a truth-value assignment on which the sentence does indeed come out true. If a branch of a tree closes, then that particular attempt at finding a truth-value assignment for the assumption has failed. It has failed because it has ended in a contradictory assignment in which a sentence has been assigned both **T** and **F** at the same time. If every branch of a tree closes, then no truth-value assignment yields the opening assumption and the assumption can't possibly be true.

A completed open branch of a tree represents a truth-value assignment that is in accord with the initial assumption. This point is easier to appreciate if you understand how to construct a truth-value assignment from an open branch. Suppose we begin by assuming:

$$[Pv(Qv{\sim}R)]\&{\sim}P$$

The resulting tree looks like this:

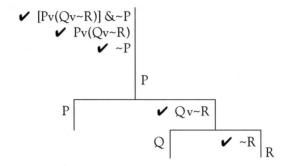

Now, in order to construct a truth-value assignment, we first find an open branch. From such a branch, we simply read off the truth-values assigned to the letters and this gives us a truth-value assignment in accord with the starting assumption. The middle branch is open. That branch shows P on the false side and Q on the truth side. R doesn't appear on this branch and so it may be assigned either true or false. This yields two truth-value assignments:

a. P = **F**   b. P = **F**
    Q = **T**   Q = **T**
    R = **T**   R = **F**

On both truth-value assignments, the sentence placed at the top of the tree computes—as assumed—true.

Thus you can see the way in which an open branch of a tree corresponds to a row of a truth-table yielding the tree's initial assumption.

A Summary of the Tree Tests   Let us say that a truth-tree constructed according to the rules of this chapter is a "TL tree." We may summarize this chapter's tree tests and introduce some new terminolgy with the following set of principles.

1. **P** is *truth-functionally true* if and only if it is the case that if **P** is placed on the false side of a TL tree, the resulting completed tree closes. A sentence that is truth-functionally true also counts as a *theorem* of truth-functional logic.

2. **P** is *truth-functionally false* if and only if it is the case that if **P** is placed on the true side of a TL tree, the resulting completed tree closes.

3. **P** is *truth-functionally contingent* if and only if it is the case that if **P** is placed on the false side of a TL tree the resulting completed tree is open and it is the case that if **P** is placed on the true side of a TL tree, the resulting completed tree is open.

4. An argument is *truth-functionally valid* if and only if it is the case that if the premises are placed on the true side of a TL tree and the conclusion

is placed on the false side, the resulting completed tree closes. An argument is *truth-functionally invalid* if and only if it is the case that if the premises are placed on the true side of a TL tree and the conclusion is placed on the false side, the resulting completed tree is open.

5. **P** and **Q** are *truth-functionally inconsistent* if and only if it is the case that if the two are placed on the true side of a TL tree, the resulting completed tree closes.

6. **P** and **Q** are *truth-functionally consistent* if and only if it is the case that if the two are placed on the true side of a TL tree, the resulting completed tree is open.

7. **P** *truth-functionally implies* **Q** if and only if it is the case that if **P** is placed on the true side of a TL tree and **Q** is placed on the false side, the resulting completed tree closes.

8. **P** and **Q** are *truth-functionally equivalent* if and only if it is the case that if the two are ever placed on opposite sides of a TL tree, the resulting completed tree closes.

Comparing Truth-Trees and Truth-Tables   The truth-tables differ from the truth-trees in a fundamental way. We begin a truth-table by assigning truth-values to the atomic components of a sentence, and then we generate the truth-value of the sentence as a whole. We begin a truth-tree the other way around, by assigning a truth-value to a sentence as a whole, and then generating truth-value assignments for the atomic components of the sentence. Truth-tables thus exhibit a "part-to-whole" strategy, and the trees reflect a "whole-to-part" strategy.

4. THE LAW OF NONCONTRADICTION

We noted in Chapter 1 that within logic, anything counts as logically possible except that which is contradictory. Contradictions are logically impossible. Recall that an explicit contradiction may be defined as any statement of the form **P&~P**. We used a truth-table in Chapter 2 to prove that any statement of the form **P&~P** is necessarily false, that is, can't possibly be true. The principle that contradictions are impossible is called the *Law of Noncontradiction* (or "LNC" for short). Now that you understand the nature of a truth-tree, you have the background to understand one important consideration in support of the LNC: If a reasoning process begins with a contradictory premise—no matter what the contradiction is—then any arbitrarily chosen conclusion validly follows. For example, consider the following argument: Pat's pet is an aardvark and it is not an aardvark. So, the Beatles disbanded in 1918. In TL, this is:

$$\frac{A\&\sim A}{B}$$

The following tree establishes the validity of this argument:

$$
\begin{array}{c}
\checkmark \ \mathrm{A\&\sim A} \quad | \quad \mathrm{B} \\
\mathrm{A} \\
\checkmark \ \sim\mathrm{A} \\
| \quad \mathrm{A} \\
\mathrm{X}
\end{array}
$$

This tree closes no matter which two sentences A and B abbreviate.

If you reason validly from a contradiction, then any and every conclusion validly follows.

So, any systematic reasoning that proceeds from a contradiction will be a piece of reasoning in which anything and everything follows. In such a case, no conclusion drawn, no matter how absurd, can possibly be the wrong conclusion to draw. However, any set of logical principles must specify some distinction between drawing the correct conclusion and drawing an incorrect conclusion. After all, that's the whole point of logical theory. After reflecting on these considerations, perhaps you will agree that if we do not presuppose the LNC, no useful system of reasoning can possibly be formulated.

PHILOSOPHY OF LOGIC

Deductive validity is the central concept of the deductive branch of logical theory. Recall that an argument is deductively valid if and only if there are no possible circumstances in which the premises are true and the conclusion is false. Most people, upon exposure to this definition, find it fairly intelligible. That is, after being shown a number of examples of deductively valid and deductively invalid arguments, most are able to classify additional arguments accurately enough, which shows that they have a working understanding of the concept.

However, the definition of deductive validity mentions some puzzling things. It contains a reference to things called "possible circumstances." It refers to things that are "true" and "false." And it speaks of things being true and false "in" a possible circumstance. Although most of us have some idea of what we mean when we speak of such things, it is rather difficult to articulate those meanings.

Since the concepts of possibility, truth, and falsity appear again and again throughout the study of logic, many philosophers believe that logical theory will make better sense if those concepts are systematically examined. Therefore, in this chapter we shall focus upon possibility, truth, and falsity.

When one reflects upon the concepts of possibility, truth, and falsity, certain philosophical questions naturally arise. These questions are addressed in the branch of philosophy known as the philosophy of logic. This chapter constitutes an introduction to that field of study.

This excursion into the philosophy of logic will serve several purposes. First, our discussion of the concepts of possibility and truth should deepen your understanding of a deductively valid argument, and it should also help you make sense of the symbolic logical languages we are studying. Furthermore, the philosophical ideas explored in this chapter have recently led to the development of a new semantical theory for the branch of logic known as *modal logic*. In the next chapter, when we take up that branch of our subject, you will see how the philosophical ideas of this chapter apply to a very interesting field of logic.

●— SOME PHILOSOPHICAL REFLECTION

Throughout the previous four chapters, we spoke of possibilities, truth, and falsity. A bit of reflection upon these raises the following *philosophical* questions.

1. When we refer to possible circumstances, what sort of thing is it that we are referring to? What *is* a possible circumstance anyway?

2. When someone says "That's true" or "Your premise is true," what is it that is said to be true? The speaker is attributing truth to something. What is this thing that is being called true? And when someone says "That's false," *what* is it that is false?

3. Pontius Pilate asked, "What is truth?" If something is true, what does its truth consist in? What constitutes truth?

4. What is truth or falsity "in" a possible circumstance? Can an explanation of this be given?

Let us now take these four questions, one by one.

2. POSSIBLE WORLDS

Consider this argument:

1. Either Jones will win or Smith will win.
2. Jones will not win.
3. So, Smith will win.

Clearly, this is a deductively valid argument, for in no possible circumstance are the premises true while the conclusion is false. One can think of lots of different possible circumstances in which sentences (1) and (2) are false and (3) is true, or in which (1), (2), and (3) are all false, or in which (1), (2), and (3) are all true. But there's no possible circumstance in which (1) and (2) are true while (3) is false. Our question is: What are these things we're calling "possible circumstances"?

In the early 1960s, several philosophers began developing an intriguing answer to this question, an answer that ushered in a new way to think about the concept of possibility. The answer they gave had first been introduced into philosophy by the philosopher and mathematician G. W. Leibniz during the seventeenth century. However, until the 1960s, Leibniz's idea was not elaborated upon or developed in any systematic way. By the 1970s, this new way to think about possibility had grown into an elaborate and fascinating theory that was being applied in many areas of philosophical research. The new theory, many philosophers argued, helps organize our thoughts about logical possibility to a degree never before attained. By the 1980s, the theory originally suggested by Leibniz had proven so illuminating in so many areas of philosophical discussion that it had become a widely accepted philosophical tool. Instead of speaking of "possible circumstances," many philosophers and logicians were now referring to entities they called "possible worlds." A whole new field of philosophical research had come into being, one called "possible worlds semantics." What do these philosophers have in mind when they speak of a possible world? We'll have to approach their idea in stages.

THE WORLD WE LIVE IN

Our planet orbits a star that sits about 30,000 light-years from the center of an enormous disk-shaped collection of stars called the Milky Way galaxy. Our galaxy is about 100,000 light-years from edge to edge and contains more than 100 billion stars, plus gigantic clouds of gas and soot that extend across millions of miles of space. This enormous quantity of matter is sailing through space at an unimaginable speed. The galaxy we find ourselves in, the Milky Way, is located near the edge of a swarm of several dozen galaxies, each with billions of stars, each trillions of miles apart. This cluster of galaxies, extending across 3 million light years of space, is called the Local Group. The Local Group lies near the edge of the Local Supercluster of galaxies—a gigantic cluster of clusters of galaxies whose radius is around 100 million light-years. Deep in the center of this supercluster sits the gigantic galaxy M87, a galaxy 30 times the size of the Milky Way. Beyond the Local Supercluster lie neighboring superclusters containing thousands and thousands of galaxies, each with billions or hundreds of billions of stars. Think of the vastness of it all. The universe—our world—it's the biggest thing in existence!

When we use the term 'our world,' let this cover everything, inclusive in space–time. That is, what we shall call 'our world' includes everything, no matter how far away in space and no matter how far away in time. Thus, nothing is too far away in space to be a part of our world, and nothing is too far away in time—past or future—to count as part of our world.

OTHER WORLDS

However, our world might have been different from the way it actually is. Instead of containing as many stars as it actually contains, it might have contained only half as many, or a fourth as many, or an eighth as many. Instead of containing protons, neutrons, and electrons, it might have contained other kinds of particles. Furthermore, it seems plausible to suppose that the laws of nature might have been different from what they are. Cosmologists plug alternative laws of nature into their computer models of the universe and come up with a picture of what the universe would have been like if the laws of physics had been different. And there are many other ways things might have been. You might have grown up in a different city, you might have been present at Woodstock when Jimi Hendrix played his version of "The Star-Spangled Banner," you might have slept an hour longer this morning, and so on.

So there are many other ways things might have been. Indeed, it is easy to see that there are an infinite number of ways things might have been. Just consider the sequence that begins 'The universe might have contained exactly 1,000 galaxies, it might have contained exactly 1,001 galaxies, it might have contained exactly 1,002 galaxies, it might have contained exactly 1,003 galaxies, . . .'

These infinitely many ways things might have been represent an infinite number of ways a world could be. Let us call each of these ways that things might have been a *possible world*. So, these infinitely many ways that things might have been constitute an infinity of possible worlds. Each of these possible worlds constitutes an alternative to the way things actually are.

Now, suppose one of these other possible worlds had been actual. Had one of these other worlds been actual, then the actual world—the way things actually are—would itself have been just another way things might have been. Consequently, the actual world is also a possible world. All the other possible worlds are *nonactual possible worlds*. The way things actually are may be called the *actual world*.[1]

●— LOGICAL SPACE

It is important to emphasize that other possible worlds are not distant planets or galaxies far off in space. Remember that anything , no matter how far away in physical space–time, counts as part of what we are calling the actual world. So other possible worlds are not located "out there" in physical space. However, we will be talking about possible worlds throughout this and the next two chapters. Where, then, do they exist? Consider how mathematicians respond to a similar question. In such branches as topology and geometry, mathematicians prove theorems about all sorts of interesting and often strange mathematical objects. Where do these objects exist? Certainly not in physical space. (No land exploration or space probe will ever discover the perfect triangle, the perfect cube, or the square root of two.) Furthermore, it seems that mathematical objects do not exist only in our heads. (When Pythagoras was thinking about the perfect right triangle, he wasn't thinking about an object that is situated at this moment inside your head.) So where do mathematical objects exist? A widely accepted answer among mathematicians is that mathematical objects exist in *mathematical space*. This is a nonphysical manifold that can be thought of but can't be physically traveled to. Now, just as mathematicians speak of a nonphysical manifold called "mathematical space," we shall speak of a similar nonphysical manifold we'll call "logical space." Let us say that the infinite realm of possible worlds constitutes *logical space*. Thus, our investigation of logical theory will take our thoughts across other possible worlds—a mental journey through logical space.

●— POSSIBLE WORLDS AS FAMILIAR OBJECTS OF THOUGHT

Possible worlds might seem to be strange entities at first exposure. However, a little reflection on the matter will show that they are actually quite familiar objects of thought. For instance, have you ever done something that you later deeply regretted having done? Did you think, "If only I had done something else instead"? If you had done something else, the world would be different from the way it actually is. When you had the "if only" thought, you were thinking of a counterfactual situation, a way

---

[1] The reasoning of this paragraph is based on an argument Robert Coburn gives in Coburn [1991], Chapter 4.

things might have been but were not. These other ways that things might have been—that you were thinking of—are simply the things we're calling "other possible worlds." Consequently, each time you have an "if only" thought, you are thinking of something that is an example of what we're calling a possible world.

Have you ever had to make a difficult decision? You probably thought long and hard about a number of alternative possibilities. These alternative ways things might be are simply the things we are calling "possible worlds."

We also entertain thoughts of possible worlds when we listen to a story, read a novel, watch a movie, or formulate a hypothesis. Each of these things describes a possible circumstance, or, as we shall now begin saying, a possible world. Our minds seem to have a natural ability to think about possible worlds.

In Chapter 1, we discussed counterexamples—possible circumstances in which the premises of a particular argument are true while its conclusion is false. We can now redescribe this in terms of the possible worlds idiom, or way of speaking: We were describing possible worlds in which the premises of a particular argument are true while the conclusion is false.

So, possible worlds are really nothing new. You have referred to them and thought in terms of them all your life.

●— DISTINGUISHING PHYSICAL POSSIBILITY and LOGICAL POSSIBILITY

Before we go any further with the idea of a possible world, we must distinguish two kinds of possibility. Could a human being swim the Atlantic Ocean—all 3,000 miles or so of it—in five minutes? It seems that such a swim would be impossible. The laws of physics won't allow anyone to swim that far that fast. At the required speed, which would be around 36,000 miles per hour, the friction alone would destroy a human body. And besides, no matter how many cans of spinach a person eats, the human body doesn't have the strength or the stored energy for such a task. So, given the laws of physics and the facts of human biology, that long-distance swim would be impossible.

So, could a person swim the Atlantic in five minutes? If you reasoned along the lines just given and answered no, then you had in mind the kind of possibility termed *physical possibility*. An event is physically possible if its occurrence would not violate a physical law of nature, and an event is physically impossible if its occurrence would violate a physical law.

Physical possibility may now be explained in terms of possible worlds. Consider just those possible worlds that have the same physical laws of nature as does the actual world. Call such worlds *physically possible worlds*. Something is a physical possibility if and only if there is a physically possible world in which it happens. So, to go back to our earlier example, it is not physically possible that someone could swim the Atlantic in five minutes, for there are no physically possible worlds in which someone swims 36,000 miles per hour.

In contrast, is it physically possible for someone to win the Washington State Lottery 10 times in a row? Well, such a thing would be extremely improbable, but it would not violate any physical laws of nature. That is to say, there are physically possible worlds in which a person wins the Washington State Lottery 10 times in a row. Thus, such an event is physically possible, although quite improbable. Notice the distinction here between *possibility* and *probability*. Some events are physically possible even though they are highly improbable.

It is therefore not physically possible that someone could swim the Atlantic in five minutes. However, if the laws of nature had been different, and if the facts of human biology had been different, wouldn't such a swim have been possible? There is a different kind of possibility, corresponding to a much broader sense of the word 'possible', according to which it is possible that someone swims the Atlantic in five minutes. Suppose the physical laws of nature had been different. And suppose human bodies were constructed in a different way. Indeed, suppose things had been so different that people could easily swim the Atlantic in five minutes. Isn't that a way things might have been? It seems plausible to suppose so. Therefore, we're speaking of a possible world. Of course, it's a possible world that is not a physically possible world. But it is a possible world nonetheless, for it's a way things might have been. More specifically, it's a way things might have been had the laws of physics been different. So, there is a possible world in which someone swims the Atlantic in five minutes. If there is even one possible world in which an event occurs, we will say the event is *logically possible*. Thus, although the Atlantic swim is not physically possible, it is nevertheless logically possible.

This distinction between logical and physical possibility is of such crucial significance in logic that we must examine it more closely.

It is traditional to use the term 'logically possible' for the broadest, most inclusive category of possibilities, the category that includes within itself all possibilities and kinds of possibilities. So, every possible world counts as a 'logically possible' world. It follows that every physically possible world, that is, every world containing the same physical laws of nature as the actual world, counts as a logically possible world as well. However, it seems plausible to suppose that some logically possible worlds are not physically possible worlds, for some possible worlds, it would seem, have laws of nature different from those we find in the actual world. So, although all physically possible worlds are also logically possible worlds, not all logically possible worlds are physically possible worlds.

The relationship between logical possibility and physical possibility may be represented visually. Let the following region represent the set of all possible worlds.

Since the region contains an infinite number of mathematical points, and since there are an infinite number of possible worlds, let each point in the region represent a unique possible world. The region thus represents all that is logically possible.

One set is a *proper subset* of a second set if and only if every member of the first set is a member of the second set, but not every member of the second set is a member of the first set. For example, the set of human beings constitutes a proper subset of the set of mammals. Every world belonging to the set of physically possible worlds is a member of the set of all logically possible worlds, but not all the members of the set of logically possible worlds belong to the set of physically possible worlds. The set of all physically possible worlds thus constitutes a proper subset of the set of all logically possible worlds. The set of physically possible worlds may be represented as a region within the region of the logically possible:

```
┌─────────────────────────────────┐
│                                 │
│  Logical Possibility            │
│        ┌──────────────┐         │
│        │  Physical    │         │
│        │  Possibility │         │
│        └──────────────┘         │
│                                 │
└─────────────────────────────────┘
```

And so, to sum this up: Is it possible that someone could swim the Atlantic Ocean in five minutes? It's not physically possible, for in no physically possible world does someone swim the Atlantic in five minutes. But such a thing is logically possible, simply because in some possible world, someone can swim the Atlantic in five minutes. From here on, we will use the term 'possible world' in place of 'logically possible world' to refer to this most inclusive kind of possibility.

●— THE POSSIBLE WORLDS SENSE OF 'IN'

Some readers might be puzzled by our talking of events happening "in" various possible worlds. Does such talk really make any sense? What do we mean when we speak of something happening "in" a possible world? Perhaps the following will help make sense of such talk. We readily understand sentences such as

> In the novel *Robinson Crusoe*, a man is shipwrecked on a deserted island.

and

> In the story *Hansel and Gretel*, two children get lost in the woods.

And so it seems that we understand what it is for an event to happen "in" a novel or a story. But a novel or story is essentially a description of a possibility. As such, a novel or story describes a possible world. Therefore, let us think of an event happening "in" a possible world in much the way we think of an event happening "in" a novel, a story, or a movie. The two senses of 'in'—the possible worlds 'in' and the literary 'in'—seem to be closely analogous, if not identical.[2]

---

[2] The analogy can be made even more explicit. Consider the set of all novels having the same physical laws of nature as does the actual world. Call such novels 'physically possible novels'. These form a proper subset of the set of all novels. There are no physically possible novels in which someone swims the Atlantic in five minutes or flies unaided to the moon. That should make sense to you. But then, similarly, it should make sense when we say that there are no physically possible worlds in which either of these things happen.

— ARE THERE LIMITS TO WHAT IS POSSIBLE?

When it comes to possible worlds, does anything one can think of qualify as possible? That is, will anything anyone suggests, no matter how absurd, count as at least logically possible? Or are there limits to what is possible? Are there some suggestions that count as logically impossible? What forms the boundary of logical space? There are a number of interesting disputes in philosophy, as well as in other fields, concerning whether a particular collection of sentences describes a real possibility or not. For instance, suppose you climb into a time machine, turn some dials, and visit an event from your childhood. Is such a thing even logically possible? Or, to take a different issue, is it possible for an object to exist in two places at the same time? There have been a number of disagreements about whether to count this or that as a "real" possibility. We won't enter into any of those disputes here. But there is one thing we will base much of our discussion on: that which is self-contradictory is a paradigm case of the impossible. The principle that contradictions are impossible, often called the *law of noncontradiction*, was discussed in Chapters 3 and 4. Our discussion of possible worlds will therefore presuppose the law of noncontradiction.

Recall from Chapters 1 and 2 that an explicit self-contradiction is a conjunction in which one conjunct is the negation of the other conjunct. Here are some examples of self-contradictory sentences.

1. Pat is retired and it is not the case that Pat is retired.

2. Rita is taller than Fred and it is not the case that Rita is taller than Fred.

3. Jim weighs 150 pounds and it is not the case that Jim weighs 150 pounds.

It is extremely important that you understand the following: In order for a sentence, in a certain context, to qualify as an explicit self-contradiction, the sentence component that has a negation operator attached to it must have one definite meaning, and the other component sentence conjoined to it must have exactly the same meaning. For example, in sentence 3 above, the sentence 'Jim weighs 150 pounds' must have exactly the same meaning in both of its occurrences in that conjunction. So, in order for the sentence 'Jim weighs 150 pounds and it is not the case that Jim weighs 150 pounds' to count as a self-contradiction, each occurrence in that sentence of 'Jim weighs 150 pounds' must refer to the same Jim, the same scale, the same moment in time, and so on.

Why should we count that which is self-contradictory as impossible? Why shouldn't we suppose that some possible worlds include or constitute self-contradictory states of affairs? Three sets of considerations support the law of noncontradiction. First, if you will reflect upon the nature of a self-contradiction, you will see that when one utters a self-contradiction, one conjunct asserts one thing, and then the other conjunct takes back or denies precisely that which the first conjunct asserted. The net result is that nothing is asserted. The one conjunct cancels out what was said by the other conjunct, leaving nothing. No possible situation is described by a self-contradiction because a self-contradiction doesn't describe anything.

Second, Chapter 3 demonstrated that if a natural deduction proof begins with a contradictory premise, then any and every conclusion follows. Chapter 4 showed that if a truth-tree is constructed for an argument that has a contradictory premise, then any and every conclusion follows. In other words, if someone reasons from a contradiction, then no distinction can be made between correct and incorrect reasoning, for any and every conclusion follows. However, any set of logical principles must draw some sort of distinction between correct and incorrect reasoning. (That's one of the fundamental purposes of logic.) So, unless we agree that contradictions are impossible, no distinction between correct and incorrect reasoning can be formulated and no system of logic can be developed.

Third, it seems that effective human communication itself rests upon the law of noncontradiction. In order to help you think about this on your own, consider what you would do in the following situation. You have accidentally swallowed poison and will die in 15 minutes unless you get to a hospital. You know there's a hospital within four blocks, but you don't know which direction to head. Fortunately, you see a doctor and you ask, "Where's the hospital?" Now, suppose she says: "It's two streets north and it's not two streets north." Which direction do you go? Has she really communicated anything of value? Has she communicated anything at all?

Consequently, the one limit to the possible that we shall base our discussion on is the self-contradictory. Surely nothing that counts as self-contradictory will also count as logically possible. Anything self-contradictory, then, counts as logically impossible. If we agree that a self-contradiction is an impossibility, then there are no possible worlds in which, for example, a person is 66 and yet is not 66 at the same time, or in which an only child plays with her brother.

•— QUESTIONING THE POSSIBLE WORLDS FRAMEWORK

The concept of a possible world seems extremely simple, almost silly at first. However, initial appearances are, in this case, deceptive. In the next chapter, as noted previously, you will see how the idea of a possible world has given birth to a new semantical theory for the branch of logic known as modal logic. This branch of logic is currently the scene of some interesting and significant philosophical research.

In this section, you have been introduced to the rudiments of the possible worlds framework of ideas. Although the possible worlds framework is becoming widely accepted among philosophers, there are many philosophers who raise strong philosophical objections to the entire possible worlds approach. They argue that if we accept the framework of ideas associated with the concept of a possible world, we must also accept certain logical consequences of that framework, consequences that seem either absurd or at least extremely implausible. Since it is not reasonable to accept a theory that has absurd or highly implausible logical consequences, they argue that the possible worlds approach should be rejected. As you might expect, philosophers who favor the possible worlds framework have developed arguments that reply to these criticisms.

It would be nice if we could examine some of the interesting arguments for and against the possible worlds framework at this point. However, those arguments presuppose a significant amount of philosophical and logical theory. Philosophy students therefore typically study those arguments in upper-level philosophy classes.

EXERCISE 5.1

Part I Read the following sentences carefully. Give each the standard meaning it carries within our linguistic community. In each case, is there a logically possible world in which the statement is true? Is there a physically possible world in which the statement is true?

1. Someone swam the Panama Canal in one minute.

2. This statement is false.

3. Eddie Murphy is the President of the United States.

4. Pat spoke the truth when he said, "I cannot speak."

5. $2 + 2 = 456$

6. A scientist who built a time machine traveled back to her childhood and watched her younger self come home from kindergarten.

7. Water is naturally purple.

8. There exists a perfectly square house, and all of its sides face north.

9. An only child is swimming with her two sisters.

10. A supernova destroyed the Earth in 1990.

11. Ed is taller than Betty, Betty is taller than Pat, and Pat is taller than Ed.

12. In Hibbing, Minnesota, there lives a person who is older than all persons.

13. Richard Nixon was the lead guitarist for the Rolling Stones on their first album.

14. Someone is 2 million years old.

15. On an island in the South Pacific there are two brothers, and each is older than the other.

16. God exists.

17. Someone is 16 and someone is not 16.

18. Someone is 16 and also not 16.

19. James spoke the truth when he said, "I'm lying."

20. There exists a mountain higher than all other mountains.

21. There exists a greatest positive integer.

22. There exists a being so perfect that nothing could possibly be more perfect, not even the being itself.

23. The Summer of Love occurred during the fall of 1967.

24. "One of themselves, a prophet of their own, said, 'Cretans are always liars, evil beasts, lazy gluttons.' This testimony is true. . . ." St. Paul, *Titus* I, 12–13.

Part II Give an example of a logically possible but physically impossible event that occurred in a novel, movie, or play.

3. THE BEARERS OF TRUTH

Suppose someone walks up to you and for no apparent reason says, "That's purple." You might naturally respond, "What is purple?" In other words, what is the thing that is supposedly purple? Now, imagine that someone says, "The moon has craters," and you reply, "That's true." When you say "That's true" or "That's false," *what* is it that is true or false? What are you attributing truth (or falsity) to? Truth and falsity seem to be characteristics *of* something. Of what? In other words, what are the "bearers" of truth and falsity?

In this section, we will examine a series of theories on the nature of the bearer of truth. In the case of each theory, opposing arguments will also be discussed. As we move from one theory to the next, the discussion will be dialectical in nature. As you think about each theory and the arguments against it, it will be up to you to decide which theory provides the most reasonable answer to the question under discussion.

● THE BELIEF THEORY

What is it that is true when something is true? One suggestion that might seem plausible initially may be called the *belief theory*: It is the beliefs in people's minds that are the bearers of truth and falsity. On this view, if Pat says, "The moon has mountains," and someone replies, "That's true," the thing that is true is Pat's belief. According to the belief theory, it is Pat's belief—a psychological state or component of his mind—that is the thing that is true. This belief arose in Pat's mind at a certain time, was caused by certain things he saw and heard, and it will cease to exist at some moment in time.

The belief theory seems, at first glance, to provide a plausible answer to our question. After all, we do sometimes say things like "Her belief turned out to be true." However, the belief theory seems to conflict with several things we know about belief and truth. Consider the following argument.

> We learn in elementary mathematics that there are an infinite number of positive whole numbers. Mathematicians study the properties of these numbers, prove theorems about them, and so on. However, corresponding to each of these numbers is the truth that that particular number has a successor, and corresponding to each there is also the falsehood that that particular number has no successor. Since there are an infinite number of positive whole numbers, and since each has associated with it at least one truth and at least one falsehood, it follows that there are an infinite number of truths and also an infinite number of falsehoods.

AN INFINITE SERIES OF TRUTHS:	AN INFINITE SERIES OF FALSEHOODS:
1 has a successor	1 has no successor
2 has a successor	2 has no successor
3 has a successor	3 has no successor
.	.
. . . and so on	. . . and so on

Unfortunately, this conclusion seems fatal to the belief theory, for in all of history, only a finite number of persons have lived, and each has had a finite number of beliefs. Even if each person acquires a billion beliefs every billionth of a second, each person still acquires, in the course of a lifetime, a finite number of beliefs. Which means that a finite number of belief states—the entities referred to by the belief theory—are available to serve as the bearers of truth and falsity. Since there exist an infinite number of truths and an infinite number of falsehoods, it follows that there are more truths and falsehoods than there are belief states. But beliefs can't be the things that are true and false if there are more truths and falsehoods than there are beliefs. There are simply not enough beliefs to serve as truth-bearers. Belief states, therefore, cannot be that which is true or false.[3]

Another argument also seems to count against the belief theory. Billions of years ago, before anyone was around to have any beliefs, it was true that water is composed of hydrogen and oxygen. It simply makes no sense to suppose that it

---

[3] Compare this piece of reasoning with the following. Suppose that someone is thinking of something, you have to guess what it is, and you can ask only three questions. Suppose you first ask, "How many of these things are there in this room?" The answer is, let us say, 100. If you look around and see that the room contains only three lamps, then you know one thing for sure: The person is not thinking of a lamp.

became true that water is composed of hydrogen and oxygen only when, in 1793, Henry Cavendish made his famous discovery and became the first human being to believe that water is composed of hydrogen and oxygen. Rather, this truth was discovered in 1793, when it was first believed. It was true that water is composed of hydrogen and oxygen long before it was believed to be true. Similarly, billions of years ago, before anyone had any beliefs about the matter, it was false that water freezes at 10,000 degrees Celsius. It makes no sense to suppose that this became false only when the corresponding mistaken belief formed in someone's mind. So it seems that truth and falsity may exist independently of, or in the absence of, human belief. This gives us further reason to reject the belief theory, for if truths and falsehoods may exist in the absence of any human beliefs, then that which is true must be something other than a belief.

The reasoning against the belief theory can be generalized to support the conclusion that the bearer of truth cannot be any sort of mental or psychological entity existing in the mind. Consider the following argument:

    a.   There are an infinite number of truths and falsehoods.

    b.   Only a finite number of people have ever lived, and each has had only a finite number of beliefs, thoughts, and other mental states.

    c.   Therefore, the bearers of truth and falsity are not beliefs, thoughts, or any other mental states.

    d.   So, the things that are true and false are entities that exist independently of—separate from—the human mind.

If this argument is sound, then the truth and falsity bearers must be nonmental, nonpsychological entities of some sort.

If you are convinced that the bearer of truth is a belief, or some kind of mental state, and you wish to defend your theory, there are several ways you might argue. You might try producing an argument for the claim that there does not exist an infinity of truths. Your argument would attack a crucial premise in the reasoning against the belief theory. Alternatively, you might try producing an argument that there does exist an infinity of beliefs. This argument would attack a different premise in the reasoning against the belief theory. Or you might develop an argument for the claim that there were no truths in existence before human believers existed. Here you would have to explain, for example, in what sense it was not true 5 million years ago that water is H_2O, and you would want to explain how something like this became true when the relevant belief formed in someone's mind.

●— THE SENTENCE TOKEN THEORY

If the truth-bearer is not a belief or a mental entity, what is it? What is that which is true? According to another proposal, the truth-bearer is a *sentence*. This suggestion seems plausible at first glance. We naturally say things such as "That sentence is true" or "That sentence seems false." However, this suggestion faces a problem. The word 'sentence' is ambiguous; two senses of the word must be distinguished. Consider the box below.

```
┌─────────────────────────┐
│                         │
│                         │
│                         │
│     Snow is white.      │
│                         │
│                         │
│     Snow is white.      │
│                         │
│                         │
│                         │
└─────────────────────────┘
```

How many sentences does the box contain? If you answered that it contains two sentences, then you were thinking of what philosophers call *sentence tokens*. These are the individual physical sentences in front of you, the marks on paper, composed of ink, with a certain size, shape, location, and so on. If you answered that the box contains just one sentence that has been written twice, you had in mind what philosophers call the *sentence type*. The box "contains" just one sentence type, although it contains two instances or tokens of that type. The relation between a sentence type and a sentence token is analogous to the relation between a sentence form and its instances.

Here is another example to help you grasp the distinction between a sentence token and a sentence type. Suppose 100 people are each asked to write down their favorite sentence on a scrap of paper. Suppose they each wrote down the same sentence. In this case, we would have 100 scraps of paper and 100 sentence tokens, but only one sentence type. The 100 written sentences are all the same sentence in virtue of being 100 tokens of the same type.

A *physical entity* is an entity that is composed of matter, that occupies a volume of space, and that has a position in space and time. We detect physical entities with our five physical senses and with physical instruments such as microscopes and telescopes. The sentence token is a physical entity. It is composed of ink or some other writing material, it occupies a volume of space, it reflects light, it has a physical location, and so on. The sentence type, however, is not a physical entity. The type is what philosophers call an *abstract entity*. This is an entity that is not composed of material particles, does not occupy a volume of space, does not have a location in space and time, and is not detected by the five physical senses. Although they cannot be detected using the five physical senses, abstract entities are such that they can be thought of and understood by the mind.

Are there such things as abstract entities? Do abstract entities really exist? *Absolute materialism* is the view that absolutely nothing exists except particles of matter and objects composed out of particles of matter. If you hold this view, then you won't accept the existence of abstract entities such as sentence types. If you agree with absolute materialism and deny that abstract entities exist, and if you also suppose that the truth-bearer is a sentence of some sort, then you will want to postulate that it is the sentence token that is the bearer of truth. Call this suggestion the *sentence token theory*.

Are sentence tokens the bearers of truth? This suggestion seems plausible at first. After all, we do sometimes say things like "That sentence is true" or "That sentence is false" while pointing at a sentence token. However, the sentence token theory faces a major objection. We have already argued that there exist an infinite number of truths as well as an infinite number of falsehoods. The truth-bearers, whatever they are, must therefore be available in infinite number. But only a finite number of sentence tokens have ever been written down, printed, or produced in any way, counting all cultures, all language systems, and all times. Even if each person in the world had been writing a billion sentence tokens every billionth of a second, that would still add up to a finite number of sentence tokens. That implies that there are more truths and falsehoods than there are sentence tokens. Consequently, whatever the bearers of truth and falsehood are, it seems they are not sentence tokens.

However, if you are convinced that the bearer of truth is a physical sentence token and you wish to defend the sentence token theory, you might argue that only a finite number of truths exist. This would undermine a key premise of the argument against the sentence token theory. Also, you might try explaining the sense in which the first production of a sentence token brought the first truth into existence, and how it could be that there would be no truths at all if we were to erase or destroy all sentence tokens.

The argument given above against the sentence token theory can be generalized to support the conclusion that the bearer of truth is not any sort of physical entity. Consider the following argument:

    a.   An infinite number of truths and falsehoods exist.

    b.   No physical object that could plausibly be construed to be the truth-bearer and falsity-bearer exists in infinite numbers.

    c.   Therefore, the bearers of truth and falsity are not any sort of physical object.

As a result of reasoning along these lines, many philosophers argue that we have good reason to suppose that the truth-bearers are neither belief states nor sentence tokens. What, then, are the bearers of truth and falsity? From here, philosophers divide into two schools of thought. Some argue that the bearers of truth and falsity are sentence types. Others argue that the bearers of truth and falsity constitute a unique kind of abstract object, a kind of abstract entity that is not a sentence type and indeed is not sentential or linguistic in nature at all. Let's look briefly at the reasoning for these two schools of thought.

●— THE SENTENCE TYPE THEORY

According to some philosophers, although the truth-bearer is obviously not in any case a belief in the mind or a physical sentence token, it does make sense to say that the *sentence type* is that which is true. However, a problem immediately arises if we simply say that the sentence type, by itself, is the truth-bearer. Suppose that at noon, Pat says,

I am hungry.

Abstracted from the context of utterance, the sentence type in this case seems to have no truth-value at all. This suggests that a sentence type by itself—that is, abstracted from its context of utterance—cannot plausibly be said to be that which is true or false.

In response to this line of argument, proponents of the sentence type theory amend their account by supposing that it is not the pure sentence type alone that is the thing that is true or false. Rather, it is the sentence type *with its context of utterance built into it* that is true or false. The suggestion is that the sentence type's context must be built into it in such a way that the sentence can be fully understood when abstracted from its context of utterance. The following example should illustrate the procedure. Assuming Pat's last name is Smith, Pat's sentence, with context built in, becomes:

"Pat Smith is hungry at noon on August 17, 1982."

Interpreted in this way, the sentence type in this case can plausibly be said to be true. It is such "context-free" sentence types—sentences with their context built in—that are proposed, then, as the bearers of truth and falsity. Call this account the *sentence type theory*. One virtue of this theory is that attributions of truth and falsity are explained in terms of familiar entities, namely, context-free sentences, which are themselves parts of a language.

● THE SUI GENERIS VIEW

Some philosophers reject the previous theories we have surveyed and argue that the truth-bearer is an entity that can be expressed by a sentence but that is neither mental, physical, nor sentential in nature. On this view, the truth-bearer is a *sui generis* abstract entity—an entity that is like nothing else on the list of proposed truth-bearers. What reasons exist for such a view? We will look at two lines of reasoning.

First of all, three sentences, drawn from three different languages, can express the same (one) truth. For example, consider the following three sentences, drawn from three different languages, and suppose each an accurate translation of the other two:

1. The Moon has craters.
2. La luna tiene cráteres.
3. Maraming maliliit na bulkau sa buwan.

One thing seems obvious. Between the three sentences, only one truth is expressed. But what is this one thing that is true here? That which is true cannot, it seems, be the individual sentences themselves, for there are three sentences and one truth.

And the truth-bearer cannot be a part of one language either, for our example displays three language systems but only one truth. This is puzzling. Perhaps we can make sense of the example if we suppose that:

  i. All three sentences express, in different ways, one and the same truth;
  ii. This truth-bearer, although it can be expressed within each language, does not itself belong to any one language.

If this is correct, then it seems that the thing that is true can't be a component of a language. Therefore, it seems plausible to suppose that the truth-bearer must be a nonsentential and indeed a nonlinguistic entity of some sort.

Another line of reasoning for the claim that truth-bearers are nonliguistic entities begins with some thoughts about our prelinguistic ancestors who lived tens of thousands of years ago. Lacking a language, they did not write, speak, or understand any sentences of any language. This is because a sentence is a part of a language, and if one is to understand or use a sentence, one must understand the grammar and semantics of the language the sentence belongs to. However, although our prelinguistic ancestors understood no sentences, they must have had beliefs. For in order to survive, they must have believed certain things about the world. And sometimes, that which they believed was true and other times that which they believed was false. Now, this argument continues, that which was believed—whether true or false—couldn't have been sentences of any kind. Remember that these individuals had no language and so were incapable of understanding any sentences, token or type. It seems to follow that whatever the objects of their beliefs were, they were not sentences or components of a language. This is another reason to suppose that the truth-bearer is a nonsentential and nonlinguistic entity.

Someone might object to this line of reasoning as follows. A being could not possibly have a belief without also having a language. Consequently, our prelinguistic ancestors had no beliefs and therefore knew no truths. But this objection faces a problem. If our prelinguistic ancestors had no beliefs, then how and why would they go on to eventually develop the communication systems we call "languages"? Specifically, why and how would they finally come to use a sentence that we would translate as, say, "Mastodons live by the river," unless they first believed that mastodons live by the river?

A DIFFICULT QUESTION

Does belief precede language or does language precede belief? Or is neither position correct?

PROPOSITIONS

Which of the theories canvassed above makes the most sense to you? In order to accommodate as many viewpoints as possible on the nature of the truth-bearer, and in order to avoid as much as possible taking sides on a controversial issue, let us at this point introduce a special terminological convention. Let us use the term 'proposition' to stand for whatever it is that is the bearer of truth. So, a proponent of the belief theory may interpret the word 'proposition' as another term for 'belief'; an advocate of the sentence type theory may treat the word 'proposition' as just another term for 'sentence type'; and so on.

Propositions Understood as Abstract Entities   The last two theories surveyed—the sentence type theory and the sui generis theory—both hold that the truth-bearer is an abstract entity. The difference between the two is that according to the sentence type theory, the truth-bearer is a sentential abstract entity while the sui generis theory holds that the truth-bearer is a nonsentential abstract entity. Propositions, understood as abstract entities, seem to be a mysterious type of object. Although they are not physical and consequently occupy no location in physical space, our minds nevertheless comprehend them. After all, we think about various propositions, we believe them true or false, and we form various other judgments about them. Which raises the question: How do our minds become aware of or "grasp" abstract entities? How are propositions accessible to our thought? Furthermore, how do such nonphysical objects fit into the modern scientific world-view, a view that seems to see everything as composed of purely physical parts? Anyone who believes in the existence of abstract entities faces this difficult line of questioning. Some philosophers argue that this difficulty makes it very hard to believe that there really are such things as abstract entities. Can we really accept a theory that posits such mysterious objects of thought? Let's explore this question for a moment.

●— THE EXISTENCE OF THEORETICAL ENTITIES

Philosophers who hold that a proposition is an abstract entity consider propositions to be *theoretical entities*. A theoretical entity is an object that cannot be seen or directly detected, but that, for theoretical reasons, we suppose must exist. Everyone has heard of the theoretical entities that physicists study, particles such as the electron, the proton, the neutron, and the quark. Consider, for example, the particles physicists call "quarks." Physicists tell us that inside each proton and neutron are three quarks. What is the nature of a quark? Nobody knows for sure. What do they look like? They can't be seen or directly detected. Physicists still know very little about the sorts of properties they supposedly possess. Quarks are at this point mysterious objects. How do we know they even exist? Our best theories of the proton and neutron tell us there must be three particles inside each proton and neutron. These particles have been named "quarks." In other words, quarks are referred to in our explanations of certain things. So, although they can't be seen, reference to them helps explain various phenomena observed in the laboratory. According to most physicists, this is good reason to suppose that such objects truly exist.

In general, if reference to a theoretical object helps make sense of some facet of the world, this constitutes a reason to suppose the theoretical object exists. Thus, one reason we suppose that electrons exist, to take another example from physics, is that although no one has ever seen one, we need to refer to electrons if we are to explain such things as the operation of a battery, the flash of a lightning bolt, or the effect of a magnet.

When we use the term 'proposition', we are simply assigning a name to the theoretical objects, whatever they are, that are true or false, just as the physicist assigns the term 'quarks' to the mysterious objects inside protons and neutrons. We assign names to theoretical entities so that we can refer to them. Philosophers who believe propositions are abstract entities argue that we need to refer to propositions, conceived as abstract entities distinct from sentence tokens, in order to explain or make sense of certain of our linguistic activities, specifically our attributions of truth and falsity. If reference to abstract objects helps make sense of our world, then this is good reason to suppose such objects exist. So, one reason for supposing that there are such things as propositions—understood as abstract objects—is that reference to them helps explain or make sense of our attributions of truth and falsity. Physicists and philosophers both posit theoretical entities in order to make sense of the world.

Propositions, understood as abstract entities, may be compared to a similar type of abstract entity studied by mathematicians: numbers. Perhaps the comparison will prove illuminating. We all think in terms of these abstract mathematical entities, and we refer to them daily. But just what is a number?

Mathematicians distinguish *numbers* from *numerals*. Numerals, like sentence tokens, are physical marks on paper, blackboards, computer terminals and so on. They are composed out of molecules of graphite, ink, chalk, or other writing compounds. Since it is physical, a numeral has a size, a shape, a location, an age, a weight, and other physical properties. Numbers, on the other hand, are simply not physical objects. The number three, for instance, which has been studied since ancient times, does not reflect light, is not located at some point in space, is not composed of molecules of some substance, and has no size, shape, weight, or age. If someone were to claim that the number three is a physical object, we could ask such questions as: "Where is it located?" "How much does it weigh?" "How tall is it?" or "How old is it?". There is general agreement among mathematicians that the number three is not located somewhere on Earth or in outer space.

Numerals are said to *express* numbers. For example, the numerals '3' and 'III' each express the number three. The number itself is distinct from the numerals that express it, for in this example, there are two numerals but only one number is expressed.

Mathematicians distinguish the number—an abstract entity—from the physical numerals that express it. Similarly, philosophers who believe propositions are abstract entities distinguish between the proposition and the sentence token that expresses the proposition. The distinction between numbers and numerals is therefore analogous to the distinction between propositions and sentence tokens, and the relationship between numbers and the numerals that express them is analogous to the relationship between propositions and the sentence tokens that express them.

Philosophers who accept the existence of propositional abstract objects address the following question to those who doubt the existence of such objects: If we can accept the existence of an infinity of abstract objects in mathematics called "numbers," then why should we balk at accepting an infinity of similar abstract objects in logic, objects we're calling "propositions."

Nevertheless, not all philosophers think of propositions as abstract entities. If you are convinced that a proposition is not an abstract entity, then interpret the term 'proposition' in the light of your favored account.

Although our discussion will not officially take sides on the question of the nature of a proposition, our wording will sometimes distinguish between sentence tokens and propositions. The reader who believes that sentence tokens are the bearers of truth and falsity may simply ignore the distinction.

EXERCISE 5.2

Part I Fill in the blanks with either 'numeral' or 'number' in such a way that the sentences all express truths:

1. The _____ on the computer screen is a pretty shade of blue.

2. We painted a new _____ on the airplane.

3. Twice the _____ four is the _____ eight.

4. That _____ five is shorter than that _____ seven.

5. The _____ 10 is half the _____ 20.

6. You write a _____ '21' by placing a _____ '2' beside a _____ '1'.

7. Many symbols can be used to refer to the _____ three.

8. The cookie is in the shape of a _____ two.

Part II Fill in the blanks with either 'sentence token' or 'proposition' in such a way that the sentences all express truths:

1. Your _____ is hard to read.

2. The first _____ you wrote is messier than the second one.

3. That _____ is very hard to believe.

4. '2 + 2 = 4' is a _____ of English and expresses a true _____.

5. The _____ 'Pat is older than Jan' means the same as the 'Jan is younger than Pat'.

6. It is possible to see a _____, but nobody can see a _____.

7. That _____ is written in yellow.

8. A _____ has a shape but a _____ has no shape.

9. That _____ is exactly two feet long.

4. TRUTH

●— THE CORRESPONDENCE THEORY

We opened this chapter with four philosophical questions. We have discussed two so far, which brings us to the third question: What is it for a proposition to be true? What characteristic does a true proposition have that makes it a true proposition?

Let's begin by looking at a specific example. Suppose someone says, "Sue owns two cars." What do we mean when we say that the proposition expressed by this sentence is true? Or false? What is it for the proposition to be true or false? Well, if Sue actually does own two cars, then the proposition expressed by 'Sue owns two cars' is a true proposition. And if Sue does not actually own two cars, then that proposition is a false proposition. More precisely:

> The proposition expressed by 'Sue owns two cars' is true if and only if Sue owns two cars. That proposition is false if and only if Sue does not own two cars.

Reflecting upon the matter, it looks as if truth involves a relation between two separate things. On the one hand, there is a proposition. A proposition in effect makes a claim about the way the world is, it specifies a possible way the world might be. On the other hand, there is the way the world really is, independent of or separate from the proposition. When the proposition's claim or specification of the world corresponds to the way the world is, the proposition is true. When the proposition's claim or specification of the world does not correspond to the way the world is, the proposition is false.

Here are some examples of sentences expressing propositions that "correspond" to or "specify" the way things are:

The Moon has more than one crater.

$$1 + 1 = 2$$

And here are some examples of sentences expressing propositions that do not "correspond" to or "specify" the way things are:

The Sun is smaller in diameter than the Earth.

$$1 + 2 = 406$$

After thinking about these rather obvious examples, do you see the sense in which a proposition may "correspond" or fail to correspond to the way the world is?

If you are puzzled by the relationship between a proposition and the world, consider for a moment the relation between a map and the world. A map also specifies a way the world is, it makes a claim about the world. The map is "correct" just in case the way the world is corresponds to or fits the map's claim or specification; the map is incorrect otherwise. In this way, maps are somewhat like propositions.

And so, according to these considerations, truth is a property that a proposition has just in case it corresponds to the way the world is. Truth, in short, is correspondence with reality. This account of truth goes back at least to Aristotle and is known as the *correspondence theory of truth*. It has been accepted by most philosophers throughout history as providing the best explanation of the matter.[4]

Objections to the Correspondence Theory   Not all philosophers find the correspondence theory satisfactory. Some object by raising questions that they claim the theory has no answers to. For instance, what is the nature of the correspondence relation between a proposition and the world? It can't be a physical relationship, for propositions are supposedly not situated in space–time and so are not physically related to anything. And the correspondence relation isn't pictorial, as when a map corresponds to some portion of the world by picturing that portion of the world, for

---

[4]   If you reflect upon the correspondence theory, you will notice that it seems to presuppose several things. Or at least, several assumptions seem to go along with the theory quite naturally. For instance, the theory speaks of the correspondence between two separate things: a proposition and the world. It is natural to interpret this as follows. There exists a world of things out beyond our minds and distinct from propositions, such things as rocks, trees, mountains, stars, and so forth. That is, an "external world," a mind-independent reality, exists. A proposition may, or may not, correspond to this independent reality. So, reality has its character independently of the propositions we express about it. For example, someone might say, "The Earth is flat." However, the Earth is round. The Earth has the shape it has independently of what this person says. The search for truth, on the correspondence view, is the attempt to bring our thoughts into correspondence with a mind-independent, proposition-independent reality.

The view that material objects exist externally to us and independently of our thoughts about them is termed *metaphysical realism*. So, the correspondence theory of truth presupposes metaphysical realism. Some people claim that there is no such thing as "reality." That is, they do not believe that a mind-independent reality exists. Such individuals will favor an alternative theory of truth. An alternative to metaphysical realism is a view called *metaphysical idealism*. Roughly, according to this view, no objects exist apart from our thoughts. Things exist, in other words, only as ideas in minds. One who accepts metaphysical idealism would probably reject the correspondence theory of truth.

propositions surely don't picture what they correspond to. In what sense, then, does a proposition correspond?

Here is another question for the correspondence theory. Consider the proposition expressed by 'The Moon has mountains'. To what does this proposition correspond? To the world as a whole? Or to just the Moon? Or only to the Moon's mountains?

One more question, which you can puzzle over on your own. Consider a negative proposition:

> Santa Claus does not exist.

To what does this proposition correspond?

●— ALTERNATIVE THEORIES OF TRUTH

Although most philosophers have found the correspondence theory of truth to be "self-evident" or in some way intellectually indispensable, two alternative theories have won some support in certain parts of the philosophical world. According to the *coherence theory* of truth, advanced by philosophers of the idealist school of thought, to say that a proposition is true is to say that it belongs to a coherent system of propositions. Very roughly, a system of propositions is said to be coherent if the propositions that belong to the system are consistent and stand in certain explanatory relations to one another. Thus, for coherence theorists, truth is a relation between propositions, and not, as in the correspondence theory, a relation between a proposition and the world.

The coherence theory of truth has its roots in the view—held by the philosophers in the idealist school of thought—that reality consists of one coherent system. It would be desirable to explain further why some philosophers hold a coherence theory of truth. However, the issue is deep and difficult and simply cannot be summed up in a few words. Any fair attempt at a summary would take us beyond the scope of an introductory logic text.

The coherence theory has always faced an objection that many philosophers find decisive. It is possible to specify two opposing systems of propositions, such that each system seems to be coherent, yet such that one contradicts the other. Since the two systems are contradictory, they cannot both be true. If this is correct, a system of propositions, it seems, can be coherent yet fail to be true. The conclusion of this line of reasoning is that truth is not coherence.

A second alternative to the correspondence theory is the *pragmatic theory of truth*. According to this theory, advanced by philosophers in the school of thought known as *pragmatism*, to say that a proposition is true is to say that it is useful in some way. Truth, on this view, is identified with usefulness. Very roughly, a proposition is useful if our knowledge of it can serve a human interest.

The pragmatic theory of truth has its roots in the pragmatic theory of meaning. That theory ties meaning and truth to use. It would be desirable to explore the pragmatic theory further here. Unfortunately, that would take us beyond the scope of an introductory symbolic logic book.

The pragmatic definition of truth also faces an objection that many philosophers find decisive. There are some propositions that, it seems, are useful to some people although they are clearly false. For instance, Hitler's racial theories were useful to him in his quest for power, yet those theories were most certainly false. Thus, many conclude that truth and usefulness are two separate concepts.

One final note. We have been speaking here of the nature of truth itself, not of how people determine what the truth is. These are two separate things. The correspondence theory is a theory of the nature of truth; it tells us what truth consists in. If one wants to know how to go about finding the truth, one is asking a question about evidence, about what sorts of things we use to help us determine what the truth is. It is important to keep these two things separate. We have been discussing the nature of truth, not how one finds the truth. A theory of the nature of truth belongs to the branch of philosophy called *metaphysics*, and a theory of evidence belongs to the branch of philosophy called *epistemology*.

●— TRUTH-IN-A-WORLD

Our task in this chapter was to explain four ideas that appear in the definition of deductive validity. We are now finished with the first three notions: "possible circumstance," "sentence," and "truth." And so we turn to the final item, "true in a possible circumstance." Some philosophers have suggested that the idea of something being true "in" a possible world doesn't make any sense. Let's see if any sense can be made of the possible worlds sense of "in."

We earlier spoke of things happening "in" various other possible worlds. We spoke, for instance, of worlds where someone swims the Atlantic in five minutes, of worlds where the laws of nature differ from those in our world, and so on. In order to help make sense of such talk, let us consider a case in which we say that a certain proposition is true "in" a particular world. Suppose someone says

> In world W, Jones is an athlete.

What does this mean? One way to make sense of this claim is the following:

> To say 'In world W, Jones is an athlete' is simply to say, 'If world W were to be actual, then Jones would exist and would be an athlete'.

Now, this sentence should make sense to you. We ordinarily have no trouble understanding similar expressions, expressions such as:

> "I didn't actually buy Microsoft Stock, but *if I had actually bought it, then I'd be rich today*."

> "I didn't actually take that job, but *if I had actually taken it, then I'd be better off today*."

In general, according to this account, to say that in a particular world a particular proposition is true is just to say that if the specified world were to be actual then the specified proposition would be true.

There are other ways to make sense of a claim that a particular proposition is true in a particular world, and not all philosophers would explain the concept as it is explained here. However, the issue is extremely complex, and any further exploration of the question would take us beyond the scope of an introductory logic text.

HISTORICAL NOTE: LEIBNIZ'S ACCOUNT OF POSSIBLE WORLDS

Leibniz, the German mathematician and philosopher (1646–1716), introduced the notion of a possible world into philosophy. He originally explained possible worlds as different ways God could have created the universe. Possible worlds, on Leibniz's interpretation, are alternative universes God could have created. For surely, Leibniz argued, God could have created different individuals and things from the ones God did actually create; different, that is, from the ones that actually populate this universe. And surely, Leibniz argued, God could have instituted different laws of nature from the ones that regulate this universe. So, God, in creating the universe, considered an infinite number of possibilities, and called into being—actualized—just one. This one.

Of course, as the term 'possible world' is used within modern logic, and as the concept was explained in this chapter, no theistic presuppositions are required. It is possible to be perfectly comfortable, logically speaking, with the idea of a possible world whether one is a theist, an atheist, or an agnostic.

In Conclusion In this chapter, we examined four concepts:

1. The concept of a possible world.
2. The concept of a proposition.
3. The concept of truth.
4. The concept of truth-in-a-world.

In the next chapter, we will put this conceptual material to work as we explore what is known as modal logic.

6

INTRODUCTORY MODAL LOGIC

1. INTRODUCTION

The basic concepts of logic were introduced using ordinary English in Chapter 1. In Chapter 2, an artificial language and mechanical procedures were developed for the evaluation of truth-functional arguments: arguments built out of truth-functional operators and the sentential components these attach to. In Chapter 5, we stepped back from the formal logical theory under construction and reflected philosophically upon the key concepts of the first four chapters. It is now time to put those philosophical reflections to work. Using the philosophical ideas generated in Chapter 5, this chapter will extend formal logic so that it covers an additional type of argumentation: the modal argument.

A *modal* argument is an argument built out of modal operators and the components these attach to. The two most commonly used modal operators are 'necessarily' and 'possibly'. If we begin with a sentence such as

    i.   It's true that $2 + 2 = 4$

and prefix to it the monadic sentence operator 'necessarily', we produce the compound sentence

    ii. Necessarily, it's true that $2 + 2 = 4$

According to (ii), it is necessarily true that $2 + 2 = 4$, which is to say that this mathematical truth could not possibly have been false.

Similarly, if we begin with a sentence such as

    iii. Someone is 200 years old

and prefix to it the monadic sentence operator 'possibly', we produce the compound sentence

    iv.   Possibly, someone is 200 years old.

'Necessarily' and 'possibly' are called 'modal' operators because they specify what are called *modes* of truth. We will begin examining the modes of truth a little further on in this chapter.

Many of the most intriguing and important arguments in the history of philosophy—arguments about such things as God's existence, free will, time, and the nature of the mind—are modal in nature and can only be properly evaluated using principles of modal logic. For example, consider the following argument:

1.   If God—as conceived within traditional theism—exists at all, then *necessarily* God exists.

2.   If God does not exist, then *necessarily* God does not exist.

3.   It is *possible* that God exists.

4.   Therefore, God exists.

Notice the argument's arrangement of modal operators. Is the argument valid or is it invalid? The techniques of truth-functional logic are of no help here, for whether the argument is valid or invalid depends on the arrangement of its modal operators, and the procedures of truth-functional logic do not apply to modal operators. Our goal now is to do for modal argumentation what we did for truth-functional argumentation, namely, to develop a formal language and formal procedures that will allow us to evaluate modal arguments precisely and systematically.

At first, this sounds like an easy task. Simply specify a truth-function for each modal operator and then develop truth-table techniques for the evaluation of modal arguments. Unfortunately, the task of developing a semantical theory for modal logic is not that simple, for the modal operators are not truth-functional in nature. This will be shown later in this chapter. Indeed, the goal of developing a semantical theory for modal logic has proven so elusive that until the late 1950s, a promising theory had not been found. However, beginning with the pioneering work of Saul Kripke in 1959, and thanks to the efforts of a number of philosophers working during the 1960s, a semantics for modal logic was developed that has only recently come into widespread use.

If we can't use truth-tables to give the modal operators a semantical theory, how do we proceed? What is going to take the place of truth-tables? The philosophical ideas generated in Chapter 5.

●— THE LAW OF EXCLUDED MIDDLE

A proposition is true, according to the correspondence theory of truth, if it corresponds to or specifies the way things are, and it is false otherwise. For any proposition one might pick, that proposition either corresponds to the way things are or it does not. There does not seem to be any middle alternative. And so we will assume a principle first formulated in ancient times, the *Law of Excluded Middle*: Every proposition is either true or, if not true, false. The Philosophical Interlude briefly takes up the question: What are the logical consequences if we reject the Law of Excluded Middle?

As we saw in the last chapter, propositions can be true or false in various possible circumstances, that is, in various possible worlds. In any possible circumstance one cares to pick, it seems plausible to suppose that any proposition one picks will either correspond to the way things are or it will not. So no matter which proposition we pick, and no matter which possible world we pick, that proposition is either true or false in that world. This is the "possible worlds" version of the Law of Excluded Middle.

2. MODAL PROPERTIES OF PROPOSITIONS

Besides being true or false in a single world, propositions have properties called modal properties. These are properties that reflect the way a proposition's truth-values are distributed across the infinite set of possible worlds. We shall consider five ways in which a proposition's truth-values might be distributed across the infinite set of possible worlds and the five modal properties corresponding to those five distribution patterns.

1. If the truth-values of a proposition are distributed in such a way that the proposition is true in at least one world, the proposition is *possibly true*.

2. If the truth-values of a proposition are distributed in such a way that the proposition is false in at least one world, the proposition is *possibly false*.

3. If the truth-values of a proposition are distributed in such a way that the proposition is true in at least one world and false in at least one world, the proposition is *contingent*.

4. If the truth-values of a proposition are distributed in such a way that the proposition is true in all worlds, the proposition is *necessarily true*.

5. If the truth-values of a proposition are distributed in such a way that the proposition is false in all worlds, the proposition is *necessarily false*.

The question in the case of each of these five patterns is: Are there any such propositions?

●— PATTERN ONE: POSSIBLE TRUTHS

A proposition is possibly true just in case it is true in at least one world. Such a proposition may be called a *possible truth*; it represents a logical possibility. Consider the following examples:

1. There are craters on the Moon.
2. George Washington was the first President of the United States.
3. $1 + 1 = 2$
4. All triangles are three-sided.

Since each of these propositions is true in the actual world, it follows that each is true in at least one world, which makes each a possible truth.

Now consider the following:

5. There are no craters on the Moon.
6. George Washington was the fortieth President of the United States.
7. Bill Cosby is the Governor of California.
8. Water always runs uphill.

Notice that each of these propositions also qualifies as a possible truth, for each is true in at least one world. Of course, none of (5) through (8) is true in this world, but that doesn't change the fact that each is possibly true.

●— PATTERN TWO: POSSIBLE FALSEHOODS

A proposition is possibly false just in case it is false in at least one world. Such a proposition is called a *possible falsehood*. Consider the following:

9. George Washington was the thirtieth U.S. President.
10. Saturn has no rings.
11. Mount Everest is 59,000 feet high.

Since each of these propositions is false in this world, it follows that each is false in at least one world, which makes each a possible falsehood.

Now consider the following:

12. George Washington was the first President of the United States.
13. The Moon has craters.
14. Mount Everest is over 20,000 feet high.

Notice that each of these propositions also counts as a possible falsehood, for each is false in at least one world. Of course, none of (12) through (14) is false in this world, but that doesn't change the fact that each is possibly false.

Incidentally, a proposition that is true in the actual world is said to be *actually true*, and a proposition that is false in the actual world is said to be *actually false*. Which of the following are actual truths and which are actual falsehoods?

15. $1 + 2 = 3$
16. $2 + 4 = 8$

17. Abraham Lincoln lived during the nineteenth century.

18. Carbon is an element.

19. Michelangelo was born in 350 B.C.

The answers are terribly obvious, of course. But this just goes to illustrate how naturally we apply terms such as 'actual truth' and 'actual falsehood'. Here is a slightly less obvious question: Which of (15) through (19) are possibly true and which are possibly false?

Some Consequences of These Definitions   Let us draw some consequences from these definitions. Any proposition that is actually true is true in this world and so is true in at least one world. Therefore, any proposition that is actually true is also possibly true. And any proposition that is actually false is false in this world and so is false in at least one world. Therefore, any proposition that is actually false is also possibly false. However, not all possible truths are actually true, and not all possible falsehoods are actually false. For example, (19) is possibly true but is not actually true, and (17) is possibly false but is not actually false.

So, the set of all actual truths forms a proper subset of the set of all possible truths and the set of all actual falsehoods forms a proper subset of the set of all possible falsehoods.

Distinguishing Epistemic Possibility from Logical Possibility   At this point, we must make an important distinction between two senses of the expression 'it's possible that'. Within the context of modal logic, the expression 'it's possible that' is short for 'it's logically possible that'. So, when we say "it's possible that P", this is just to say "There's at least one possible world in which P is true." For example, consider the following proposition:

20. Groucho Marx was King of England.

Surely this proposition would have been true if things had been sufficiently different. So, there is a possible world in which (20) is true. Therefore, since there is a possible world in which (20) is true, (20) qualifies as possibly true.

There also exists a different sense of 'possibility', a different use for the expression 'it's possible that . . .'. In some cases, if someone were to say "It's possible that (20)", they would mean "Based on what we know, (20) might be true." A proposition is said to be *epistemically possible* if it is possible, on the basis of what we know, that the proposition is true. So, in this type of case, the person is saying that (20) is *epistemically* possible. ('Episteme' is from the Greek word for knowledge and something is "epistemic" if it pertains to knowledge.) Of course, in the epistemic sense of 'it's possible that', (20) does not qualify as possibly true, because, based on what we know, (20) is not true.

These two senses of 'it's possible that'—the logical and the epistemic senses—are easily confused. However, they must be kept distinct or else serious logical confusion will result. So, before going on, make sure you understand why (20) is logically possible but not epistemically possible.

So, 'It's logically possible that (20)' means 'There's a possible world in which (20) is true', and 'It's epistemically possible that (20)' means 'It's possible, based on what we know, that (20) is true'. From now on, when we use the word 'possible', unless we specify otherwise, let it be understood that we mean 'logically possible'.

●— PATTERN THREE: CONTINGENCY

Consider the proposition

> **21.** Winston Churchill was the Prime Minister of Britain during World War II.

Are there worlds in which this is true? Are there worlds in which it is false? Clearly, this proposition is both possibly true and possibly false, for it is true in at least one world and it is false in at least one world. Thus, proposition (21) is *contingent*. Here are some examples of other contingent propositions:

> **22.** The Beatles first appeared on the Ed Sullivan show in February 1964.
>
> **23.** World War I ended in 1918.
>
> **24.** John F. Kennedy grew up in China.
>
> **25.** Abraham Lincoln was the King of Romania.

Notice that although each of (22) through (25) is contingent, not all are actually true. (22) and (23) are actually true, and (24) and (25) are actually false. Thus, when we say that some proposition is contingent, we are not saying that it is actually true. We are saying just that the proposition is true in some worlds, false in others.

As noted previously, in everyday discourse, we often use the word 'contingent' to mean 'dependent'. For example, someone might say, "The sale of the house is contingent on the sale of the buyer's house." One way to keep in mind the logical sense of 'contingent' is to think of it this way: If a proposition is contingent, then whether it is true or false *depends* on which world you are talking about, for a contingent proposition's truth-value varies from world to world.

If we say "(23) is contingent" or "It is contingent that (23)," we mean only that there are some worlds in which (23) is true and some in which it is false. These expressions make no reference to the actual world. However, suppose that some proposition is both contingent and actually true. Then we shall say that it is *contingently true.* Notice that in the expression 'contingently true', the word 'contingently'—with its adverbial ending—modifies 'true'. So, when we say that a proposition is contingently true, we are saying first of all that the proposition is true, and secondarily, we are adding that it is contingent.

Therefore, if someone were to say "P is contingently true" or "it is contingently true that P," she or he would be asserting two things:

    i. P is true;

    ii. P is true in a contingent way, that is, P is contingent.

A contingently true proposition is thus one that is actually true but that might have been false whereas a contingent proposition is simply one that is true in some worlds, false in others.

● NONCONTINGENT PROPOSITIONS

If a proposition is contingent, then its truth-value varies from world to world, that is, in some worlds it is true and in some worlds it is false. Is there an alternative to contingency? Are there any noncontingent propositions? Before we answer this, let's work out what modal features noncontingent propositions would have. If a proposition were to be noncontingent, then it would not be contingent, which is to say that its truth-value would not vary from world to world. There are only two ways in which the truth-value of a proposition might be invariant:

    i. If a proposition were to be true in all worlds and false in no world, then its truth-value would not vary from world to world and it would therefore be a noncontingent proposition.

    ii. If a proposition were to be false in all worlds and true in no world, then its truth-value would not vary from world to world and it would therefore be a noncontingent proposition.

Let us now see if there are any such propositions.

Pattern Four: Necessary Truth A necessary truth, if one existed, would be a proposition that is true in every possible world, false in none. That is, such a proposition would be true and could not possibly have been false. But are there any such propositions? Consider the following and judge for yourself:

    a. $2 + 2 = 4$

    b. The interior angles of a Euclidean triangle sum to 180 degrees.

    c. $\sqrt{2}$ is an irrational number.

    d. 9 is an odd number.

    e. All bachelors are unmarried.

    f. A sister is a female sibling.

    g. All triangles have three sides.

    h. All material objects occupy a volume of space.

i. Either Sue is 21 or it is not the case that Sue is 21.

j. If Pat is 16, then Pat is 16.

k. Nothing is both red all over and green all over at the same time.

l. If something is red, then it is colored.

m. If one person is taller than a second person, and the second person is taller than a third, then the first is taller than the third.

n. It is necessarily true that $2 + 2 = 4$.

o. For any sentences **P** and **Q**, if **P** is true, and if **P** implies **Q**, then **Q** must be true.

Given the standard meaning that our linguistic community attaches to each of the above *sentences*, each expresses a unique *proposition*. Are there any possible circumstances in which any of those propositions are false? To answer this, try the following thought experiment. In the case of each sentence, reflect carefully on the standard meaning the sentence possesses. Make sure you understand that meaning. Now, consider the proposition that is expressed. (Obviously, in each case, the sentence itself will have different meanings in different worlds, but that has nothing to do with the truth or falsity of the unique proposition expressed by the sentence when that sentence is given the standard meaning we give it.) Finally, once you grasp the proposition actually expressed by the sentence, try to write down a description of a possible world in which that proposition would be false. If you try this for any of (a) through (o), you will find that your description will be self-contradictory. In each of the cases, you cannot describe a possible circumstance in which the proposition would be false without falling into a self-contradiction.

Let's briefly look at some of the propositions expressed above. Propositions (a) through (d) are mathematical truths. Mathematicians, from the earliest days of the discipline, realized that the truths they were discovering were in some sense necessary. Mathematical truths, they believed, couldn't possibly be false. This is part of the reason mathematical truths were viewed with awe and even reverence when first discovered. Pythagoras, one of the first mathematicians, is said to have offered a great sacrifice to the gods upon discovering the necessary truth we call the Pythagorean Theorem.

Propositions (e) through (h) are called *analytic* truths because they have the following feature. If a certain type of logical analysis is applied to the concept expressed by the subject term of the sentence, one can see that if the subject concept applies to an individual, then the predicate concept must also apply. Some philosophers express this point by saying that the subject concept "contains within itself" the predicate concept. By logically analyzing the subject, the predicate automatically follows. Thus, through a process of logical analysis, one can see that the propositions expressed by these sentences are necessarily true. Consequently, such sentences are said to be analytically true.

The propositions expressed by (i) and (j) are tautologies. In truth-functional logic, we can translate such sentences into symbols and then with the aid of a truth-table prove that each expresses a necessary truth.

Propositions (k) through (o) don't fit into any easily characterized category. Many philosophers consider these to be *synthetic necessary* truths. Roughly, synthetic truths are truths that are not analytic—their truth can't be determined just by logical analysis. Synthetic necessary truths would be truths that are necessary yet not analytic. There is considerable debate within philosophy as to whether any synthetic necessary truths exist. Although the question is interesting, it is also complex, and an investigation of the issue would take us beyond the scope of this introductory chapter.

An Objection to the Idea of Necessary Truth   We must turn now to a certain objection that often arises when someone is introduced to the idea of a necessary truth. One often hears the following argument, or one very much like it:

> It seems that there are no necessary truths; any truth could have been false. Take '1 + 1 = 2' as an example. There are worlds where mathematicians speak the truth when they say "It's false that 1 + 1 = 2." There are worlds where people speak the truth when they say "Not all triangles are three-sided." And similarly for each of the so-called "necessary truths" you have listed. Each could be false. Therefore, necessary truth does not exist.

In response to this argument, let us first try to understand the "possibilities" referred to. So we ask:

> How do you make sense of the possibility in which people speak the truth when they say, "It's false that 1 + 1 = 2" or "Not all triangles are three-sided"?

And the usual reply to this question goes along the following lines:

> Well, I'm thinking of worlds in which the sentence '1 + 1 = 2' means something different from what it means in our world, and the term 'triangle' has a meaning that is different from its meaning in our world. That is why in those other worlds they say things like "It's false that 1 + 1 = 2" and "Not all triangles are three-sided."

The error in this line of thought is not difficult to detect. The argument under consideration confuses sentences with the propositions those sentences express. When we say that it is necessarily true that 1 + 1 = 2, we do not mean to say that the sentence '1 + 1 = 2' is a necessary truth. Nor do we mean to say that the sentence, '1 + 1 = 2', has the same meaning in every world. Surely, that sentence has different meanings in different worlds. But this just means that the sentence '1 + 1 = 2' is used in different worlds to stand for different propositions. This does not imply that the proposition we express with the sentence has different truth-values in different worlds. When we say that it is necessarily true that 1 + 1 = 2, we mean that the unique proposition—the proposition expressed by the sentence '1 + 1 = 2' as we use that sentence—is necessarily true. Which unique proposition is that? It's the one we typically have in mind when we use that sentence with its standard meaning, the meaning established by the linguistic practices and conventions of our linguistic community. Once given a specific meaning, our sentence '1 + 1 = 2' expresses a specific proposition, a proposition about numbers. Now, it is that proposition that is true in every possible circumstance or world.

How then do we make sense of worlds where mathematicians speak the truth when they say "1 + 1 = 2 is false"? The only way to make sense of such a possibility is to suppose that these are situations—worlds—in which the sentence '1 + 1 = 2' has a different meaning from the meaning we typically give it and so expresses a different proposition, a proposition that is false. Thus, in the worlds under consideration, those mathematicians are not attributing falsity to the proposition we have in mind when we say "1 + 1 = 2." Rather, they are attributing falsity to the proposition they have in mind when they use those words.

What proposition are they thinking of? In order to answer this, we would have to be able to translate their sentences into our sentences. Suppose that when they say "1 + 1 = 2," they are expressing the same proposition we express with '2 + 2 = 7'. Then, assuming they mean by 'it's false that' what we mean when we say "it's false that," it follows that when they say "it's false that 1 + 1 = 2" they are attributing falsity to the proposition we refer to when we say "2 + 2 = 7." And we're in agreement on the falsity of that proposition! None of this contradicts the plain fact that as we use the sentence, '1 + 1 = 2' expresses a necessarily true proposition.

The Linguocentric Proviso   When we communicate with others, when we try to understand what it is they are saying, we must—if we are going to communicate anything at all—assume what has been termed the *linguocentric proviso*:

> *Unless we specify otherwise, the sentences we put forward carry the standard meanings established within our linguistic community. That is, unless otherwise specified, our sentences are assumed to express the propositions our linguistic group typically expresses with those sentences.*

Of course, we can redefine terms and coin new meanings, but our listener won't understand us unless we use the standard meanings as we construct our new definitions.

If we are going to successfully communicate, one more assumption is necessary. The meanings of the words we use must be given a consistent meaning throughout our discourse. For instance, if 'circular' has the standard meaning at its first usage, then unless we say otherwise, each additional usage of the term within the same context should carry the same meaning. If you are going to communicate effectively, you should not use one meaning and then, without signalling the change, switch to a different meaning.

Pattern Five: Necessary Falsehoods   We have spent a long time introducing the first category of noncontingent propositions, that of necessary truth. Let us now turn to the second category of noncontingency. Here we have propositions that are possibly false but not possibly true; such propositions would be false in all worlds, true in none. Propositions that are false in every possible world are called 'noncontingent falsehoods', 'necessarily false propositions', or simply 'necessary falsehoods'. Are there any such propositions? Judge for yourself. Take each of the sentences (a) through (o) used above to express necessary truths and prefix 'it is not the case that' to each of those sentences. The result will be the negation of each sentence. Now, are the propositions expressed by the negations of those sentences necessarily false propositions?

Some Logical Consequences of Our Definitions   Let us draw out a few of the implications of some of our definitions. First, every necessarily true proposition is possibly true as well. Here is why. A possibly true proposition is one that is true in at least one world. A necessary truth is a proposition that is true in all worlds. Since a necessary truth is true in all worlds, it is true in at least one world, and so every necessary truth counts as possibly true as well as necessarily true. And it follows that every necessarily true proposition is actually true as well. The argument for this is obvious. A proposition is actually true if it is true in this world. Since a necessary truth is true in all worlds, it is true in this world. Therefore, every necessary truth is actually true as well. However, not all possible truths are necessarily true, and not all actual truths are necessarily true either. For instance, although it's actually true, and possibly true, that Los Angeles has a population of over 1 million people, that proposition is not necessarily true, for it's not true in every possible world.

Turning to necessary falsehoods, we may observe that each necessarily false proposition is both possibly false and actually false as well. A possibly false proposition is one that is false in at least one world. A necessary falsehood is a proposition that is false in all worlds. Therefore, any necessary falsehood is false in at least one world and so counts as possibly false as well as necessarily false. However, not every possible falsehood is necessarily false, and not every actual falsehood is necessarily false. Can you state an actual falsehood that is not necessarily false? Such a proposition would also be possibly true, of course.

●— CONTRASTING LOGICAL and PHYSICAL NECESSITY

Necessary truths are also called 'logically necessary truths' or 'logical necessities'. It is important not to confuse logically necessary truths with physically necessary truths. Consider the set of all possible worlds that have the same laws of nature as does the actual world. This is the set of all physically possible worlds. Physical necessity may be defined as follows:

> A proposition is physically necessary if and only if (i) the proposition is contingent; and (ii) the proposition is true in every physically possible world.

The physical laws of nature that are catalogued in science textbooks are physically necessary. Examples of such truths include:

> Protons repel each other.
> Copper conducts electricity.
> The speed of light is constant in all reference frames.

If you will compare the two kinds of necessity you will notice that logical necessity is the broadest sort of necessity and physical necessity is a narrower or more restricted kind, at least in the sense that logical necessities characterize all possible worlds and physical necessities characterize only a proper subset of the set of all possible worlds.

3. THE LANGUAGE ML

We have already used symbolism in the course of this study. For instance, each word you have read is a symbolic element of an enormously complicated symbolic system called the English language. And if you have learned how to read music, you have learned a symbolic language that represents the musical structure of a piece of music. And, of course, logical symbolism was introduced in Chapter 2. It is time to extend our logical symbolism in order to help us represent the principles of modal logic. The symbolism we shall introduce will greatly simplify the way we express certain fairly complex matters.

The Necessity Operator Suppose we begin by considering the proposition expressed by

26. $2 + 2 = 4$.

This proposition is obviously true. But more than this, the proposition is necessarily true as well. We can express this point by prefixing to sentence (26) the expression 'it is necessarily true that':

27. It is necessarily true that $2 + 2 = 4$.

Sentence (27) expresses a new proposition.

The expression 'it is necessarily true that' is a monadic sentence operator, for when it is prefixed to a sentence, it "operates" on the sentence in such a way that a new, compound sentence is formed. Since 'it is necessarily true that' serves to attribute necessary truth to a proposition, let us call this operator a *necessity operator* and say that a necessity operator is any operator that is used to attribute necessary truth to a proposition.

There are many alternative ways to attribute necessary truth to a proposition. For instance, instead of saying 'it is necessarily true that P', one may say 'P is necessarily true', 'P is true in all worlds', 'P must be true', 'P is a necessary truth', and so on. Each of these sentences contains an operator that attributes necessary truth and so each contains a necessity operator.

Let us now introduce some of the symbolism we will be using. Let P abbreviate $2 + 2 = 4$. We will use the symbol '$\Box$', called the *box*, when we abbreviate a necessity operator. Now we can symbolize (26) simply as:

P

and we can symbolize (27) as:

$$\Box P$$

We may use $\Box P$ to abbreviate 'It is necessarily true that P', or 'P is necessarily true', or 'P is true in all worlds', or 'P must be true', or 'P is a necessary truth', and so on.

We saw that 'it is necessarily true that' is a monadic sentence operator. We will also call '$\Box$' a sentence operator, since it represents an English sentence operator. And since '$\Box$' reflects a modal property, we will call the box a "modal" operator.

The Possibility Operator   Consider next the sentence

28.   The Earth is struck by a comet.

Surely, the proposition expressed here is logically possible. We can express this point by prefixing to (28) the expression 'it is possibly true that', which produces the sentence

29.   It is possibly true that the Earth is struck by a comet.

The expression 'it's possibly true that' is a monadic sentence operator, for when it is prefixed to a sentence it operates on the sentence in such a way that a compound sentence is formed. Since 'it is possibly true that' serves to attribute possible truth to a proposition, let us call this operator a *possibility operator* and let us say that a possibility operator is any operator that is used to attribute possible truth to a proposition.

There are many ways to attribute possible truth to a proposition. For instance, instead of saying 'It is possibly true that P', one may say 'P is possibly true', 'P is true in some worlds', 'P might be true', 'P is a possible truth', and so on. Each of these sentences contains an operator that attributes possible truth and so each contains a possibilty operator.

We will use the symbol '$\Diamond$', called the *diamond*, when we abbreviate a possibility operator. Thus, '$\Diamond P$' abbreviates 'it is possible that P', or 'P is possibly true', or 'There's at least one possible world where P is true', and so on.

Now, if we abbreviate proposition (28) as

$$Q$$

then we can abbreviate (29) as:

$$\Diamond Q$$

We have seen that 'it is possibly true that' is a monadic sentence operator. We will also call '$\Diamond$' a monadic sentence operator, since it represents an English sentence operator. And, since '$\Diamond$' concerns a modal property, the diamond is a modal operator.

The Contingency Operator   Consider the sentence

30.   Dick's Drive-in has had the same menu for 40 years.

Surely this sentence expresses a contingent proposition. Dick's *could have* changed its menu in the last 40 years. Therefore we may say:

31.   It is contingent that Dick's Drive-in has had the same menu for 40 years.

The expression 'it is contingent that' is a monadic sentence operator, for when it is prefixed to a sentence, it operates on the sentence in such a way that a compound sentence is formed. Since 'it is contingent that' is used to attribute the modal property of contingency to a proposition, let us call this operator a *contingency* operator and let us say that a contingency operator is any operator that attributes contingency to a proposition.

If we let the symbol '∇', called *nabla*, abbreviate any contingency operator, and if we let D abbreviate the sentence 'Dick's Drive-in has had the same menu for 40 years', then sentence (30) may be symbolized as

$$D$$

and (31) may be symbolized as

$$\nabla D$$

●— THE LANGUAGE ML

We may now specify the syntax for a formal language for modal logic. The language will be called ML (for 'Modal Logic').

THE VOCABULARY FOR ML

    i. Sentence constants with or without subscripts:
      A, B, . . . Z, A_1, A_2, . . B_1, B_2, . . . etc.
    ii. Sentence operators:
      Monadic operators: ~, □, ∇, ◇
      Dyadic operators: &, v, ⊃, ≡, o, φ, →, ↔
    iii. Parenthetical devices:
      Parentheses: ()
      Brackets: [ ]
      Braces: { }

THE GRAMMAR FOR ML

M1. Any constant, with or without a subscript, is a sentence of ML.

M2. If **P** is a sentence of ML then ~**P**, □**P**, ∇**P**, ◇**P** are sentences of ML.

M3. If **P** and **Q** are sentences of ML, then (**P&Q**), (**PvQ**), (**P⊃Q**), (**P≡Q**), (**PoQ**), (**PφQ**), (**P→Q**), (**P↔Q**) are sentences of ML.

Any expression that contains only items drawn from the vocabulary of ML and that can be constructed by a finite number of applications of the rules of grammar M1 through M3 is a *sentence* or *well-formed formula* of ML. Nothing else counts as a sentence of ML.

You have not yet been introduced to the dyadic modal operators contained within ML's vocabulary. Those operators will be discussed shortly, after further groundwork has been laid.

If you will examine ML's syntax, you will notice that it is exactly the same as the syntax for TL with the exception of one minor change: the modal operators have been added to the lists of monadic and dyadic operators. Thus, the language ML "contains" the language TL in the sense that any sentence that is a wff of TL will automatically count as a wff of ML, although it's not the case that any sentence that is a wff of ML will count as a wff of TL.

In order to simplify things, we will continue the following practice:

> You may leave off the outer parentheses in cases in which no operator applies to those parentheses.

Recall the distinction between object language and metalanguage. In the context of this chapter, the object language will be ML and the metalanguage will be English. Be careful to keep the two languages distinct. Also, recall the distinction between the use of a linguistic expression and the mention of an expression. In contexts where the distinction is clear, we will sometimes simplify things by omitting the quotes on mentioned expressions.

• **QUESTION** •

Give an example of a wff of ML that is also a wff of TL. Give an example of a wff of ML that is not a wff of TL.

●— LINKING OPERATORS

Our three monadic modal operators can be applied to each other or "concatenated"—from the Latin for "linked together"—as in the following examples:

> '◇~P' is translated as 'It's possible that it's false that P', or 'There's at least one world in which P is false', or 'There's a world in which P is false'.

'□~P' is translated as 'It's necessarily true that it's false that P', or 'In every world, P is false', or 'P is necessarily false'.

'~◇P' is translated as 'It's not the case that it's possible that P', or simply as 'It's not the case that there's even one possible world in which P is true', or 'P is false in every possible world'.

'~◇~P' is translated 'It's not the case that there's even one possible world in which P is false', which is to say 'P is true in every world'.

'~□P' is translated 'It's not the case that P is necessarily true' or 'P is not a necessary truth'.

'~□~P' is translated as 'It's not the case that in every world P is false' or 'P is not a necessary falsehood'.

In these cases, one operator "operates" on another, generating a new compound sentence in the process.

●— TRANSLATING BOXES INTO DIAMONDS

Our symbols allow us to express the concepts of possibility and necessity in terms of each other. For instance, to say 'In every world, P is true' is just to say 'There's not even one world in which P is false'. Thus:

$$\text{'□P' means the same as '~◇~P'}$$

If one person says, 'There's a possible world in which P is true' and another says, 'It's not the case that in every world P is false', then the two have said the same thing. Therefore:

$$\text{'◇P' means the same as '~□~P'}$$

To say 'There's no world in which P is true' is just to say 'In every world P is false'. So,

$$\text{'~◇P' means the same as '□~P'}$$

Finally, the expression 'It's not the case that in every world P is true' is equivalent to 'There's at least one world in which P is false'. Consequently:

$$\text{'~□P' means the same as '◇~P'}$$

There is a simple rule for translating formulas containing boxes (or diamonds) into formulas containing no boxes (or no diamonds). The rule has three parts:

1. Add a tilde to each side of the box or diamond.
2. Change the box to the diamond or the diamond to the box.
3. Cancel out any double negatives that result.

For example, suppose we want to translate ~□P into a formula containing no boxes. Here are the three steps:

1. Add a tilde to each side of the box or diamond:

$$\sim\sim\square\sim P$$

2. Change the box to the diamond or the diamond to the box:

$$\sim\sim\Diamond\sim P$$

3. Cancel out any double negatives that result:

$$\Diamond\sim P$$

SYMBOLIZING MODAL SENTENCES

The Scope of a Modal Operator When symbolizing sentences containing modal operators, it is important to pay attention to the scope of the modal operators. The box, diamond, and nabla are monadic operators. The scope of one of these three operators will always be exactly what the scope would be if the operator were changed to a tilde. In the following examples, the scopes of the tildes are marked by brackets:

$$\sim (A \ \& \ B) \qquad \sim A \ v \ \sim B \qquad \sim P \supset Q$$

In the following, the scopes of the boxes are marked by brackets:

$$\square(A \ \& \ B) \qquad \square A \ v \ \square B \qquad \square P \supset Q$$

The scopes of the diamonds are marked with brackets below:

$$\Diamond (A \ \& \ B) \qquad \Diamond A \ v \ \Diamond B \qquad \Diamond P \supset Q$$

The scopes of the nablas are marked off with brackets in the following formulas:

$$\triangledown (A \ \& \ B) \qquad \triangledown A \ v \ \triangledown B \qquad \triangledown P \supset Q$$

If you are uncertain about the scope of a box or diamond, figure the scope as if the operator were a tilde, and you will have your answer.

Recall that the main connective of a sentence is the operator of greatest scope. In each of the following, the main connective is the box:

$$\Box P \qquad\qquad \Box(PvQ) \qquad\qquad \Box(P\supset Q)$$

And the main connective is the diamond in each of these:

$$\Diamond(P \; v \; Q) \qquad\qquad \Diamond P \qquad\qquad \Diamond\sim R$$

However, in each of the following, the main connective is a truth-functional operator:

$$\Box P \; v \; \Box Q \qquad\qquad \Diamond P \supset \Box Q \qquad\qquad \Diamond P \; \& \; \Box Q$$

In the following examples, various truths are expressed first in English and then using our new modal symbols. In these examples, suppose that P abbreviates your favorite proposition-expressing sentence. It won't matter which proposition-expressing sentence that is, because what we will say about P will apply equally to any sentence.

1. If P is necessarily true, then P is actually true as well.

$$\Box P \supset P$$

2. If P is necessarily true, then P is possibly true as well.

$$\Box P \supset \Diamond P$$

3. If P is actually true, then P is possibly true as well.

$$P \supset \Diamond P$$

4. If P is contingent, then there are worlds in which P is true and there are worlds in which P is false.

$$\nabla P \supset (\Diamond P \& \Diamond\sim P)$$

5. If P is noncontingent, then either P is necessarily true or P is necessarily false.

$$\sim\nabla P \supset (\Box P v \Box\sim P)$$

Notice the way the English translates into boxes, diamonds, and nablas.

●— EVALUATING SENTENCES CONTAINING MULTIPLE MODAL OPERATORS

Sentences containing several concatenated operators can be difficult to evaluate. For instance, what is the truth-value of the following sentence?

It is noncontingent that it is false that 2+2=4.

When dealing with complicated sentences such as this, it is best to begin at the right and figure out the truth-value of the primary sentential component. Then move one step left and apply to the primary sentence unit the operator that sits to its left. Then move another step left and apply the next operator to this. Keep applying operators in this way until the final operator is reached.

For example, the sentence above can be evaluated in three steps.

Step 1 Determine the truth-value and modal status of the proposition expressed by the primary sentence component. That component is '2+2=4'. It is true that 2+2=4. Furthermore, it is necessarily true and therefore noncontingent that 2+2=4:

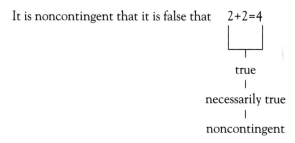

Step 2 Now, move one step to the left and apply the operator to the primary sentence. If someone asserts that it is false that 2+2=4, what they say is false. How false is their statement? It is necessarily false. Since it is necessarily false, it is also noncontingent. Thus:

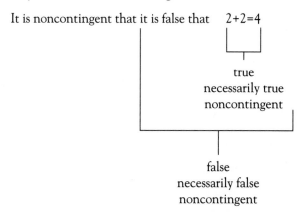

Step 3 The next operator happens to be the final operator. Our analysis so far has disclosed that the part of the sentence saying that it is false that

2+2=4 is itself the expression of a noncontingency. The last operator attributes noncontingency to this. Consequently, the sentence as a whole expresses a true proposition.

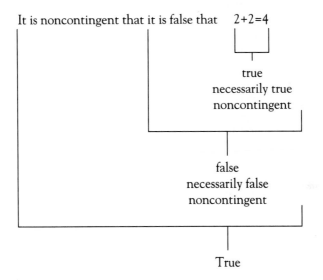

It is noncontingent that it is false that   2+2=4

true
necessarily true
noncontingent

false
necessarily false
noncontingent

True

• EXERCISES •

Exercise 6.1

1. From the following list, pair together formulas that have the same meaning.

   1. $\Box P$ 5. $\Box \sim P$

   2. $\sim \Diamond P$ 6. $\Diamond P$

   3. $\Diamond \sim P$ 7. $\sim \Box \sim P$

   4. $\sim \Diamond \sim P$ 8. $\sim \Box P$

2. If we wish to express 'it is contingent that P' in symbols, we write '∇P'. However, if we wish to symbolize 'P is contingently true' we must write '$\nabla P \,\&\, P$'. Explain why.

3. 'Noncontingently true' is just another way of saying 'necessarily true', and 'noncontingently false' is just another way of saying 'necessarily false'. Explain why.

4. Explain the difference between the following two statements:

   a. It is noncontingent that it is false that $1 + 1 = 2$.

   b. It is noncontingently false that $1 + 1 = 2$.

5. A. True or false?

    1. Every possibly true proposition is also possibly false.

    2. Every necessarily true proposition is possibly true.

    3. Every necessarily false proposition is possibly false.

    4. Every possibly true proposition is necessarily true.

    5. Every necessary truth is possibly false.

    6. Every contingent proposition is possibly false.

  B. For each question in part A, explain the reason for your answer.

6. The second page of this chapter contains an argument concerning the existence of God. Symbolize that argument in ML.

Exercise 6.2

1. Suppose that the sentence constant in the following formulas abbreviates the sentence 'Jupiter has three moons'. Translate each formula into more or less ordinary English.

a.	J	f.	$\Box$J	k.	∇J
b.	~J	g.	~$\Box$J	l.	~∇J
c.	$\Diamond$J	h.	$\Box$~J	m.	∇~J
d.	~$\Diamond$J	i.	~$\Diamond$~J	n.	~∇~J
e.	$\Diamond$~J	j.	~$\Box$~J		

2. For each English sentence in the problem above, decide whether it expresses a true or a false proposition. (Jupiter does possess three moons.)

3. In problems (a) through (n) above, let 'J' abbreviate the sentence 'Two times two equals four'. Decide whether each formula represents a true or a false proposition.

4. In problems (a) through (n) above, let 'J' abbreviate the sentence 'Some circles have corners'. Decide whether each formula represents a true or a false proposition.

5. True or false?

  a. It is contingent that FDR's presidency ended in 1992.

  b. It is contingently true that FDR's presidency ended in 1992.

  c. It is contingent that it is false that the Moon has craters.

  d. It is contingently false that the Moon has craters.

  e. It is noncontingent that all squares have four sides.

    f.   It is noncontingent that some circles have corners.

    g.   It is noncontingently true that all squares have four sides.

    h.   It is noncontingently false that some circles have corners.

    i.   It is noncontingently true that the Earth has only one moon.

    j.   It is noncontingent that it is false that $1 + 1 = 2$.

    k.   It is noncontingent that it is false that $1 + 1 = 8$.

    l.   It is noncontingently false that $1 + 1 = 2$.

    m.   It is noncontingently false that $1 + 1 = 8$.

Exercise 6.3

    1.   Why would it be incorrect to translate '∇P' as 'P is contingently true'?

    2.   Why would it be incorrect to translate '~∇P' as 'P is noncontingently true'?

4. MODAL RELATIONS

A modal property such as necessary truth, necessary falsity, or contingency reflects the way in which the truth-values of a single proposition are distributed across the set of possible worlds. We shall now introduce four *modal relations*. Modal relations reflect the way in which the truth-values of pairs of propositions are distributed across the set of possible worlds.

— INCONSISTENCY

Why does the prosecuting attorney ask the defense witness so many questions? For one thing, if the witness is weaving together a false story, it is hoped that he or she will say something that is inconsistent with something he or she said earlier. The more details that are put on the table and the longer the testimony, the greater is the chance of an inconsistency. But why is an inconsistency so important here? Two propositions are inconsistent just in case they cannot both be true. If the witness's story is inconsistent, it must contain one or more falsehoods. The concept of inconsistency was introduced in Chapter 1. Within the context of modal logic, inconsistency may be defined in terms of possible worlds:

> Two propositions are inconsistent if and only if there is no possible world in which both are true.

For example, consider the following two propositions:

    a.   Pat is 18.
    b.   Pat is not 18.

There is no possible world in which (a) and (b) are both true, so (a) and (b) constitute a pair of inconsistent propositions.

Consider next:

    c.   Pat is 18.

    d.   Pat is 40.

There is no possibility that (c) and (d) are both true, so (c) and (d) are inconsistent as well.

●— CONSISTENCY

Two propositions are *consistent* with each other if and only if they are *not inconsistent*. (Compare: Two people are related if and only if they are not unrelated.) But two propositions are inconsistent just in case there is no possible world in which they are both true. So it follows that two propositions are consistent just in case there is a possible world in which both are true.

Examples of Pairs of Consistent Propositions

    P: George Washington was the first President.
    Q: John F. Kennedy was assassinated.

    P: Mao Tse Tung became President of the United States.
    Q: George Washington was China's first President.

    P: There are only three planets.
    Q: The Earth has no moon.

Notice that two actually false propositions may, in some cases, be consistent. For example, it's actually false that Kennedy is President and it's actually false that Michael Jackson is Vice President. However, there is a possible world in which these two propositions are both true, and so the two propositions are consistent. Of course, some pairs of actually false propositions are not consistent.

————————● QUESTIONS ●————————

In each case, explain your answer:

1. Suppose two propositions are necessarily false. What relation obtains between them?

2. Suppose one member of a pair of propositions is necessarily false and the other is contingent. What relation obtains in this case?

3. Suppose one member of a pair of propositions is necessarily true and the other is necessarily false. What relation holds in this case?

4. Suppose two propositions are (both) actually false. Does this imply that they are inconsistent?

5. Any two actually true propositions are consistent. Explain why.

6. Give an example of two actually false propositions such that the two are inconsistent.

● IMPLICATION

The relation of implication was introduced in Chapter 1. In many respects, implication lies at the heart of logical theory. Within the context of modal logic, we may define this relation in two different but equivalent ways.

a. One proposition implies a second proposition if and only if there is no possible world in which the first is true while, in that world, the second is false.

b. One proposition implies a second proposition if and only if the second is true in all those possible worlds—if any—in which the first is true.

In each of these examples, the first proposition implies the second:

P: Susan is 15.
Q: Susan is a teen.

P: Jim is taller than Fred.
Q: Fred is not taller than Jim.

P: World War II began in 1939.
Q: World War II began before 1960.

If one proposition implies a second proposition, then the second is *an implication of* the first. Thus, in each case above, the second proposition is an implication of the first. Incidentally, does the second imply the first in any of the three cases above?

The Implication Test Suppose you are considering two propositions, and you aren't sure whether or not the first implies the second. Here is a test you can apply. Ask yourself this question: Are there any worlds in which: (i) the first is true and (ii) the second is false?

If the answer is *no*, then: the first implies the second.

If the answer is *yes*, then: the first does not imply the second.

Implication and Our Pursuit of Truth   An understanding of implication and an ability to "see" the relationship—to follow it from proposition to proposition— can help us enormously in our quest for truth. For example, if you begin with a true proposition and then, by a process of reasoning, trace the logical implications of that proposition, you will always arrive at additional true propositions. For if some proposition is true and implies a second proposition, then the second proposition must be true as well. All the propositions implied by a true proposition must themselves also be true. So, if we begin with propositions already known to be true and reflect upon the implications of those truths, we will acquire new truths and enlarge our knowledge of the world.

Suppose we do not know the truth-value of some proposition. However, suppose we trace the implications of the proposition and discover that it implies a second proposition. Suppose we discover that this second proposition is false. If one proposition implies another proposition, and the second is known to be false, then the first must be false as well, for a truth can never imply a falsity. Thus, using our understanding of implication, we can in some cases discover the truth-values of propositions whose truth-values we are unsure of. Scientists often test a theory by deriving a logical implication of the theory and determining the truth-value of the implication. If the implication of the theory is false, then at least some part of the theory must be false.

• SOME QUESTIONS •

1.   If one of the implications of a proposition turns out to be true, that fact alone doesn't prove that the proposition is itself true. Explain why.

2.   Give an example of two propositions that are such that the first implies the second, the first is false, yet the second is true.

Paradoxes of Implication   Our definition of implication makes perfectly good sense when we apply it to contingent propositions. For instance, most find it entirely natural to say that

Pat is 16

implies that:

Pat is a teenager.

However, things are different when we apply the definition of implication to non-contingent propositions. When we do, we get genuinely paradoxical results. You can see this for yourself by performing the following simple logical experiment. On one line of a sheet of paper, on the left side, write any sentence expressing a necessary falsehood. On the right side of the line, write any proposition-expressing sentence of your choice. Call the proposition expressed by the sentence on the left 'A' and the proposition expressed by the sentence on the right 'B':

A	B

Now, ask yourself this question:

> Is there any world in which: A is true and B is false?

If you think about it, you will see that your answer is "no." But then A implies B. And if you think about the matter further, you will see that your answer would be the same no matter which necessarily false proposition you represented on the left and no matter what proposition you represented on the right. The general principle in this case is paradoxical:

A necessary falsehood implies any and all other propositions.

That is, any necessarily false proposition implies every proposition. Thus, the proposition that 2+2=88 implies the proposition that Kennedy is President. The proposition that 2+2=88 also implies the proposition that 1+1=2, and so on.

Here is a second logical experiment that will uncover a second paradox. On one line of a sheet of paper, on the left side, write any proposition-expressing sentence. On the right side of the line, write any necessary truth of your choice. Call the proposition expressed by the sentence on the left 'A' and the proposition expressed by the sentence on the right 'B'. Now, ask yourself:

> Is there any world in which: A is true and B is false?

The answer is again "no." But then A implies B. Further reflection shows that the answer is the same no matter which necessary truth you choose and no matter what the other proposition is. The general principle in this case is also paradoxical:

A necessary truth is implied by any proposition.

That is, any and every proposition implies a necessary truth. For example, the proposition that Carter is President implies the proposition that all triangles have three sides. The proposition that 2+2=1,003 implies the proposition that 1+1=2, and so on.

Most people find these two principles paradoxical, yet the two paradoxes follow from our definition of implication. Consequently, they are called the *paradoxes of implication*.

In everyday life, we normally look for the implications of contingent propositions. This is the context in which our intuitions about implication are formed. With contingent propositions, when one proposition implies another, the two have something to do with each other; the two are relevant to each other. For example, that Chris is 50 implies that Chris is an adult. Notice that both are about Chris and both are about age. But when we extend the definition of implication to cases of noncontingent propositions, we get the paradoxical consequences in which one proposition can imply another proposition even though the first is not about the second, nor is relevant to it in any obvious way. Perhaps this consequence shouldn't be terribly surprising. After all, noncontingent propositions are very different in nature from contingent ones. Why expect the logic of the one to be just like that of the other?

However, the paradoxes are bothersome. Why isn't the definition of implication altered so as to eliminate these paradoxical consequences? Logicians have found that the definition of implication cannot be altered without resulting in other paradoxical logical results. Why not restrict implication to contingent propositions only? The problem with this suggestion is that logic makes extensive use of the implication relations obtaining between noncontingent propositions.

Incidentally, logic is not the only field containing paradoxes. Modern physics contains a number of puzzling and even bizarre paradoxes, including the "twin paradox" and the "clock paradox" within relativity theory and the "Schrodinger's cat paradox" within the theory of quantum mechanics.

Softening the Paradoxes We saw in Chapter 2 that if the premises of an argument are contradictory, then the argument must be valid, no matter what its conclusion is. But this is just another way of saying that a contradiction implies any conclusion. In Chapters 3 and 4 we saw that if we apply a tree or natural deduction proof to an argument that has contradictory premises, then any conclusion whatsoever follows. But this too is just another way of saying that a contradiction implies any and every proposition.

In Chapter 2, we also saw that if the conclusion of an argument is a necessary truth, then the argument must be valid, no matter what its premises are. But this is just to say that any proposition implies a necessary truth. In Chapter 3, we saw that we can prove a proposition to be tautological or necessarily true using a premise-free proof. In such a case, no matter what premises we add to the argument, the argument remains valid, for we can prove it valid without even using the premises. But this is another way of saying that any and every proposition implies a necessary truth.

Perhaps in the light of these considerations, the paradoxes of implication will not seem quite so paradoxical.

● EQUIVALENCE

One final modal relation is left to be defined. You were introduced to the concept of equivalence in Chapter 1. In terms of possible worlds, equivalence may be defined this way:

Two propositions are equivalent if and only if their truth-values match in every world, that is, in every world the two have matching truth-values.

Thus, if one proposition is equivalent to a second proposition, then it follows that:

a. In any world in which one of the two is true, the other is also true;

b. In any world in which one of the two is false, the other is also false;

Reflection upon the concept of equivalence reveals that if one proposition is equivalent to a second proposition, then the first implies the second and the second implies the first. This is so because if the first is equivalent to the second, then there is no world in which the first is true while the second is false and there's no world in which the second is true while the first is false, which implies that the first and the second imply each other.

Equivalence Is not Identity When we say that two propositions are equivalent, we are not saying that they are identical. That is, we are not saying that they are the same proposition. We are simply saying that they imply each other. Here are examples of pairs of equivalent propositions:

A. Jim is older than Pat.
Pat is younger than Jim.

B. Sue is married to Sam.
Sam is married to Sue.

C. $1 + 1 = 2$
All triangles are three-sided.

D. $2 + 3 = 8$
Some triangles have seven sides.

EXERCISE 6.4

1. In order to test your understanding of the "possible worlds" account of consistency, rework exercise 1.2 in Chapter 1 using the possible worlds definition of consistency.

2. In order to test your understanding of the "possible worlds" account of implication, rework exercise 1.3 in Chapter 1 using the possible worlds definition of implication.

3. In order to test your understanding of the "possible worlds" account of equivalence, rework exercise 1.4 in Chapter 1 using the possible worlds definition of equivalence.

SOME ADDITIONAL SYMBOLISM

We can now extend our symbolism as follows.

1. The symbol 'o', called the *consistency circle* (or "circle" for short), abbreviates 'is consistent with'.
2. The symbol 'φ', called the *inconsistency sign*, abbreviates 'is inconsistent with'.
3. The symbol '→', called the *fishhook* (or "hook" for short), abbreviates 'implies'.
4. The symbol '↔', called the *double fishhook* (or "double hook" for short), abbreviates 'is equivalent to'.

Thus, if P abbreviates 'Mars has two moons' and Q abbreviates 'Pluto has no satellite', then:

'PoQ' abbreviates "'Mars has two moons' is consistent with 'Pluto has no satellite'".

'PφQ' abbreviates "'Mars has two moons' is inconsistent with 'Pluto has no satellite'".

'P→Q' abbreviates "'Mars has two moons' implies 'Pluto has no satellite'".

'P↔Q' abbreviates "'Mars has two moons' is equivalent to 'Pluto has no satellite'".

The concept of implication defined in modal logic, represented above by the fishhook, is different from the truth-function associated with the horseshoe operator. In order to distinguish the two, the truth-function for the horseshoe is named 'material implication' and the modal concept associated with the fishhook is termed 'strict implication'.

Each of these four new modal operators is a dyadic operator, since each is used to join two components into a compound. As you work with these operators, you must be able to recognize the scopes of the operators. The scope of a dyadic modal operator is always going to be exactly what the scope would be if the operator were to be a dyadic truth-functional operator such as an ampersand or wedge. In the following examples, the scopes of the wedges are marked off with brackets:

A v B   A v (B&C)   (A v B) ⊃ E   ~[A v (S & H)]

In these examples, the scopes of the consistency circles are marked with brackets:

A o B   A o (B&C)   (A o B) ⊃ E   ~[A o(S & H)]

In these examples, the scopes of the fishhooks are marked with brackets:

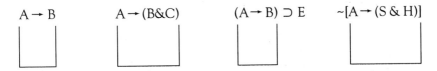

$$A \rightarrow B \qquad A \rightarrow (B\&C) \qquad (A \rightarrow B) \supset E \qquad \sim[A \rightarrow (S \& H)]$$

If you are ever uncertain as to the scope of a circle, fishhook, or double fishhook, figure the scope as if the operator were a wedge.

Recall that the main connective of a sentence is the operator of greatest scope. Thus, in the following, the main connective is the fishhook:

$$(A\&B) \rightarrow G \qquad \sim H \rightarrow (T\&O)$$

In these, the main connective is the consistency circle:

$$A \text{ o } (\sim B\&E) \qquad [(\sim FvD) \supset S] \text{ o } (J\&M)$$

●— SYMBOLIZING SOME MODAL TRUTHS

Using these new symbols, plus symbols introduced earlier, we can now abbreviate additional truths. In the following examples, let P and Q stand for any two proposition-expressing sentences you choose. It won't matter which proposition-expressing sentences P and Q symbolize, because what we will say about P and Q would be equally true if said about any proposition-expressing sentence.

1. Remember that two propositions are inconsistent just in case there's no world in which both are true. Suppose P and Q are both necessarily false. It follows that each is false in every world. Therefore, there's no world in which both are true. But if that's the case, then the two must be inconsistent. So, if P and Q are both necessarily false, then they are inconsistent. That is,

$$(\Box \sim P \& \Box \sim Q) \rightarrow (P \phi Q)$$

In general: any two necessary falsehoods are inconsistent.

2. If P is necessarily true and Q is necessarily false, then there is obviously no world in which the two are both true. That is:

$$(\Box P \& \Box \sim Q) \rightarrow (P \phi Q)$$

In general, any necessary truth is inconsistent with any necessary falsehood.

3. Remember that two propositions are consistent if and only if there's at least one world in which both are true. If P and Q are both necessarily

true, then there must be at least one world in which both are true, since both are true in every world. In that case, the two must be consistent. So, if P and Q are both necessarily true, then P and Q are consistent. That is,

$$(\Box P \,\&\, \Box Q) \to (P \text{ o } Q)$$

4. If P is a necessary truth and Q is contingent, then the two must be consistent. For if P is necessarily true, then P is true in every world, and if Q is contingent, then Q is true in one or more worlds, which implies that there must be at least one world in which both are true. Thus:

$$(\Box P \,\&\, \triangledown Q) \to (P \text{ o } Q)$$

5. Recall the first paradox of implication:

A necessary falsehood implies any other proposition.

If P is a necessary falsehood, then P implies Q no matter what proposition Q happens to be:

$$\Box {\sim} P \to (P \to Q)$$

6. The second paradox of implication is the thesis that:

A necessary truth is implied by any proposition.

If P is necessarily true, then Q implies P no matter what proposition Q happens to be:

$$\Box P \to (Q \to P)$$

7. Suppose you know that the negation or opposite of P is inconsistent with itself. Can you tell just from this information what P's modal status must be? Might P be necessarily false? Or contingent? Or necessarily true? If you think about the various possibilities, you will see that in such a case, P can only be a necessary truth. The only type of proposition that is not consistent with itself is a necessary falsehood. Therefore, the negation of P must be necessarily false, which makes P a necessary truth. Thus:

$$({\sim}P \,\phi\, {\sim}P) \to \Box P$$

8. Remember that two propositions are equivalent if and only if the truth-values of the two match in every world. Suppose P and Q are both necessarily false. What follows? Both are false in every world. But in that case, they have matching truth-values in every world, and this implies

that they are equivalent. So, if P and Q are both necessarily false, then P is equivalent to Q:

$$(\Box\text{-}P \,\&\, \Box\text{-}Q) \rightarrow (P \leftrightarrow Q)$$

Of course, if P and Q are both necessarily true, then they are both true in every world and so their truth-values match in every world. Consequently, if P and Q are necessarily true, they are equivalent as well:

$$(\Box P \,\&\, \Box Q) \rightarrow (P \leftrightarrow Q)$$

EXERCISE 6.5

Consider the following statements:

A: $1 + 1 = 2$

B: Ringo plays drums.

C: Some circles are square.

D: Exactly 7 million people live in New York City.

1: Ringo owns a drum set.

2: All triangles have three sides.

3: $2 + 2 = 463$

4: Less than 20 million people live in New York City.

5: More than 8 million people live in New York City.

6: Between 6,999,999 and 7,000,001 people live in New York City.

7: More than 1 million people live in New York City.

Notice that A is consistent with the following numbered propositions: 1, 2, 4, 5, 6, 7. We can express this point by writing: Ao 1, 2, 4, 5, 6, 7. In a similar fashion, fill in the following blanks with the appropriate numbers:

Ao _____	Aφ _____	A→ _____	A↔ _____
Bo _____	Bφ _____	B→ _____	B↔ _____
Co _____	Cφ _____	C→ _____	C↔ _____
Do _____	Dφ _____	D→ _____	D↔ _____

5. SEMANTICS

Modal Operators Are Nontruth-Functional   At the start of this chapter, we referred to the difficulty philosophers have encountered developing a semantical theory for modal logic. Although the semantics for truth-functional logic was developed during the first half of this century, a semantical theory for modal logic wasn't developed until the late 1950s and early 1960s. If the modal operators had been truth-functional in nature, the semantics of modality would have been as easy as drawing up a few new truth-tables. Unfortunately, as noted previously, the semantics for the modal operators cannot be stated in terms of truth-functions, for the fundamental modal concepts are not truth-functional in nature. We can now explain why this is so. Essentially, the reason is that the truth-value of a sentence whose main operator is a modal operator is not a function of the truth-value (or values) of its component (or components). That is, when the main operator is a modal operator, there is no determinate or functional relationship between the truth-value of the components and the truth-value of the compound as a whole. For example, suppose all we know is that some sentence **P** has the truth-value true. Given just this information, what is the truth-value of □**P**? There's no way to tell. In some cases, a sentence **P** will be true while □**P** is true, and in other cases **P** will be true while □**P** is false. So, given that **P** is true, nothing follows regarding □**P**. Therefore, we cannot fill in the first row of a truth-table for □**P**.

P	□P
T	?
F	

Thus, the truth-value of □**P** is not a function of the truth-value of **P**.

Suppose some sentence **P** is false. Given just this information, what is the truth-value of ◇**P**? There's no way to tell. If **P** is false, the truth-value of ◇**P** is not thereby determined. Thus, if we were to try to write a truth-table for the diamond operator, we could not fill in row 2:

P	◇P
T	
F	?

If we try to specify truth-tables for the other modal operators, in each case we find rows that cannot be filled in. For example:

P Q	P → Q		P Q	P ↔ Q		P Q	P ∘ Q
T T	?		T T	?		T T	T
T F	F		T F	F		T F	?
F T	?		F T	F		F T	?
F F	?		F F	?		F F	?

Thus, the modal operators are not truth-functional in nature. No truth-functions exist that can be associated with the modal operators. So how shall we specify the semantics for ML and its modal operators? In place of the truth-tables with their finite arrays of columns of T's and F's, we will refer to the infinite array of possible worlds.

A Semantics for ML   There are a number of alternative sets of semantical principles for the modal operators and philosophers are not in agreement on which set is the "correct" set. The following set of semantical principles is widely accepted. In the following, 'iff' abbreviates 'if and only if':

1. ~**P** is true in a given world iff **P** is false in that world.
2. (**P&Q**) is true in a given world iff **P** is true in that world and **Q** is true in that world.
3. (**PvQ**) is true in a given world iff **P** is true in that world or **Q** is true in that world or both **P** and **Q** are true in that world.
4. (**P⊃Q**) is true in a given world iff it's not the case that **P** is true and **Q** false in that world.
5. (**P≡Q**) is true in a given world iff either **P** and **Q** are both true in that world or **P** and **Q** are both false in that world.
6. □**P** is true in a given world iff **P** is true in every world.
7. ◇**P** is true in a given world iff **P** is true in at least one world.
8. ∇**P** is true in a given world iff **P** is true in at least one world and **P** is false in at least one world.
9. **P o Q** is true in a given world iff **P** and **Q** are both true in at least one world.
10. **P ф Q** is true in a given world iff there is no world in which **P** and **Q** are both true.
11. **P → Q** is true in a given world iff there is no world in which **P** is true and **Q** false.
12. **P ↔ Q** is true in a given world iff there is no world in which **P** and **Q** have differing truth-values.

Recall that ML "contains" TL. Consequently, the first five clauses simply summarize the semantical principles of truth-functional logic. In other words, clauses 1 through 5 above summarize the truth-tables for the five truth-functions we studied in Chapter 2. Clauses 6 through 12 go on to specify, in terms of possible worlds, a semantics for the modal operators.

Let us for a moment compare modal logic and truth-functional logic. The semantical principles of truth-functional logic are specified in terms of rows on truth-tables. Each of those tables has only a finite number of rows. The semantical principles of modal logic are defined in terms of a manifold comprising an infinite number of possibilities, possibilities called 'possible worlds'. So, the truth-functions are defined against a finite manifold while the modal operators are defined against an infinite manifold.

●— ALTERNATIVE SYSTEMS OF MODAL LOGIC

A *system* of modal logic consists of two parts: (i) a formal language to be used for the symbolization of modal arguments; (ii) a set of principles to be used for the evaluation of modal arguments. A number of different systems of modal logic have been developed in recent years. The semantical principles summarized in the 12 clauses above, combined with the language ML, constitute the basis of the system of modal logic known as the system S5. Using S5—the language ML and the 12 semantical principles—arguments can be symbolized and tested for validity, sentences can be symbolized and the logical properties of the corresponding propositions investigated, and so on.

Philosophers are not in total agreement on clauses 6 through 12. Those clauses reflect a particular understanding of the modal operators, an understanding shared by many but not by all philosophers. There are other ways the modal operators can be interpreted and, accordingly, there are alternative sets of semantical principles. Philosophers who favor alternative sets of semantical principles have developed alternative systems of modal logic. A large number of alternative systems of modal logic have been developed in recent years. Many of these systems carry deep philosophical consequences. System S5 is developed here because there is widespread agreement among philosophers that S5 best represents the standard concepts of logical necessity and possibility that we have been studying in this chapter. However, in Chapter 7, we will discuss other systems of modal logic and see how they differ from S5. Then we will briefly look into the question of which system best represents the concepts of logical necessity and possibility.

6. EPISTEMOLOGY and LOGIC

In this chapter, you have been introduced to the idea of a necessarily true proposition, a truth that cannot even possibly be false. If you have reflected upon the matter, perhaps the following philosophical question has arisen:

> How do we know that a particular truth is necessarily true, true in every world? What's our claim to knowledge based upon? In everyday life, we acquire information about the world through our five senses. In the physical sciences, we investigate the world by observing it and probing it with our senses. Thus, if a scientist claims to know that a proposition P is true, he or she can establish P on the basis of evidence presented by the senses, that is, on the basis of what is termed 'sense-experience'. But what evidence can the philosopher point to in order to prove that some truth is a necessary truth? It can't be based on observation; nobody can scan with their eyes that infinite manifold of possibilities called the set of all possible worlds. So, what sort of evidence is a claim of necessary truth based upon?

In order to examine the issue this raises, we must raise a more general question. How do we come to acquire knowledge? Since ancient times, most philosophers have distinguished two separate sources of knowledge. One is reason. The other is sense-experience. Let's take each in turn.

● THE SOURCES OF KNOWLEDGE

Sense Experience as a Source of Knowledge   At least some of our knowledge is based upon the input we receive from the world through our five senses. That is, we have knowledge that is justified on the basis of or by appeal to our sensory experience of the world. Such knowledge is called *experiential knowledge*. For an example of experiential knowledge, consider the claim that Mount Everest rises 29,000 feet above sea level. How do we know this? Our knowledge is based upon observations and measurements people have made, which is to say that this knowledge is ultimately justified in terms of things seen, felt, or otherwise detected by the senses. Our physical senses of sight, hearing, touch, smell, and taste constantly supply us with new knowledge of our world. Other things we know on the basis of sense observation include:

> The Moon has craters.
> Lemons taste sour.
> Snow is usually white.

Reason as a Source of Knowledge   Many philosophers throughout history have held that there exists a second source of knowledge. Some truths, it is claimed, are known purely on the basis of an inner process of reasoning and not on the basis of information provided by our senses. This type of knowledge is called *ratiocinative knowledge*. Thus, reason has traditionally been understood as a second source of knowledge. If some of our knowledge is ratiocinative knowledge, then our minds have the power to apprehend some truths about the world by a process of pure reasoning, without using input from the senses in the process. Ratiocinative knowledge is said to arise when the mind directly grasps—independently of the senses—a truth.

Those philosophers who hold that some knowledge is ratiocinative might point to the following example. Surely we know that:

> **a.**   Either Abraham Lincoln ate an apple on his sixth birthday or he did not eat an apple on his sixth birthday.

But this knowledge does not seem to be justified in terms of sense-experience. It seems to be based solely on a process of logical reasoning that one who had no sense-experience of the world could still think through. Of course, we must use our senses to learn the language and to read the sentence, sentence (a). But that process just gives us an understanding of the meaning of the sentence. It's one thing to understand the meaning of a sentence; it's another thing entirely to come to know that the sentence expresses a truth. Once we understand the meaning of sentence (a) we then grasp the proposition it expresses. Then, by an inner process of reasoning alone, we come to see directly that the proposition is itself necessarily true, true in every world. No experiments or observations or sense data of any kind, it seems, could possibly justify the claim that the proposition expressed by (a) is true. We simply reflect on the proposition and it is self-evident that it is true in every world.

Philosophers who hold that some knowledge is ratiocinative might point to another example. We know it is necessarily true that

The decimal expansion of π is nonrepeating and goes on to infinity.

Yet how could this item of knowledge possibly be based on experiments, or observations, or on any sort of sense-experience? How could we possibly observe every number in π's decimal expansion? This mathematical truth seems to be something that is known by reason alone, by rational reflection, operating independently of the senses.

Rationalism and Empiricism Epistemology is the branch of philosophy concerned with exploring the nature of knowledge. In epistemology, there are two broad schools of thought concerning the sources of knowledge. Some philosophers argue that there is no such thing as ratiocinative knowledge. These philosophers therefore deny that we have any ratiocinative knowledge whatsoever. All knowledge, according to this position, is based ultimately only on evidence provided by the senses. Philosophers who hold this position are called *empiricists*. Thus, empiricists hold that our knowledge of the world ultimately comes from one source, namely, our senses.

Some philosophers claim that we do have at least some ratiocinative knowledge. Philosophers who hold this position are called *rationalists*. Since few rationalists wish to deny that some knowledge is also based upon information received from the senses, most rationalists add that some knowledge is also experiential knowledge. Thus, rationalists typically claim that our knowledge comes from two sources: (i) sense-experience, and (ii) reason operating independently of the senses.

Arguments Against Rationalism The rationalist claims that some of our knowledge is based on pure reasoning alone operating independently of the senses. For example, according to the rationalist we know by reasoning alone that

√2 is irrational

is true. Someone might object to the rationalist along the following lines. Don't we use our senses—our eyes—to read the sentence? And don't we use our senses to learn the language that the sentence belongs to? Then doesn't it follow that all knowledge—even what's called ratiocinative knowledge—is really based only upon the senses?

A rationalist might answer as follows: Yes, we do use inputs from our senses to learn a language and to learn the meaning of a sentence such as '√2 is irrational'. However, this doesn't mean that our knowledge of the truth of this particular necessary truth is itself based on the evidence of the senses. It is one thing to learn the meaning of a sentence; it is another thing altogether to come to know that the sentence expresses a truth. Truth and meaning are two different things. We use our senses to learn the meaning of the sentence. Once we understand the meaning of the sentence, we can grasp the proposition it expresses. But from there, it takes an act of pure reasoning to know that the proposition is necessarily true. The senses provide no help at this stage. And it is this knowledge—that the proposition in question is necessarily true—that is ratiocinative. One simply "sees," by rational reflection alone, that the proposition is necessarily true.

Rationalists argue that our minds have the ability to directly grasp certain truths by reason alone. The standard examples of ratiocinative truths are typically necessary truths. No sense observation of the world, it seems, could possibly justify such knowledge. So the rationalist is claiming that our minds have the ability to attain a direct intellectual insight into some of the necessary aspects of reality. This is an insight into an extralinguistic, extraconceptual reality, an insight into the way some things necessarily are in themselves.

The empiricist denies that such an ability exists. How, the empiricist asks, could our minds possibly know something without using the physical senses? An ability to know certain truths without using the physical senses would be, it seems, a supernatural ability. Ratiocinative knowledge, the empiricist argues, is inherently mysterious and cannot be fit into a scientific view of the world. The empiricist concludes that our physical senses are the only conduits through which information enters our minds.

Objections to Empiricism   Rationalists challenge the empiricist by producing examples of truths that, they claim, could not possibly be known solely on the basis of input from the senses. For example, we know that $\sqrt{2}$ is irrational. But how could we possibly know the truth of this proposition on the basis of observation or other sense inputs? How could our senses of sight, touch, smell, taste, or hearing prove that the decimal expansion of $\sqrt{2}$ goes on forever in a never-repeating sequence? (Nobody has ever seen or computed $\sqrt{2}$'s decimal expansion in its entirety.) An inner process of reasoning—a process that could be performed by someone who had lost the use of all five senses—is, it seems, all that proves that $\sqrt{2}$ is irrational. Thus, suppose a person knows the meaning of the various principles of mathematics. That person could be put in a sensory-deprivation tank and—without any sense-experience of the world—still think through the reasoning for the irrationality of $\sqrt{2}$. This person would then know, on the basis of reasoning alone, that $\sqrt{2}$ is irrational.

Another example that rationalists cite is the following necessary truth:

> Nothing in the entire universe is red all over and green all over at the same time.

Surely this is true. Yet how could it possibly be that we know this to be true on the basis of information our senses have provided? What sense-experience could possibly prove that nothing in the entire universe is red all over and green all over at the same time?

The "Further Readings" section at the back of this book contains references for those who wish to look more deeply into some of the issues raised here.

SOME QUESTIONS

1. Are there two sources of knowledge? Or is there only one source of knowledge? Give an argument for your answer.

2. If you hold that the senses are the only source of knowledge, then what is that claim based upon? Is your claim based on your sense observation of the world? In what way?

APPENDIX: HORSESHOES, FISHHOOKS, and THE HISTORY OF LOGIC

The horseshoe and the fishhook represent different logical concepts, namely the concepts of *material* implication and *strict* implication respectively. Material implication is, of course, a truth-function defined on a (finite) four-row truth-table. Strict implication, on the other hand, is a modal concept specified in this text in terms of an infinity of possible worlds. The horseshoe and the fishhook thus represent two ways to formalize the "if, then" or conditional operator, and philosophers are not in agreement regarding which way best captures the semantics of conditional sentences.

Shakespeare once observed that there is nothing new under the sun. It is interesting that the contemporary debate between proponents of the truth-functional analysis of the conditional sentence and advocates of the modal analysis actually goes back to ancient times. A brief examination of the origins of this debate will introduce you to some of the characters who played a role in the early development of logical theory and help you see that, as abstract as logic may be, it was and is the concern of real live human beings.

In the ancient Greek world, two major schools of logical theory developed, each associated with a major school of philosophical thought. The *Peripatetic* school of philosophy consisted of those philosophers teaching and writing in the tradition started by Aristotle (384–322 B.C.), the founder of formal logic. The logical theory associated with this school of thought is known as Aristoteleon logic. The *Megarian* school of thought was founded by Euclides, a student of Socrates (469–399 B.C.) and a contemporary of Plato (428–348 B.C.). Euclides taught either Stilpo or Stilpo's teacher Thrasymachus of Corinth. Stilpo taught Zeno of Citium (ca. 336–264 B.C.), the founder of the important school of thought known as Stoicism.

In the third century B.C., the Megarian philosophers Diodorus Cronus (d. 307 B.C.) and his pupil Philo engaged in a spirited and famous debate over the logical nature of the conditional sentence, with Diodorus advocating the semantics we now associate with the fishhook and Philo advocating the semantics we now associate with the horseshoe. Reasoned debate—dialectic—generally leads to increased understanding and the expansion of knowledge. The logical theory that grew out of the debate between Diodorus and Philo became the logical theory associated with the Megarian and Stoic schools of philosophy.

Since the actual writings of Philo and Diodorus were not preserved, our source for their debate is the writings of a third-century philosopher, Sextus Empiricus. In the *Outlines of Phyrronism* (II, 110), Sextus contrasts the views of Philo and Diodorus as follows:

> Philo says that a sound [true] conditional is one that does not begin with a truth and end with a falsehood, e.g., when it is day and I am conversing the statement "If it is day, then I am conversing." But Diodorus says it is one that neither could nor can begin with a truth and end with a falsehood. According to him, the conditional statement just quoted seems to be false, since when it is day and I have become silent, it will begin with a truth and end with a falsehood.[1]

This passage makes it clear that Philo advocated the truth-table analysis of the conditional operator and that Diodorus advocated the modal analysis.

The following passage, from another of Sextus' works, makes it clear that Philo advocates the analysis represented by the modern day truth-table for the horseshoe. In *Against the Mathematicians*, Sextus presents Philo's account as follows:

> Since, then, there are four possible combinations of the parts of a conditional—true antecedent and true consequent, false antecedent and false consequent, false and true, or conversely true and false—they say that in the first three cases the conditional is true (i.e., if the antecedent is true and the consequent is true, it is true; if false and false, it again is true; likewise for false and true); but in one case only is it false, namely, whenever the antecedent is true and the consequent is false.[2]

It seems that quite a number of philosophers joined the debate over the semantics of the conditional. Indeed, so many philosophers entered the fray and the discussion became so intense that Callimachus is reported to have written:

> Even the crows on the roofs caw about the nature of conditionals.[3]

And this was more than 2,300 years ago!

Some of the ancient logicians were interesting characters. For instance, Zeno, who was from Cyprus, did not make himself very popular when, upon arriving in Athens, he proposed to reform the Greek language before he had even learned to speak it! Zeno is said to have died at the age of 98 by holding his breath. According to legend, Diodorus finally committed suicide because he couldn't solve a logical

---

[1] This passage appears in Kneale and Kneale [1984] p. 128.
[2] This passage is quoted in Mates [1972] p. 214.
[3] Quoted in Kneale and Kneale [1984] p. 128.

puzzle Stilpo had given him. (Now that's taking logic seriously!) Chrysippus (ca. 280–205 B.C.) the founder of Stoic logic, is supposed to have written to Cleanthes (a fellow Stoic), "Just send me the theorems; I'll find the proofs for myself." One of the old sayings was, "If there is any logic in Heaven, it is that of Chrysippus." According to Diogenes, Chrysippus wrote 311 books on logic. Incidentally, Cleanthes has been described as a "poverty-stricken prize fighter who came to Athens and entered Zeno's school, became its head, transmitted Zeno's doctrines without change, and eventually starved himself to death at the age of 99."[4]

In addition to being the first to formulate a truth-functional analysis of the conditional operator, the Stoic logicians were the first, as far as we know, to formulate what amount to truth-functional analyses of the 'and', 'or', and 'not' operators. For instance, Sextus Empiricus described the Stoic truth-functional account of conjunction in the following way:

> A conjunction is [true] when it has everything in it true . . . , false when it has one thing false.[5]

This is clearly a summary of the ampersand's truth-table.

The Stoics were also the first logicians, as far as we know, to formulate natural deduction inference rules. (Natural deduction was covered in Chapter 3.) They used numbers instead of variables for sentences. Following are several Stoic inference rules:

1. If the first, then the second;

   The first;

   Therefore, the second.

2. If the first, then the second;

   Not the second;

   Therefore, not the first.

3. The first or the second;

   Not the first;

   Therefore, the second.

4. Not both the first and the second;

   The first;

   Therefore, not the second.[6]

---

[4] See Mates [1972] pp. 212–213.

[5] Quoted in Rist [1978] p. 17.

[6] These rules appear in Mates [1972] pp. 214–216. See also Kneale and Kneale [1984] p. 163.

SOME QUESTIONS

Part I   The following questions are for those who covered Chapter 3:

1.   Which modern truth-functional natural deduction rule corresponds to the first Stoic rule above? Which corresponds to the second? Which to the third?

2.   State the Hypothetical Syllogism rule in the Stoic fashion. State the Constructive Dilemma rule in Stoic form.

Part II   The following question is for those who covered either Chapter 3 or Chapter 4: Which valid argument form corresponds to the first Stoic rule above? Which corresponds to the second? Which to the third?

PHILOSOPHICAL INTERLUDE

What's the point of studying all this abstract logical theory? What could it possibly be used for? These are questions students often raise. Logical theory actually has a number of significant uses. First, symbolic logical languages are an indispensable part of computer science. For this reason, symbolic logic has played an important role in the development of computer technology.

Second, you are bombarded with arguments every day of your life. These arguments come from friends, from people you work with, from politicians, and from many other sources. The study of symbolic logic can deepen your understanding of reasoning, which in turn can contribute to your ability to evaluate the arguments others present to you.

Third, logical terminology, principles, and procedures are employed extensively in the field of philosophy. In this interlude, you will have an opportunity to think through some fundamental philosophical issues. These are issues that have been of concern to reflective persons throughout history. As you explore each issue, and the reasoning for and against each of the various philosophical viewpoints, you will see that the evaluation of the arguments requires an understanding of many of the concepts and principles of formal logic. The arguments in this chapter will thus give you a chance to apply some of the logical principles you are learning.

Some people believe that forming one's own philosophical view is a waste of time. However, consider that people's lives are shaped by the philosophical views they hold on religious and ethical questions. One obvious example is the question of abortion. The decision to have or not to have an abortion is ultimately based on one's views concerning what constitutes a human life, what constitutes a justified taking of life, and so on. In addition, our political, legal, economic, and social systems reflect certain fundamental philosophical viewpoints.[1] Whether we like it or

---

[1] For an introduction to the abortion issue, see Glover [1977]. The authors of the Declaration of Independence and the United States Constitution held philosophical views on human nature, human rights, the purpose of govern-

not, our lives are deeply affected by the philosophical views we and others hold. Philosophy seems to be an inescapable part of human existence.

This philosophical interlude introduces and explores possible answers to three philosophical questions:

1. Does God exist?
2. Do we possess free will?
3. What is the nature of the mind?

As you consider the arguments presented in this interlude, try to adopt a philosophical attitude. Fundamentally, this involves three things. First, try to set aside your opinions and convictions and approach each argument with as open a mind as possible. Second, try to evaluate each argument on the basis of reasons rather than emotion. Third, the first time you read an argument, try to understand the argument before you begin objecting to it. Often, people will begin objecting to a philosophical argument as soon as they read or hear the first premise—before they have even the slightest understanding of the argument.

1. DOES GOD EXIST?

●— THE CONCEPT OF GOD WITHIN TRADITIONAL THEISM

A theist is a person who believes that the universe has a creator. An atheist is one who believes that the universe does not have a creator. *Traditional theism* is the view that there exists a transcendent being, sometimes called God, who created the entire universe and who governs it from moment to moment. Traditional theism includes orthodox believers within Judaism, Islam, and Christianity.

Theology is the study of God. *Philosophical theology*, the subject we are about to enter, is the application of philosophical methods to the field of theology. The fundamental goal of philosophical theology is to think about God using only our natural reasoning abilities—the abilities we use when we reason about philosophical, scientific, historical, and other matters.

Theists sometimes say that God is the "Supreme Being". Among philosophical theologians, there is widespread agreement that the core idea of theism—the idea shared by orthodox Jews, Muslims, and Christians—is the claim that God is a being so perfect that nothing could even possibly be more perfect. To put it another way, God, according to traditional theism, is the greatest being possible.

---

ment, and the nature of social justice. For an excellent account of those views, see White [1987 and 1978]. Remember that if you wish to explore on your own any of the philosophical issues raised in this text, a "Further Reading" section, organized by topics, appears at the end of the book.

Of course, the big question is: Does such a being exist? The concept of God presents us with an interesting possibility. But is there actually such a being? Before we consider this question, however, a more fundamental question must be asked: If a greatest possible being were to exist, what would its nature be? What would it be like? Before we can decide whether a particular entity exists, we must first characterize the properties the entity supposedly possesses. If you are trying to decide whether a certain sort of thing exists, and you lack a characterization of the properties it supposedly has, you really won't know what it is you are supposed to be considering the existence of. For instance, if you are sent to Loch Ness and told to look for a cadborasaurus, you won't know what to look for unless someone explains to you the characteristics a cadborasaurus (the so-called Loch Ness monster) is supposed to possess.

So, let's ask a hypothetical question: If the greatest being possible were to exist, what would its nature be? According to traditional theism, the greatest being possible—if it were to exist—would be a being that is absolutely perfect in every way. This may be put as follows:

> The greatest being possible would be a being possessing the greatest possible set of consistent perfections.

But what is a perfection? And what is a set of consistent perfections?[2]

A *perfection* may be characterized as the highest possible degree of a property that (a) can be possessed in various degrees, and (b) is such that possession of the property adds to the *intrinsic* goodness of that which possesses the property. The concept of an intrinsic good is best introduced by contrasting it with the concept of an *extrinsic* good. Some things are good only insofar as they lead to or are used to bring about something else that is itself good. Apart from what they lead to or bring about, they are not in themselves good. In other words, they are good only in the sense that they are a means to an end. Other things are good in themselves, independently of the goodness of anything else. In other words, they are an end rather than a means to an end. Roughly, an extrinsic good is something that is good only because of some further thing that it leads to or is used to bring about, while an intrinsic good is something that is good in itself and not because of something it leads to. For example, a trip to the doctor is good only in that it leads to the good we call "health," and a car is good only in that it is used to transport you from one point to another. Cars and visits to doctors are extrinsic goods. On the other hand, love, health, friendship, knowledge, and aesthetic beauty are generally agreed to be intrinsic goods.[3]

A set of properties is consistent if it is possible that all the properties in the set are possessed by one individual thing. For example, the property of being a teenager and the property of being an octogenarian are not consistent properties, while the property of being a cousin is consistent with the property of being an adult.

---

[2] In my discussion of the nature of perfection, I am indebted to the accounts in Morris [1987b] and Feldman [1986].

[3] For further discussion see Finnis [1982].

The question that arises next is: Which properties are perfections? To answer this, we must consult our value intuitions—our intuitions about what is intrinsically good and what is not. Let us begin with knowledge. Being knowledgeable is a property that can be possessed in varying degrees, for some people are more knowledgeable than others. Furthermore, knowledge is intrinsically good, for it is better to have knowledge than to lack it, no matter what one's external circumstances are. Perfection with respect to knowledge would be the possession of the highest degree of knowledge it is possible for anything to have. This perfection has traditionally been called *omniscience*, and a being having perfect knowledge would be an omniscient being.

Power, understood as the ability to make things happen, also comes in degrees and seems to be an intrinsic good. Perfection with respect to power would be the possession of the highest degree of power anything could possibly have. This perfection has traditionally been termed *omnipotence*, and a being possessing the highest possible degree of power would be an omnipotent being.

Moral goodness is a quality that comes in degrees and is surely an intrinsic good. The highest possible degree of moral goodness anything could possibly possess is called *omnibenevolence*. An omnibenevolent being would be a being who is morally good to the highest possible degree.

Creativity comes in degrees and also seems to be an intrinsic good. The highest degree of creativity anything could possibly exercise would be to create everything else in the universe. Let us say that a being possessing the highest degree of creativity possible is an *omnigenic* being.

These are some of the most important perfections studied within philosophical theology. Since traditional theism holds that God possesses all of these perfections, traditional theism holds that omniscience, omnipotence, omnibenevolence, and omnigenesis constitute a consistent set. Putting this all together, traditional theism claims that:

> The greatest possible being would be a being possessing the greatest possible set of consistent perfections, including the perfections of omnipotence, omniscience, omnibenevolence, and omnigenesis.

Traditional theism may thus be summed up as the view that there exists an omniscient, omnipotent, omnibenevolent creator of the universe.

We now have a basic account of the traditional concept of God, a characterization of the nature a greatest possible would have if such a being were to exist. The next question is: Does such a being exist?

•— ST. ANSELM'S ONTOLOGICAL ARGUMENT FOR GOD'S EXISTENCE

An *ontological* argument for God's existence is an argument that claims to deduce the truth of the proposition that God exists from nothing more than a logical analysis of the concept of God. Such an argument relies on no empirical evidence whatsoever, that is, it does not use any evidence based upon our sense observation of the physical world. Rather, the argument relies only upon logical reasoning to reach the conclusion that God must exist. Philosophers who argue for the soundness of an ontological argument for God's existence hold that the proposition that God exists is a *ratiocinative* truth. (Ratiocinative truth is discussed in Chapter 6.)

St. Anselm, a scholar and monk who lived from 1033 to 1109, formulated the very first ontological argument for God's existence. Anselm's argument is an example of an indirect proof. In order to understand Anselm's reasoning, one must first understand the logical nature of an indirect proof. In Chapters 3 and 4, we examined the indirect proof of the irrationality of $\sqrt{2}$. That proof began with the assumption that $\sqrt{2}$ is a rational number. The argument then demonstrated that this assumption implies a contradiction. Since only that which is necessarily false can ever imply a contradiction, it followed that the assumption was itself necessarily false, which proved that the opposite of the assumption must be true. This implied that $\sqrt{2}$ must be irrational. In general, when you construct an indirect proof, you first assume the logical opposite of that which you aim to prove. You then demonstrate that this assumption implies a contradiction, which proves that the assumption is itself contradictory. It follows that the opposite of your assumption must be true. It is recommended that you keep this general format in mind as you think through Anselm's reasoning.

Philosophers disagree over how best to express Anselm's argument. The following line of reasoning represents one interpretation of Anselm's argument.[4]

It was noted earlier that if you are going to argue for the existence of something, it is important that you begin by defining just what it is that you are claiming exists. In general, when putting forth an argument, if you do not define your fundamental terms, your listeners won't know what it is you are arguing about. Consequently, in proper logical fashion, Anselm's reasoning begins with a definition of his most important term:

> 1. By the term 'God' let us mean 'that than which a greater is not possible'.

Once the definition of the term 'God' is understood, Anselm and his intended audience can at least agree on what it is they are arguing about.

The next premise seems to be one that can be granted without conceding anything to Anselm:

> 2. It is *possible* that that than which a greater is not possible exists.

Here Anselm merely asserts the possibility of God's existence. It remains to be seen if Anselm can prove God actually exists.

Why suppose this second premise is true? Perhaps the following reasoning might make (2) plausible. Even the atheist understands and can make sense of the

---

4 My interpretation of Anselm is influenced by the discussion in Plantinga [1974]. For further discussions of the ontological argument, see Plantinga [1965] and Hick and McGill [1967].

definition of the term 'God'. But if we can understand and make sense of the concept of a thing, then there is a strong presumption in favor of the logical possibility of such a thing. Of course, this is not to suppose the thing actually exists; there's a big gap between the mere possibility of a thing and its actual existence. Consequently, if we can make sense of the concept of the greatest possible being, then it is reasonable to suppose that such a being is logically possible. Nevertheless, the question still remains: Does such a being actually exist?

Recall that every indirect proof contains an assumption. Specifically, in an indirect proof, one assumes the opposite of what one is trying to prove. Anselm aims to prove that God exists. Consequently, the assumption Anselm makes in this argument is simply that God, understood as the greatest possible being, does not exist:

3. That than which a greater is not possible does not exist. (Assumed premise)

Here Anselm has assumed the opposite of that which he seeks to prove. His goal now is to demonstrate that this assumption implies a contradiction. If he accomplishes this, he will have proven that the opposite of his assumption must be true, from which it will follow that there does actually exist a greatest possible being.

Anselm's next premise will make more sense to you if, before you read it, you first reflect on the relation between greatness, possibility, and actuality. Suppose someone describes something and as the entity's various properties are described, the entity described seems to be very great in some respect. You want to encounter whatever it is that is being described. Now imagine being told that the entity doesn't actually exist. It is a mere possibility that is being described. Wouldn't you be disappointed? Wouldn't you feel let down? And wouldn't that which you are considering be greater if it were actual rather than merely possible? At least in the case of something good, to be actual seems greater than to be merely possible.

After reflecting upon these thoughts, Anselm's next premise should at least make sense to you:

4. For any individual thing x, if it is possible that x exists, but x does not exist, then x could be greater than x is.

This is a controversial claim. Many objections to Anselm's argument focus on this premise.

The next step in Anselm's argument is implied by the fourth premise:

5. If it is possible that that than which a greater is not possible exists, and that than which a greater is not possible does not exist, then that than which a greater is not possible could be greater.

Now, (2), (3), and (5) imply:

6. That than which a greater is not possible could be greater.

Think about (6) for a moment. Does it sound contradictory to you? How could that which is the greatest possible being even possibly be greater? It wouldn't be the greatest possible being if it could possibly be greater. There's no possibility of a level of greatness above the level of greatness of that which is the greatest possible being. Yet (6) seems to imply that there is such a level. If that implication sounds contra-dictory, then you should agree with Anselm's next premise, which is:

> 7. It is not possible that that than which a greater is not possible could be greater.

Notice that steps (6) and (7) seem to form a contradiction. If a contradiction is derived in an indirect proof, it follows that the assumption implies a contradiction. Since only that which is contradictory can imply a contradiction, it then follows that the assumption itself must be contradictory or necessarily false. If the assump-tion is contradictory, then the opposite of the assumption must be true. The as-sumption was:

> 3. That than which a greater is not possible does not exist. (Assumed premise)

The conclusion, which is the opposite of the assumption, is therefore:

> 8. That than which a greater is not possible actually exists.

In sum, the assumption that God does not exist implies a contradiction, from which it follows that God must exist. What is intriguing about Anselm's argument is that it begins simply with a logical analysis of the concept of God as that concept is found within traditional theism, and from just that, Anselm claims to prove that such a being necessarily must exist. Once we understand the mere concept of God and its logical implications, Anselm claims, we can see that God must exist.

Since St. Anselm first put forward his argument, a number of alternative onto-logical arguments for God's existence have been developed. Ontological arguments have historically been favored by mathematically inclined philosophers because ontological arguments typically employ a method of reasoning frequently used in mathematics, the method of indirect proof. For example, Descartes, who discovered analytic geometry, and Leibniz, who codiscovered calculus, both produced and de-fended their own versions of Anselm's ontological argument.[5]

●— GUANILO'S CRITICISM OF ANSELM'S ARGUMENT

Shortly after Anselm put forward his argument, a monk named Guanilo produced a criticism of the argument. Guanilo's criticism took the form of a *parody* of Anselm's argument. A parody of an argument is an argument that, using the same reasoning

---

[5] For the arguments by Descartes and Leibniz, see Plantinga [1965].

as the original argument, reaches a ridiculous conclusion, thereby purporting to show that the original argument must itself be logically flawed in some way. Guanilo's parody may be summarized as follows.

• Guanilo's Lost Island Parody

Let us use the name 'Lost Island' to refer to the greatest possible island. Suppose that Lost Island doesn't exist. But existence—according to Anselm—adds to the greatness of a thing. Lost Island would be greater if it were to exist. So, if we suppose Lost Island doesn't exist, then the greatest possible island could have been greater than it is. But this is contradictory, for it is impossible that the greatest possible island be greater than it is. So, the assumption that Lost Island doesn't exist is contradictory. Therefore, Lost Island must exist.[6]

Guanilo went on to claim that Anselm's reasoning could be adapted to form proofs for the existence of all sorts of other bizarre entities, entities we know don't exist. The obvious conclusion to draw from this is that Anselm's reasoning must be seriously flawed.

One reply to Guanilo is this. Islands are brought into being by geological processes, they eventually pass out of existence, and they are composed of perishable matter. Any island you pick would not have existed had the geological processes that brought it into existence not occurred. Thus, any particular island, by its very nature, must be a *contingent* entity, which is to say that it exists in some possible worlds and fails to exist in other possible worlds. It follows from this that even if we suppose there's a possible world containing a greatest possible island, it does not follow that such an island exists in the actual world, for since such an island exists in some worlds but not in all worlds, it might not exist in this world.

PLANTINGA'S CRITICISM OF ANSELM'S ARGUMENT

Alvin Plantinga, a contemporary philosopher who has made important contributions to the fields of modal logic, epistemology, and the philosophy of religion, has developed an interesting criticism of Anselm's reasoning.[7] The following is a summary of Plantinga's criticism. In his ontological argument, Anselm compares the level of greatness possessed by an entity in one world with the level of greatness possessed by that entity in other worlds. In order to consider Anselm's point, we will have to think about greatness for a moment.

Ordinary people have different degrees of greatness in different worlds. For instance, Socrates attained a degree of greatness in his life that still inspires us today. However, under different circumstances, he might have attained far less greatness. Of course, this is just to say that he attained a high degree of greatness in the actual

---

6   For Guanilo's parody, see Plantinga [1965].
7   For Plantinga's critique of Anselm, see Plantinga [1974] Chapter X.

world, but in other worlds he attained far less greatness. The degree of greatness an entity possesses seems to be a quality that varies in degree from world to world. One's height, weight, and hair color, it would seem, are further examples of qualities that vary from world to world.

With the variability of greatness in mind, let us return to Anselm's argument. At premise (2), Anselm asks us to consider the possibility of a being so great nothing could possibly be greater. If we let 'G' stand for a being so great that nothing could possibly be greater, then premise (2) may be rewritten as:

> **2.*** There is a possible world w and there is a being G in w, and G is so great in w that nothing in any world could possibly be greater.

Of course, at this point in the argument, we have no reason to think that w is the actual world or that G exists in the actual world.

At premise (3), Anselm makes the assumption that the being G does not actually exist. If we let 'α' stand for the actual world, this premise may be rewritten as:

> **3.*** G does not exist in α.

Anselm thinks he has derived a contradiction at step (6). That premise, you will recall, was:

> **6.** That than which a greater is not possible could be greater.

This certainly sounds contradictory at first. It seems impossible that that than which a greater is not possible could be greater than it is. However, a closer look at this premise reveals a problem. The phrase 'could be greater', which appears in that premise, is ambiguous. There are at least two alternative ways to interpret step (6)'s 'could be greater'. Using 'G' in place of 'that than which a greater is not possible', the two interpretations of (6) are:

> **6a.** G could be greater than G is in α.
> **6b.** G could be greater than G is in w.

Which of these two interpretations of premise (6) does Anselm intend?

If we give premise (6) interpretation (6a), then (6a) follows logically from the previous steps. Remember that by hypothesis, G doesn't even exist in α. So, surely in worlds where G exists, G attains a level of greatness that is greater than the level of greatness G has in α.

However, if you will reflect on the matter, interpretation (6a) is simply not contradictory. It is not contradictory to suppose that G possesses a level of greatness in some worlds that is greater than the level of greatness that G possesses in α, for in α the being G—by hypothesis—doesn't exist. So, on this interpretation of (6), the assumption of the argument does not imply a contradiction and Anselm's argument is invalid.

Suppose we interpret (6) in terms of interpretation (6b). Interpreted in this way, the premise is contradictory, for (6b) is contradictory. Nothing could have a degree of greatness exceeding the greatness G has in w, for at w, G has—by hypothesis—the highest possible degree of greatness that anything could have in any world. However, when (6) is interpreted as (6b), it no longer follows from the previous steps, for nothing in the argument implies (6b). Anselm's argument is therefore invalid on this interpretation.

Thus, whether premise (6) is given interpretation (6a) or interpretation (6b), the argument is invalid. No matter which way we turn, the argument is not a valid argument.

A REVISED VERSION OF ANSELM'S ARGUMENT

Plantinga has put forward a revised version of Anselm's argument, arguing that this revised ontological argument is both valid and sound.[8] In this section, Plantinga's revision of Anselm's argument will be summarized. First, some new terminology must be introduced. Let us say that a *maximally excellent* being would be a being so great that nothing could possibly be greater. Such a being, if it existed, would possess the greatest possible array of perfections anything could possibly possess. Therefore, a maximally excellent being would be, among other things, omniscient, omnipotent, and omnibenevolent. That is, a maximally excellent being would possess the properties of omniscience, omnipotence, and omnibenevolence.

An entity possesses a property *essentially* just in case it possesses the property and could not have failed to possess the property, which is to say that the entity could not have existed without possessing the property. In terms of possible worlds, an entity possesses a property essentially just in case it possesses the property in every world in which it exists. In contrast, an entity possesses a property *contingently* just in case it possesses the property but it might not have possessed the property, which is to say that the entity possesses the property in some but not in all of the worlds in which it exists. For example, it seems reasonable to suppose that Abraham Lincoln possessed a rational nature essentially but possessed a beard only contingently. In terms of possible worlds, Lincoln possessed a rational nature in every world in which he existed, although he possessed a beard in only some of those worlds.

Upon reflection, it seems plausible to suppose that it is greater to have a perfection essentially rather than to have it contingently. So, any perfection possessed by a greatest possible being would be an essential property of that being. A maximally excellent being, then, wouldn't merely be contingently omnipotent. Rather, it would be essentially omnipotent—omnipotent in every world in which it exists. A maximally excellent being wouldn't merely be contingently omniscient. Rather, it would be essentially omniscient. Similarly, a maximally excellent being would not merely be contingently omnibenevolent; it would be essentially omnibenevolent.

With this as background, the first several steps of Plantinga's revision of Anselm's ontological argument may be summarized as follows.

---

[8] For Plantinga's revision of Anselm's argument, see Plantinga [1974], Chapter X. A simplified version of the argument is presented in Plantinga [1975].

1. If a being were to be maximally excellent, it would have all the perfections it is possible for a being to have and it would have those perfections essentially.

2. Therefore, a maximally excellent being would be essentially omnipotent, omniscient, and omnibenevolent.

3. A maximally excellent being is at least a logical possibility.

4. So, there is a possible world, call it w*, and a being in that world, call that being g, such that g is maximally excellent in w*.

5. Thus, in w* it's true that g is omnipotent, omniscient, and omnibenevolent, and furthermore g is omnipotent, omniscient, and omnibenevolent in every world in which g exists.

Notice that at this point in the argument, we have no explicit reason to suppose w* is the actual world and we have no explicit reason to suppose that g exists in the actual world. Our reasoning concerns a mere logical possibility. That is, we are merely reasoning at this point about a possible world.

We earlier examined the concept of perfection. Is there such a thing as perfection with respect to *existence*? Consider human existence for a moment. A human being has a precarious hold on existence. You exist, but had things been only slightly different, you might never have been born. A *contingent being* is a being that exists but that might not have existed had things been different. Each human being is therefore a contingent being, for a typical human being exists but might not have existed. In terms of possible worlds, something exists contingently just in case it exists in some worlds but not in every world. In contrast to a contingent being, a *necessary being*—if one were to exist—would be a being that could not have failed to exist. No matter how things might have been, this being would have existed. In terms of possible worlds, a necessary being would be a being that exists in every possible world. Wouldn't it be greater to exist necessarily than to exist merely contingently?

Now, what would perfection with respect to existence be? A plausible suggestion is this: a perfectly existent being would exist in all possible worlds, that is, it would be a necessary being. Its existence would be so secure that it could not possibly have failed to exist. Necessary existence—existence in all worlds—seems to be the highest degree of existence anything could possibly have.

We can sum these considerations up in the following premise:

6. If a being has the property of maximal excellence, it must also have the property of necessary existence.

In other words, if a being were to be maximally excellent, then it would also be necessarily existent. So:

7. At w*, it is true that g exists at every world.

If in one possible world it is true that a particular proposition is true in all worlds, then that particular proposition must be true in all worlds. (This is a principle of the

S5 system of modal logic.) According to 7, at w* it is true that the proposition that g exists is true in every world. Therefore:

8.   In every world, it is true that g exists.

It follows that:

9.   There actually exists a maximally excellent being.

If this revised version of Anselm's argument is sound, then God—as conceived within traditional theism—actually exists.

•— OBJECTIONS TO THE ONTOLOGICAL ARGUMENT and TO THEISM IN GENERAL

There are a number of ways to object to one or more premises of the revised Anselmian argument. Some philosophers have argued that the traditional theistic conception of God is contradictory or, as some would say, *incoherent*. An example of a contradictory or incoherent concept is the concept of a square circle. Nothing could possibly be both square and circular at the same time, for a circle is a locus of points all equidistant from a single point, while a square is a figure that has four equal sides and four equal interior angles and so is not a locus of points all equidistant from a single point. If a square circle is a contradictory concept, then there is no possible world containing a square circle, for there is no such thing as a contradictory possible world. A contradiction is a paradigm case of the impossible. If it could be established that the concept of a maximally excellent being is itself contradictory, then it would follow that no possible world includes a maximally excellent being. From this it would follow that (a) the third premise of the revised version of Anselm's ontological argument is false, and (b) the argument is therefore unsound.

Is it possible that orthodox religious believers within Judaism, Islam, and Christianity have built their religious lives around a concept that is impossible or logically contradictory? There are a number of lines of argument in favor of an affirmative answer to this question. If even one of these arguments is sound, then premise (3) of the revised ontological argument must be false. The *Paradox of the Stone* is one argument for the claim that the traditional theistic concept of God is contradictory.

• The Paradox of the Stone

According to traditional theism, God is omnipotent. An omnipotent being is said to be a being who has the power to do anything. Can God make a stone so heavy God can't lift it? If the answer is no, then it follows that there is something which God can't do, namely, make such a stone. If the answer is yes, then there is something God can't do, which is to lift such a stone. The answer is either yes or no. Either way, there is something God can't do. So an omnipotent being would be a being who (i) can do any-thing, and yet (ii) cannot do everything. This is contradictory. Therefore, no being could possibly be omnipotent.[9]

• PROBLEM •

Symbolize the Paradox of the Stone and then use either natural deduction or truth-trees to show that it is valid.

If the Paradox of the Stone is a sound argument, then the concept of omnipo-tence is a contradictory concept. Since omnipotence is supposedly part of maximal excellence, if this argument is sound, then the concept of a maximally excellent be-ing—a being essentially omnipotent, omniscient, and omnibenevolent, and neces-sarily existent—is contradictory. Since contradictions are impossible, and since no possible world contains an impossibility, it would follow that a maximally excellent being is not a logical possibility. If so, then premise (3) of the revised Anselmian ar-gument is false, which means that the revised ontological argument is unsound.

How might the theist respond to this argument? In order to reject the stone paradox, the theist must revise the definition of omnipotence in such a way that the paradox doesn't apply to God. Essentially, this requires specifying the scope of God's power in such a way that the paradox of the stone doesn't arise. Let us try out this re-vision:

> A being is omnipotent just in case it can bring about anything that is logi-cally possible.

The scope of God's omnipotence, on this account, covers every possibility, but it does not extend to the impossible.

Using the definition of omnipotence put forward above, the theist's reply to the Paradox of the Stone may now be put this way:

> Omnipotence is not to be understood as the power to do literally anything. It is the power to do anything logically possible. If a being is omnipotent, then it has the power to raise any stone of any weight. Thus, the phrase 'a stone so heavy it can't be lifted by an omnipotent being' is itself a contra-diction in terms, for a stone so heavy it can't be lifted by an omnipotent be-ing would be a stone so heavy it can't be lifted by a being who can lift stones of any weight. Therefore, a stone so heavy it can't be lifted by an omnipotent being is not even a logical possibility. So, God cannot make such a stone. But this represents no limitation on God's power—it does not detract from God's omnipotence—for such a stone is impossible and

---

[9] For a number of articles on the paradox, see Urban and Walton [1978]. See also Morris [1987a and 1987b].

the power of omnipotence doesn't extend to the impossible. To say that God is omnipotent is not to claim that God can do the impossible—it is only to say that God can do anything logically possible.

However, one might reply that on this account of omnipotence, God's power still seems to be limited or restricted in some way. For if God can't bring about a contradiction, then it seems that there are things God can't do—namely, contradictory things—and so we shouldn't attribute "real" omnipotence to God.

In reply, the theist may argue as follows. Logical impossibilities shouldn't be thought of as things an omnipotent being should be able to bring about. There is no possibility of a logical impossibility being brought about. For example, an omnipotent being should not be expected to be able to make a square circle, for the words 'a square circle' constitute a contradiction in terms and hence do not describe something it makes sense to suppose exists in even one possible world. Does this imply that there is something an omnipotent being can't do? No, for the term 'square circle' doesn't refer to something that might possibly be made, since no such thing (as a square circle) is even possible.

◆ PROBLEM ◆

As we have seen, God, according to the traditional theistic conception, is essentially omnipotent, omniscient, and omnibenevolent. Some philosophers have argued that it's not possible for any entity in any possible world to possess all three of these perfections essentially. If it is not possible for a single entity to possess all of a group of properties, the properties are said to be incompatible or inconsistent. If it could be established that it's not possible for any entity in any possible world to possess essentially all three of these perfections, that is, that essential omnipotence, omnibenevolence, and omniscience are incompatible, then this would establish that the theist's conception of God is contradictory. If this were to be established, then it would again follow that premise (3) of the revised Anselmian argument is false. Below are two such arguments. If either of these two arguments is sound, it follows that essential omnipotence, omniscience, and omnibenevolence are incompatible. Evaluate each argument. Are either of the two arguments valid? Sound? Argue for your answer.

The Omnipotence–Omnibenevolence Paradox

1. If a being is omnipotent, then it can do anything logically possible.

2. A sinful action is a logical possibility.

3. So, if a being is omnipotent, then it can commit a sinful action.

4. If a being is essentially omnibenevolent, then it cannot even possibly commit a sinful action, that is, in no possible world does an essentially omnibenevolent being commit a sin.

5. Therefore, there is no possibility of a being that is essentially omnipotent and essentially omnibenevolent as well.

6. Therefore, essential omnipotence, omniscience, and omnibenevolence are not consistent properties—no being could possibly have all three of these perfections.

The Omnipotence–Omniscience Paradox

1. If a being is essentially omnipotent, then it can do anything logically possible.

2. An omnipotent being would therefore be a being who has never tried and then failed at something.

3. In order to know firsthand what failure is, one must experience failure by trying and failing at something.

4. So, an omnipotent being cannot possibly know firsthand what failure is.

5. Therefore, there is something that an omnipotent being cannot know.

6. If a being is essentially omniscient, it knows all things.

7. Thus, no being could possibly be both essentially omnipotent and essentially omniscient.[10]

ANOTHER ARGUMENT AGAINST THE EXISTENCE OF GOD: THE ARGUMENT FROM EVIL

Some philosophers have claimed that the following two propositions are inconsistent:

A. An omnipotent, omniscient, and omnibenevolent God exists.

B. The world contains suffering.

Why suppose these propositions are inconsistent? Well, if God were to be omnibenevolent, God would be opposed to all suffering. If God were to be omniscient, God would know about any suffering that might exist. If God were to be omnipotent, God would have the power to prevent any suffering from happening. Putting these qualities together, if an omnipotent, omniscient, omnibenevolent God were to exist, it would not allow any suffering whatsoever. So, if God exists, then suffering does not exist. And if suffering exists, then God does not exist. Therefore, (A) and (B) are inconsistent.

Now, one fact about the world seems beyond dispute: The world contains large amounts of suffering. A trip to a hospital or a study of the history of warfare is proof enough of this. But if the world contains large amounts of suffering, then (B) is true.

---

[10] My formulation of these paradoxes owes much to the presentations in Gale [1991].

If two propositions are inconsistent, then it is not possible that both are true. So if (A) and (B) are indeed inconsistent, then both cannot be true. Therefore, if (B) is true then (A) is false. But (B) is obviously true. It follows that (A) is false and there does not exist an omnipotent, omnibenevolent, omniscient creator of the world.

Traditionally, this line of reasoning has been called the *Problem of Evil*, for suffering is a bad thing, an "evil." And traditionally, this reasoning has been put in the form of an argument against God's existence, an argument we will call the argument from evil.

— **The Argument from Evil** —•

1. The following two propositions are inconsistent:
   a. An omnipotent, omniscient, omnibenevolent being exists.
   b. Suffering exists.
2. Proposition (b) is true.
3. Therefore, proposition (a) is false.
4. Therefore, an omnipotent, omniscient, omnibenevolent being does not exist.

If this argument is sound, then there does not exist an omnipotent, omniscient, omnibenevolent creator of the world, and traditional theism is therefore false. Furthermore, if the argument from evil is sound, then the revised ontological argument is unsound, since the conclusion of this argument contradicts the conclusion of the revised ontological argument.

•— REPLIES TO THE ARGUMENT FROM EVIL

In response to the argument from evil, theists have formulated *theodicies*. A theodicy is an explanation of why God would create a world containing suffering. A theodicy, if plausible, would help us see how it could be possible that both (a) God exists and (b) suffering exists. A theodicy, if plausible, would therefore give us reason to reject the first premise of the argument from evil. The first theodicy we will consider goes back at least to the writings of St. Augustine and is known as the *free-will theodicy*.

The Free-Will Theodicy

Freedom is an intrinsically good thing. It is good that people have free wills and the power to chart their own paths in life. The alternative is that persons be mere automatons whose every action is determined by God. It would be logically impossible for God to create creatures possessing free will and then to determine precisely how those creatures will freely behave, for if God were to determine the actions of the creatures God created, those actions wouldn't be free actions. Now, the free creatures God created have apparently used their freedom to do some bad things. Because

> free will is good, God allows people to exercise their freedom in bad ways, for the bad must be allowed if God is going to give creatures the gift—the intrinsic good—of free will.[11]

Things that are not in themselves good are sometimes morally justified if (a) they are necessary for or lead to an intrinsic good, and (b) the good outweighs the bad. According to the free-will theodicy, free will is the intrinsic good that explains why God allows some creatures to do bad things.

A theodicy helps us see why God would allow the existence of suffering; it explains how it could be that God and evil both exist. Thus, a theodicy gives us a reason to reject any premise that asserts that God's existence and the existence of evil are inconsistent. If the free-will theodicy is correct, then we have an explanation—a reason—why God would create free creatures who sometimes exercise their freedom in harmful ways. We would then have reason to suppose that premise 1 of the argument from evil is false.

A number of replies have been suggested to the free will theodicy. One reply may be put this way:

> If God is omnipotent, then God has the power to create free creatures who—though free—nevertheless never do anything bad. In other words, God should be able to create free creatures who all freely live good and saintly lives, harming no one, causing no suffering.[12]

If this is correct, then God does not have to allow wrongdoing and evil just in order to bring about the existence of free creatures. Instead, God can choose to create only those free creatures who will freely live perfectly good lives. If this is a possibility, then the free-will theodicy fails to explain why God would allow evil, for if it is indeed logically possible for God to bring about free creatures who never do wrong, surely a perfectly good God would prefer this to a world containing free creatures who cause large amounts of suffering.

According to traditional theism, God created saints and good angels who freely live lives of exemplary goodness. The question for the theist, then, is this: If God can create saints and angels, who freely live good lives, why wouldn't God create only such creatures?

Another criticism of the free-will theodicy is this. The purpose of a theodicy is to explain why God allows evil. There are at least two kinds of evil:

a. *Human evils* are evils that are caused by the actions of human beings.

b. *Natural evils* are evils such as earthquakes, hurricanes, floods, and diseases that are caused by the operation of nature and not by the actions of human beings.

---

[11] For further discussion see Hick [1968], Plantinga [1974], or Plantinga [1975].

[12] For further discussion, see the chapter on the problem of evil in Mackie [1982].

The free-will theodicy offers an explanation of only one kind of evil—human evil. However, it offers no explanation for the vast amount of natural evil our world contains. It fails to explain why God would create a world containing earthquakes, hurricanes, floods, diseases, and so on. Thus, the free-will theodicy, even if true, would explain only part of the world's evil. The free-will theodicy is therefore incomplete and consequently fails to explain why an omnipotent, omniscient, omnibenevolent being would allow all the suffering we see around us in the actual world.

The second theodicy we shall examine goes back at least to the writings of St. Irenaeus and may be called the *higher qualities theodicy*.[13] This theodicy offers an explanation for the existence of natural evil.

The Higher Qualities Theodicy Imagine a possible world, inhabited by human beings, in which there is absolutely no possibility of any kind of suffering. This would be a world in which there is:

> no disease
> no poverty
> no hunger
> no pain
> no war
> no crime
> no natural disasters (earthquakes, hurricanes, and so forth)
> no loneliness
> no hardships
> no need

Now, what would the people be like in such a world? It seems that a number of valuable character traits would not exist in such a world. Let us think about human character and the qualities of character we admire.

First of all, in a world in which nobody has ever experienced any form of need or suffering, it seems that nobody would ever have felt sympathy or compassion for another person, for one can't possibly experience sympathy or compassion unless one confronts an instance of suffering or need. Therefore, in the possible world under discussion, the character traits of sympathy and compassion would simply not exist.

Furthermore, in a world in which nobody had ever experienced suffering or need, nobody would ever have engaged in an act of charity, for one can't possibly be charitable unless one confronts someone in some sort of need. Thus, charity would not exist in the world under discussion.

It seems that nobody, in a world containing absolutely no suffering, would ever have faced danger. Consequently, nobody would ever have the opportunity to be courageous or brave, for courage and bravery develop only as one confronts dangers.

---

[13] See Hick [1968] for an extended treatment of this theodicy.

Consequently, in the world under consideration, the character traits of courage and bravery also wouldn't exist.

Let us call these admirable character traits the "higher moral qualities." It seems that if God were to make a world containing not one iota of suffering or hardship, then the inhabitants of such a world would have no opportunity to develop the higher moral qualities, for these higher qualities only develop as we confront the perils, hardships, and suffering of a world such as the actual world. According to this theodicy, God made a world containing suffering and hardships—including disease and natural disasters—in order to bring about creatures who develop, by overcoming these hardships, the higher moral qualities. In other words, moral growth and development, which is an intrinsic good, is the overall good that justifies or outweighs the suffering and hardships our world contains.

If this theodicy is correct, we have an explanation that tells us why an omnipotent, omniscient, omnibenevolent being would create a world containing hardships and suffering, and we again have a reason to suppose that the first premise of the argument from evil is false.

Several replies can be made to the higher qualities theodicy. First, it has been argued that if God is omnipotent, then God should be able to build the higher qualities into people at birth—just as God builds eye color or musical talent into a person at birth—without anyone ever having to confront hardship or suffering. If this is possible, then it is possible that people possess the higher qualities without ever suffering the type of hardships we see in this world. And surely a perfectly good God would prefer to bring about the higher qualities in the most painless way, by simply building them into people at birth.

Is this reply to the higher qualities theodicy correct? If it is, then the higher qualities theodicy fails to explain why God would allow the hardships and suffering of this world. However, a defender of that theodicy might reply that a person can't possibly develop the higher qualities unless he or she actually experiences the hardships of this life. For instance, a person can't possibly develop compassion, or even know what compassion is, without confronting and responding to a case of need. The higher moral qualities are not qualities like eye color that can simply be programmed into a person at birth. Rather, they only develop over time as a person freely confronts the perils and suffering of a world such as this. Therefore, there is no logical possibility of God bringing about the higher qualities in a world containing absolutely no suffering or need.

Another reply to the higher qualities theodicy concerns the amount of suffering and hardship our world contains. Suppose we grant the theist that hardship and suffering are necessary for moral growth and development. One question remains. Why does the world contain so much suffering? If moral development is the goal, the world seems to have more suffering and greater hardships than are necessary. A wholly good God, it seems, would not allow or cause unnecessary suffering. The conclusion of this line of reasoning is that an omnibenevolent God does not exist.

The two theodicies we have examined have the following in common. Certain evils, it is argued, are logically necessary for the realization of a greater good which outweighs them. God therefore cannot bring about the greater good without permitting the corresponding necessary evils.

• PROBLEMS •

1. Make your own list of admirable human character traits that you think couldn't possibly develop in a world containing absolutely no suffering and no need. If you believe there are no such traits, argue for your position.

2. Does the world contain unnecessary suffering and hardship? Present an argument for your answer.

2. DO WE HAVE FREE WILL?

•— FATALISM

According to the view known as *fatalism*, (a) each person's life is made up of a particular pattern or sequence of actions and events; (b) before each person is born, this pattern is already determined or predestined; and (c) the pattern of a person's life is therefore not of the person's own making and is therefore beyond the person's control. If fatalism is true, then nobody has free will.

Some theists have argued that if God is omniscient, then before each person is born God knows the life each person will lead. They then reason that if God knows in advance each person's life, then each person's life is determined or predestined. This view is called *theological fatalism*. According to the view known as *logical fatalism*, purely logical considerations alone—apart from whether or not God exists—prove that our actions are predestined. In this section, we shall examine first theological fatalism and then logical fatalism.

Theological Fatalism Consider the following line of reasoning.

1. If an omniscient God exists, then God has known for billions of years precisely what you are going to do tomorrow at 2 p.m.

Most theists hold that God knows the future.

It is natural to suppose, as far as our future actions are concerned, that we have choices to make. However, we also naturally suppose that the past is outside of our control. Past events have happened, and there is nothing we can do now to alter those events. However, if an omniscient God exists, and has known for billions of years exactly what you will do tomorrow at 2 p.m., then the fact that God has known your future action all those years seems to be one more unalterable part of the past. Therefore, it seems to follow that:

2. If God has known for billions of years exactly what you will do tomorrow at 2 p.m., then there is nothing you can do to change the fact that God has known exactly what you will do tomorrow at 2 p.m.

Traditional theism also holds that God cannot possibly have a mistaken belief. If God believes or knows that you will do a particular thing, then you definitely will do that very thing. It seems to follow from this that:

3. Tomorrow at 2 p.m., when you do what God has always known you will do, you must do that; you cannot do anything else but that.

In everyday life, if someone does something, and we discover that the person couldn't have done anything else, we suppose the person did not act freely. For instance, suppose someone falls into a beautiful garden and ruins several pretty flowers. If the person fell because someone pushed him or her into the garden, then the person could not have done otherwise. In this case, we do not suppose the person fell of the person's own free will. If someone does something of his or her own free will, this means, at least partly, that given the same circumstances the person could have done something else instead. Thus:

4. If a person performs an action, and the person could not have done otherwise, then the person does not act of his or her own free will.

So,

5. If God has known for billions of years exactly what you will do tomorrow at 2 p.m., then tomorrow at 2 p.m., when you do what you are going to do, you will not be performing that action of your own free will.

This reasoning does not apply only to what you will do tomorrow at 2 p.m.; it applies to everything each person does at each and every moment. For according to traditional theism, God knows in advance each event of each person's life. Therefore,

6. If an omniscient God exists, then nobody possesses free will.

If this argument is combined with theism, one has the view known as theological fatalism.

Replies   Most traditional theists do not accept theological fatalism. There are a number of interesting theistic replies to this argument. We have space to consider two. In order to understand them, we must look briefly at the relationship that exists, according to theism, between God and time. According to traditional theism, God is an eternal being. There is general agreement that this at least means that

God did not begin to exist at some moment and God will not cease to exist at some moment. However, there are two ways to understand God's eternity.

1. *Temporal Eternalism*: God has always existed and always will exist. God's life is thus infinite in both temporal directions. Nevertheless, God exists within the influence of time, that is, God's "life" involves movement through time from moment to moment, past to future, just as ours does.

2. *Atemporal Eternalism*: God exists outside of the passage of time, in a nontemporal realm or mode of existence. That is, God has no location in time, God's life has no duration, and God's existence does not involve temporal succession, that is, passage from moment to moment from past toward future.

If God is atemporally eternal, then it is not the case that God knows in advance precisely what you will do tomorrow at 2 p.m. For in order to know something in advance, one must exist within the stream of time, passing from moment to moment toward the future. According to the atemporal view, God does not exist within the stream of time. If God is atemporally eternal, then tomorrow at 2 p.m., when you put your choice into action, God knows—from outside the stream of time—what you do, but God's knowledge of your action is not prior to the action. On this account, God's knowledge is more like simultaneous knowledge than advance knowledge. Therefore, if God is atemporally eternal, then premise (1) of the argument for theological fatalism is false.

Another theistic reply to the theological fatalism argument involves supposing that propositions about the future are neither true nor false. Suppose that until tomorrow at 2 p.m., there is no fact of the matter about what you will do. Until you do whatever it is you will do, it is not yet true that you will do that. So, there is no true proposition—at this moment—corresponding to what you will do tomorrow at 2 p.m. Tomorrow at 2 p.m., when you do what you will do, your action makes a proposition true or brings a new truth into being. If this is right, then although an omniscient God knows all truths, God doesn't know in advance what you will do, because until tomorrow at 2 p.m. when you do what you will do, there is no truth of the matter to be known by anybody. If this reasoning is sound, then we must again deny premise (1) of the theological fatalism argument, since God does not know the future in advance.

Not all theists accept this position, because it involves denying that God knows the future, and many traditional theists hold that God knows the future.

Another problem with this position is the following. If we accept this position and deny that propositions about the future are either true or false, then we must give up the Law of Excluded Middle, for the Law of Excluded Middle holds that every proposition is either true or false. If we do away with the Law of Excluded Middle, and suppose that propositions about the future are neither true nor false, then we will have to admit a third truth-value to our list of truth-values: (1) true; (2) false; (3) indeterminate. But if we suppose there are three truth-values, then we will have to drastically alter the definitions of the key logical terms, the structure of the truth-tables, and many other parts of logic. Indeed, the move to what is called "three-valued" or "many-valued" logic requires a conceptual overhaul of the whole of logical theory. The result of this would be a considerably more complicated logic, one containing a number of conceptual difficulties. This is one reason why many philosophers are reluctant to reject the Law of Excluded Middle.[14]

An Argument for Logical Fatalism   Aristotle was one of the first philosophers to consider the arguments for logical fatalism. What follows is a version of an argument he examines in *De Interpretatione IX*.

1.   At this very moment, it is either true that tomorrow at noon you will eat a hamburger, or it is false that tomorrow at noon you will eat a hamburger.

2.   If, on the one hand, it is true now that tomorrow at noon you will eat a hamburger, then tomorrow at noon you can do nothing else but eat the hamburger, for nobody can alter the truth, and the truth is that you will eat the hamburger tomorrow.

3.   If you can do nothing else but eat the hamburger tomorrow, then you will necessarily eat the hamburger.

4.   If you will necessarily eat the hamburger, then when you do eat the hamburger, you are not choosing to eat the hamburger, you are not freely eating the hamburger.

5.   If, on the other hand, it is false now that tomorrow at noon you will eat a hamburger, then tomorrow at noon, you can do nothing else but not eat the hamburger, for nobody can alter the truth, and the truth is that you will not eat the hamburger tomorrow.

6.   If you can do nothing else but not eat the hamburger tomorrow, then you will necessarily not eat the hamburger.

7.   If you will necessarily not eat the hamburger, then when you do not eat the hamburger, you are not choosing to not eat the hamburger, you are not freely not eating the hamburger.

8.   Either way we turn, whether the proposition about tomorrow is true or whether it is false, it seems that whatever you will do tomorrow at noon is something that is predestined and beyond your control.

9.   However, it's not just tomorrow at noon that is at issue. The above reasoning applies equally to everything you do at every moment of your life. Therefore, it seems that every action you have ever performed or will ever perform is predestined and beyond your control.

[14] Rescher [1969] is a good introduction to what is known as "many valued logic."

Upon reflection, it looks as if the truth-values of propositions somehow mysteriously spell out our entire life stories in advance and thereby determine—before we are born—the lives we are fated to live. Thus, the view known as logical fatalism.

Replies There are a number of replies to this argument. One reply may be called the "noninteractionist reply":

> The truth-values of propositions do not determine or influence in any way what we shall do, for propositions and their truth values do not have the power to cause things to happen. No scientific theory attributes causal powers or causal properties to propositions. Rather, it's the other way around: It is the things we do which determine the truth-values of the relevant propositions, for those propositions merely passively reflect (or fail to reflect) the things we do. So, even though it may be true right now that tomorrow at noon you will eat a hamburger, it does not follow that tomorrow when you eat the hamburger, your behavior is determined or predestined by the truth-values of propositions in such a way that you lack free will. Thus, premises (2) and (5) of the fatalist's argument are false.[15]

According to this reply, propositions do not have the power to interact with things to determine what happens. Rather, true propositions passively reflect the way things are, and false propositions simply fail to reflect the way things are.

Let us look briefly at two fatalist responses to the noninteractionist argument. One fatalist response depends upon the correspondence theory of truth. First, if some proposition about a person's future is true now, it is presumably true now in virtue of the way the world is now. So there must be something about the world now that accounts for the truth of the proposition and that will inevitably bring it about that the person performs the corresponding action. It seems to follow from this that the person will not freely perform the action.

The fatalist has a second reply to the noninteractionist response. If the noninteractionist response is correct, then an incredible and unbelievable consequence seems to follow. It seems to follow that when you perform an action tomorrow at noon, you make it be the case that it was true billions of years ago that the corresponding proposition was true. In other words, when you perform an action, your action gives a proposition the truth-value it had billions of years ago. How can this possibly be? It's as if your action has the power to reach back billions of years and affect the proposition. But if this is right, then it seems that you have the power to affect the past—something that seems to be logically impossible. How can one of the effects of your action be transmitted to something in the past?

One answer to this involves supposing that propositions are nontemporal entities. A nontemporal entity, if one were to exist, would be an entity that does not exist within the stream of time. If a proposition is nontemporal, then it does not have an age, there is no moment of time at which it began to exist, it does not grow older

---

[15] My formulation of this argument owes much to Bradley and Swartz [1979], p. 105.

each moment, and so on. On this view, propositions timelessly reflect reality, that is, they correspond to reality from a vantage point outside of time. If this is correct, then when your action determines a proposition's truth-value, it does not thereby affect something in the past, for propositions don't exist within the temporal sequence. Rather, when you perform an action, a proposition's truth-value timelessly reflects the fact.

Does this response make sense to you? Some philosophers argue that the idea of a nontemporal entity being affected by a temporal entity makes no sense at all. They ask: How can something supposedly outside the stream of time be affected by something that is situated within the stream of time?

Another reply to the fatalist's argument involves rejecting its first premise. According to the first premise, it is either true now or it is false now that you will eat a hamburger tomorrow at noon. But why suppose this proposition is *right now* either true or false? Why not suppose instead that the proposition (that tomorrow at noon you will eat a hamburger) is neither true nor false until tomorrow at noon when you eat or fail to eat a hamburger? Tomorrow at noon, when you eat (or fail to eat) the hamburger, the proposition will become true (or false). The general principle here might be put this way:

> A proposition about a future moment does not become true or false until that moment arrives. Until then, the proposition is neither true nor false.

If this is correct, then premise (1) of the logical fatalist's argument is false.

However, premise (1) of the fatalist's argument is implied by the Law of Excluded Middle. This objection to that premise amounts to a rejection of the Law of Excluded Middle. As we saw above, a rejection of that law has serious and undesirable logical consequences. Many philosophers therefore argue that it is more reasonable to retain the Law of Excluded Middle and reject this objection to fatalism.

●— DETERMINISM

If a neurosurgeon hooks an electrode up to a particular spot on a person's brain and applies an electric current to the brain cells (neurons) at that spot, certain bodily movements will automatically occur. When the electrode is attached to a different region of the brain, the patient will report a visual experience, or a memory experience, or some other type of mental event. The discoveries of neurophysiology suggest that our bodily movements are the direct result of certain neuronal events within our brain.[16]

Suppose a group of neuroscientists has discovered a way to send some sort of signals into a person's brain in such a way that they can activate any particular neuron they choose. Suppose further that they know exactly which bodily movements and

---

[16] My discussion of determinism owes much to the discussion in Coburn [1991].

mental events are associated with each possible pattern of neuronal activation. Then this group of scientists can decide on a particular bodily behavior, send the appropriate signals into the brain, and cause the desired behavior.

Next, imagine that everything a particular person named Charlie has done during the past month has been caused by this group of scientists. That is, by sending the right signals into Charlie's brain, they have caused every one of his actions. You can picture the scientists saying things like, "Let's send the signal for a sudden burst of anger so he'll yell at his wife," or "Let's now have him eat a hamburger."

If this were to be the case, how would we judge Charlie's actions? Surely we would not hold him *morally responsible* for anything he did during that month. And the reason why he would not be morally responsible for his actions would be this: Given the signals sent by the scientists, he couldn't have done anything other than what he did. In everyday life, when someone couldn't have done otherwise, we don't hold the person morally responsible. For instance, if you are standing in front of a beautiful painting at an art exhibit and someone violently pushes you into the painting, nobody would hold you responsible for ruining the painting. And the reason why you would not be morally responsible is clear: Given the violent push from behind, you couldn't have done anything else but fall into the painting.

If someone is not responsible for his or her actions, then it follows that the person deserves neither praise nor blame for those actions. If the person did something good, he or she deserves no praise. If the person did something bad, he or she deserves no blame. So, if during Charlie's programmed month, he performed some rotten act, he deserves no blame for it. And if during that month he performed some good deed, he deserves no praise for doing that either. Indeed, it seems in this case that Charlie does not have free will. For in cases where someone could not possibly have done otherwise, it seems correct to say that the person has not acted of his or her own free will.

Let us now change the picture. Instead of imagining a group of scientists behind the scenes causing each of Charlie's actions through the activation of his brain cells, suppose instead that the principle of determinism (or 'PD' for short) is true.

The Principle of Determinism

> *Every event is caused to occur by an immediately preceding event, which was in turn caused to occur by an immediately preceding event, in a chain of cause and effect that goes back to the beginning of the universe, if the universe has a beginning, or goes back forever if the universe has no beginning.*

In order to understand the principle of determinism one must understand something about causation. In general, if an event A causes an event B to occur, then B had to occur or was necessitated once event A occurred, in the same sense that a normal cube of sugar has to melt after it is dropped into a cup of boiling water or that an ordinary wine glass has to shatter after it has fallen 10 stories onto a concrete sidewalk. If B has to occur once A has occurred, then nothing else but B can occur once A has occurred.

If the principle of determinism is true, then each and every neuronal event in Charlie's brain was caused—necessitated—by an immediately preceding event, which was itself caused by a preceding event, in a chain of cause and effect that extends back in time at least billions of years, if not forever. So, each brain event in Charlie's brain was brought about by a prior event, which was brought about by a prior event, which was brought about by a prior event, in such a way that each brain event had to occur once the prior events leading up to the brain event had occurred. It follows that, given the conditions prevailing billions of years ago, it was only a matter of time before the chain of cause and effect, like a long row of falling dominoes, would end up producing each of Charlie's brain events and consequently each of the associated bodily movements.

Suppose the principle of determinism is true and the above account is the true story behind Charlie's behavior. In this case, is Charlie morally responsible for his actions? The answer, it seems, is again no. For Charlie is not responsible for the events of billions of years ago. And those events are the ultimate causes of his actions. Once those events happened, Charlie's actions were bound to eventually follow. So it seems that he is not morally responsible for any of his actions. Whether, as in the previous case, a group of neuroscientists causes his every action, or whether, as in the present case, the ultimate causes of his actions are events that happened billions of years ago, the result seems to be the same: Charlie is not responsible for his actions. In this case, as in the previous case, it seems to follow that Charlie deserves neither praise for his good actions nor blame for his bad actions.

So, if the principle of determinism is true, then the behavior of Charlie's body does not seem to be up to him; rather, it seems more accurate to say that the motions of his body are things that just happen to him. These thoughts give rise to the following argument:

1. If the principle of determinism is true, then nobody can ever do otherwise than what they do.

2. If nobody can ever act otherwise, then nobody is ever responsible for what they do.

3. If nobody is ever responsible for what they do, then nobody deserves praise or blame for anything they do, and nobody has free will.

4. The principle of determinism is true.

5. So, nobody is responsible for what they do, nobody deserves praise or blame for anything, and nobody has free will.

Determinism will be the name of the point of view that accepts as true the principle of determinism. So, a *determinist* is anyone who accepts the principle of determinism as true, and determinism may be defined simply as the claim that the principle of determinism is true. Some determinists agree with all the premises of the argument just given and also accept its conclusion as true. Other determinists disagree with one or more of the argument's premises and consequently deny the conclusion. Thus, there are different forms of determinism, that is, different views that are alike in that they accept PD as true but differ in that they disagree on other issues.

Hard Determinism   If you agree with the premises of this argument, then you must accept the conclusion, for the argument is valid. If you believe that this argument is sound, then you accept the view known as *hard determinism*, a form of determinism that may be defined by the premises and conclusion of the above argument.

Hard determinism is a difficult view to accept (hence the "hard" in its name). Can it really be true that nobody is ever morally responsible for anything they do? Can it be that everything you have ever done was as determined and inevitable as the last solar eclipse? A hard determinist, if he or she is to be consistent, must hold that nobody ever genuinely deserves praise and nobody ever genuinely deserves blame.

Incidentally, some hard determinists do advocate bestowing praise and blame on people. A hard determinist might advocate such activity on the pragmatic grounds that praising and blaming people for various things is an effective way to influence or direct their behavior. Nevertheless, a hard determinist, if he or she is consistent, will hold that individuals never genuinely deserve or earn praise or blame.

An Argument Against Hard Determinism   Might it be, then, that hard determinism is true? Consider the following argument against hard determinism.

— **The Responsibility Argument Against Hard Determinism** —•

> If hard determinism is true, then nobody is ever morally responsible for what they have done, and nobody ever deserves praise or blame for any of their actions. So, if hard determinism is true, then there is no such thing as moral responsibility and every ascription of moral responsibility is false. But many people are in fact morally responsible for things they have done or have failed to do; moral responsibility does exist. For: (a) people sometimes do morally wrong things and are morally blameworthy for doing those things; and (b) people sometimes do morally good things and are morally praiseworthy for doing those things. In other words, there does exist a real *moral order*. Therefore, hard determinism is false.[17]

•— INDETERMINISM

Hard determinism is a difficult view to accept. Is there any alternative? Some have suggested that the hard determinist's negative conclusion about free will can be avoided if we simply deny that brain events have causes. Let us suppose that when a neuronal event associated with a free action happens, the neuronal event in the brain has no cause whatsoever. Nothing brings it about or necessitates it. The neuronal event just happens out of the blue, so to speak.

Twentieth-century physics provides one way to understand the idea of such an uncaused event. According to one interpretation of *quantum mechanics*, some types

---

[17] For an interesting discussion of this argument, see Van Inwagen [1983] Chapter 6.

of subatomic events—such as the decay of a uranium atom—happen spontaneously, which is to say that these events have no cause whatsoever. Individual "quantum events," as these events are called, happen "out of nothing," for no reason. Perhaps the neuronal events in our brains that lead to the movements of our bodies are like quantum events in the sense that they have no causes whatsoever.[18]

Quantum Indeterminism If some of our brain events are quantum events, then the principle of determinism is false, since the principle of determinism holds that every event has a cause. *Indeterminism* is the claim that the principle of determinism is false. Since the suggestion under consideration—that some of our brain events have no causes whatsoever—amounts to a denial of the principle of determinism, let us call this view an *indeterminist* view. Since there are several forms of indeterminism, that is, several different views that deny PD, let us call the view under consideration here *quantum indeterminism* in order to distinguish it from other forms of indeterminism.

If quantum indeterminism, as spelled out above, is true, does it then follow that we have free will? It doesn't seem so. As far as free will is concerned, things seem no different when we switch from the hard deterministic to the quantum indeterministic viewpoint. For if quantum indeterminism is true, then each thing you have ever done can be traced back to a neuronal event that itself had no cause and therefore happened for no reason. The neuronal event simply happened—out of the blue—and then your body moved in certain ways as a consequence. Therefore, if quantum indeterminism is true, then for anything you have ever done, you do not have the power to have done anything other than what you did, which implies that you are still not morally responsible for your actions and deserve neither praise nor blame for them. In this indeterministic view, your actions don't seem to be up to you. They seem to be merely things that happen to you.

For example, suppose that after Charlie commits some rotten deed, we are able to trace back the chain of cause and effect that produced his action. The chain of causes, let us suppose, goes back to a particular neuronal event in Charlie's brain that itself had no cause and happened for no reason. It's hard to see, if this is correct, how the action was Charlie's own free action. That is, it's hard to see how the action was under Charlie's control. If the neuronal events that trigger Charlie's actions have no causes, then they seem to be accidents that just happen to Charlie. So, if quantum indeterminism is correct, it seems to follow that nobody is ever morally responsible for what they do, nobody deserves praise or blame for anything they do, and nobody has free will.

Might it be that the quantum indeterminist view is the correct view? The moral responsibility argument against hard determinism can be rewritten so that it constitutes an argument against the quantum indeterminist view:

---

[18] A number of books dealing with quantum theory have been written for the general public. Feynman [1985] is an excellent introduction. Davies and Brown [1986] is also an excellent place to start. See also Trefle [1980] and Hawking [1988].

— The Responsibility Argument Against Quantum Indeterminism —•

If quantum indeterminism is true, then nobody is ever morally responsible for what they have done, and nobody ever deserves praise or blame for any of their actions. So, if quantum indeterminism is true, then there is no such thing as moral responsibility, and every ascription of moral responsibility is false. But moral responsibility does exist; many people are in fact morally responsible for things they have done or have failed to do. For: (a) people sometimes do morally wrong things and are morally blameworthy for doing those things; and (b) people sometimes do morally good things and are morally praiseworthy for doing those things. In other words, there does exist a real *moral order*. Therefore, quantum indeterminism is false.

The Dilemma of Determinism It seems as if we are in the middle of a dilemma. Whether we accept the indeterminist view outlined above or the determinist view given, it follows that nobody has free will. Considerations such as these have led some to argue as follows.

1. If determinism is true, then nobody has free will.
2. If indeterminism is true, then nobody has free will.
3. Either determinism or indeterminism is true.
4. Therefore, nobody has free will.

This argument is known as the *Dilemma of Determinism*. It is obviously a valid argument. If it is sound, then it is true that nobody has free will. If we could somehow come up with an alternative to the two theories canvassed above—the hard deterministic theory and the quantum indeterministic theory—then we might find a way to avoid the dilemma's conclusion.

————————————• PROBLEM •————————————

Symbolize the three arguments given above—the two "responsibility" arguments and the Dilemma of Determinism—then prove each valid with either a natural deduction proof or a truth-tree.

•— SOFT DETERMINISM

Consider the following account of the nature of a free action. When we judge that a person—call this person P—performed an action of his or her own free will, what do we normally base our judgment on? We call P's action "free" if there is no person, external constraint, or external agency of some sort forcing P to perform the action. For instance, if someone pushes P violently into a painting, or throws a rope around P and yanks P into the painting, or a gust of wind blows P into the painting, then P does not freely destroy the painting. But if no person or external constraint or agency of some sort forces P to perform the action, then we say that P performs the action "freely." Nevertheless, P's action surely must have a cause. It wouldn't be a free act if it happened "out of the blue," from no cause whatsoever. If no person or external constraint or agency forces P to perform the action, then the cause of P's action must lie within P. More specifically, the cause of P's action in such a case must be an inner act of choice—which we'll call a "volition" or "decision"—to perform the act.

So, to say P's act is free, on the account being put forward here, is just to say that (a) no external agent or constraint forced P to act in a particular way; and (b) the cause of P's act was an inner act of choice within P, that is, a decision or volition to perform the act. In short, one acts freely if one does what one wants or chooses to do. On this account, 'free' does not mean 'uncaused', and freedom is not to be contrasted with determinism. An action can qualify as a free action even though it has a cause and is determined.

Let us call the claim that we possess free will the *Doctrine of Free Will* or 'DFW' for short. If the account of a free action given above is correct, then there is no inconsistency in asserting that we possess free will and that the principle of determinism is true. That is, if the account of free will just presented is correct, then DFW and PD are consistent or, as some would say, *compatible*, for they might both be true. Thus, the account of free will just sketched is called a 'compatibilist' account of free will.

Why, assuming the compatibilist account of free will, are DFW and PD consistent? How could it be that our actions are free in the sense just outlined and yet determined in accord with the principle of determinism? The fact that an action is free—in the compatibilist sense outlined above—implies that it is not forced by an external constraint or agency. But this does not imply that the inner events leading up to the action, including all the brain events involved, are not all determined in accord with PD. It is possible, if the compatibilist account of free action is correct, that every inner event leading up to each of your actions—including the volitions that cause each action—is determined in accord with the principle of determinism. Therefore, given the compatibilist account of free will presented above, it is possible that the principle of determinism is true—all events are determined—and that we have free will.

Compatibilism is the view that the existence of free will and the truth of the principle of determinism are compatible. The compatibilist account of free will sketched above, combined with the claim that we have free will, constitutes a version of compatibilism.

If compatibilism is correct, then free will and determinism are compatible or consistent in the sense that it's possible that we have free will and yet also that all our actions are determined in accord with PD. In addition, if compatibilism is correct, then premise (1) of the argument for hard determinism is false, and we thus have a justification for rejecting the conclusion of that argument. Furthermore, if compatibilism is true, then premise (1) of the dilemma of determinism is false, and we then have a reason to reject the conclusion of that argument.

Suppose you agree with hard determinism that the principle of determinism is true. Suppose you also agree with compatibilism that free will and the principle of determinism are not inconsistent. Finally, suppose you agree that we do have free will. Since the viewpoint just specified accepts the truth of the principle of determinism while rejecting the hard determinist's difficult claim that we lack free will, the view is often called "soft determinism."

Let 'C' represent compatibilism, let 'SD' represent soft determinism, and let 'HD' represent hard determinism. The difference between the two forms of determinism may be outlined compactly as follows:

HD is equivalent to [PD & ~C & ~DFW]

SD is equivalent to [PD & C & DFW]

Why would someone be attracted to soft determinism? Soft determinism offers us a way to deny one of the premises of the dilemma of determinism, for if soft determinism is true, then premise (1) of that argument is false. This is one way to avoid the unwelcome conclusion of that argument. Soft determinism has another attractive feature. Many philosophers find the principle of determinism to be extremely plausible. It has served as an important component of the modern scientific world view. Yet these philosophers also have reason to suppose that we possess free will. Consequently, they are inclined to accept both PD and DFW. At first, it seems that both cannot be true. (We earlier looked at a good reason to suppose that if PD is true, then DFW is false.) One virtue of the doctrine of soft determinism is simply that soft determinism, with its compatibilist account of a free action, explains how it could be that PD and DFW might both be true.

An Argument Against Soft Determinism   According to soft determinism, an action of yours is free if (a) you are not forced to do it by another person or by an external force of some sort, and (b) the cause of your free action is an inner act of choice—a volition or decision to perform the act. But now let us ask: Is this inner act of choice—the volition to perform the act—caused? Remember that soft determinism also includes the claim that the principle of determinism is true. So, according to soft determinism, every event—including each of your inner acts of choice—is caused by a prior event, which in turn is caused by a prior event, and so on back billions of years. And this gives rise to the following argument against soft determinism:

> If soft determinism is true, then each event in your brain, including the inner events associated with your decisions or volitions, is caused by a prior event, which was caused by a prior event, forming a chain of cause and effect that goes back billions of years before your birth. But then given the way the universe was a billion years ago, things were set up in such a way that it was absolutely inevitable that you would make the decisions you in fact have made. For instance, given the way the universe was a billion years ago, it was inevitable that you would be reading this very sentence now and—given the truth of soft determinism—you could not be doing anything else except reading this sentence. But you are not responsible for the way things were a billion years ago, for you had no control or influence over how the world was a billion years before you were born. Yet those ancient conditions caused your actions.

Therefore, if soft determinism is true, then you are not really responsible for your inner acts of choice—your decisions—and the actions those lead to, and you therefore do not genuinely possess free will. But human beings do possess free will. Therefore, soft determinism is false.

Suppose you find hard determinism, soft determinism, and quantum indeterminism all equally implausible. Perhaps you hold that none of those theories makes sense of our moral experience, our very strong sense that sometimes we are morally responsible for the things we do. That is, none of the three theories—one might argue—accounts for the fact that persons are morally responsible for at least some of their actions. Is there any alternative theory available? One might think that there are no other alternatives, for the following reason. Either determinism is true or it is not true, which is to say that either determinism is true or indeterminism is true. But it's hard to imagine a deterministic alternative to the two versions of determinism examined above. And it's equally hard to imagine an indeterministic alternative to quantum indeterminism. Therefore, our only alternatives are the two deterministic theories—hard and soft determinism—and quantum indeterminism.

This argument is unsound, for it turns out that there is an alternative to the three views canvassed so far. We shall turn to that theory next.

● AGENT CAUSATION

There is another way to understand the origin of a free action. Suppose a person, which we will refer to as the "agent," performs an action that we will call action "x." According to *agent causation theory*, an action x—performed by an agent A—is a free action just in case:

    i. the chain of cause and effect leading up to x (consisting of physiological events within the agent's body) goes back to a first event in the chain, which will be called event "e";

    ii. event e was not necessitated or caused by a prior event;

    iii. event e did not originate out of nothing;

    iv. event e occurs because it was caused or brought about by the agent willing or deciding that e occur;

    v. The agent's inner act of choice—the willing or deciding that e occur—was not itself caused or necessitated by a prior event.[19]

Normally, causation involves one event being caused to occur by a prior event and causing in turn the occurrence of a subsequent event. This type of causation is

---

[19] My presentation of agent causation theory follows closely the account presented in Coburn [1991]. For a defense of agent causation, see the article by Chisholm in Lehrer [1966]. For a critical discussion, see Coburn [1991]. See also Taylor [1975]. An interesting discussion of the free will question may be found in Nozick [1981], Chapter 4.

investigated within the physical sciences and may be called *event causation*. However, in the case of a free act, according to agent causation theory, when the agent wills or decides to bring about e, this is not a case of event causation, for by hypothesis there is no event before e that necessitates or causes e to occur. Rather, the agent himself or herself—a person and not an event—causes e to occur, and no event causes or necessitates the agent's act of choice. Since an agent rather than a prior event causes e, this is called a case of *agent causation* to distinguish it from event causation.

Agent causation is very different from event causation. The agent's inner act of choice—the agent's willing that e occur—does not cause or bring about e in the way one event normally causes or brings about another event. In other words, the causal relation between the agent and the first physiological event e (presumably in the brain) that starts the chain of cause and effect leading to the subsequent action is not the same as the causal relation we normally find between one event and another. For when an event A causes an event B, some event before A caused A to occur. Event A is merely a link in an ongoing process of cause and effect. But when an agent causes the event e, according to agent causation theory nothing causes the agent to cause e. The agent initiates a chain of cause and effect; the agent's act of choice is the beginning of the chain. Therefore, an agent-caused event is a special kind of event, unlike the type of events studied within the physical sciences.

According to agent causation theory, free will is the power to *initiate* chains of cause and effect. The agent initiates a chain of cause and effect when the agent brings about the first event (e) in the chain.

If agent causation theory is the correct account of free will, then some events are not caused by prior events, for they are caused by agents—persons exercising free will. Thus, if agent causation is correct, the principle of determinism is false, for according to that principle, every event is caused by a prior event. Thus, agent causation theory is a form of indeterminism.

Objections Puzzling questions arise. For instance, how, according to agent causation theory, does the agent bring about the first event e? Agent causation theorists have no theory that explains how agent causation supposedly works. Because of this, critics charge, agent causation theory is an utterly mysterious theory and for this reason is unacceptable.

The proponent of agent causation answers that in order to make sense of human action we must *hypothesize* that the agent has a special power, the power to *agent-cause* an event such as e, which will produce a pattern of bodily behavior. The agent exercises this power when the agent brings about an event such as e, that is, when the agent puts its will into effect. The fact that we do not yet have a theory explaining how this happens does not mean that agent causation theory is false. There are many phenomena in this world that are unexplained.

This way of understanding a free action faces a number of additional objections. One further criticism of agent causation is this. When we trace things back from an action to its ultimate origination, where, according to the agency theory, does the chain end? At the agent. What causes the agent to bring about the first event? Nothing, for according to agent causation theory, there is no prior event that causes the agent to agent-cause the event e. But then the agent's willing that e occur seems to be the sort of thing that just happens out of nothing. If this is so, then it's hard to see how agent causation differs from the quantum indeterminism examined earlier. It looks as if both theories ultimately trace things back to the same starting point, namely a point at which something occurs or comes to be out of absolutely nothing. But then the same objections that applied to quantum indeterminism apply to agent causation theory.

The challenge for the defender of agent causation is to explain how agent causation differs from the quantum indeterministic theory in this respect. One suggestion is this. On the quantum indeterminist view, a free action traces back to an uncaused event that itself has no explanation. Hence, on the quantum indeterminist view, how a free action arises is an utter mystery and seems to makes no sense at all. However, on the agent causation view, a free action traces back to the agent's willing that the action occur. This is a wholly different type of source. We each will things to occur, and each of us knows by experience what it is to put our will into effect. This source of an action is no mystery at all.

In reply, the critic of agent causation might charge at this point that the agent's willing that an action occur is a phenomenon that—according to agent causation—has no cause. But then agent causation again seems to make no more sense than the quantum events posited by quantum indeterminism.

The agency theorist has a reply to this. Granted, the agent's willing that the action occur has no cause. But this does not imply that the agent's act of willing has no explanation. A human agent's choices are explained perfectly well in terms of the reasons the agent has for his or her actions. And reasons are not the same as causes. Thus, if we want to know why Charlie walked up to the store, and are told that he wanted to get a quart of milk for tomorrow's breakfast, we have a perfectly good explanation of his decision to go to the store. And this is so even if we do not have an account of a chain of causes leading to his decision to walk to the store.

Another criticism of agent causation is the following. Event causation is a type of causation that can be investigated by science. Scientific theories may be invoked to explain how one event causes another. However, agent causation, that is, the relation between the agent and the first event e in the chain of cause and effect leading to the action, does not seem to be a type of causation that can be investigated by science. If agent causation occurs, it seems that it is something beyond the reach of science. Thus, if we adopt the agent causation theory, it looks as if we must suppose that a free action is a scientifically inexplicable phenomenon.

But many philosophers argue that a human action is in principle scientifically explicable. On the basis of the past success of science and its rapid progress in recent years, they believe it is only a matter of time before we develop a complete scientific explanation of human action. A scientific theory of human action is one of the goals of cognitive science. Some philosophers therefore reject agent causation theory on the grounds that if we accept agent causation theory, it seems we must give up hope that a scientific theory of human action will ever be discovered.

Agent causation theorists admit that the relation between the agent and the first event the agent brings about is so unlike the relation between one event and the event it causes that agent causation probably cannot be analyzed scientifically. Agent causation, its defenders claim, is a *sui generis* type of causation—a type of causation unlike any other type. Those who favor the agent causation view ask: Why suppose that the nature of human free will is scientifically explicable? Perhaps there are things in this world that science will never be able to explain, and perhaps free will is one of those things. And why suppose that scientific theories are the only acceptable theories? Not all acceptable explanations are scientific explanations. Perhaps the nature of a free action is something that will be explained in terms other than scientific ones.

Why do some philosophers suppose that a human action is in principle scientifically explicable? *Materialism* is the view that everything in the universe is either a particle of matter or is composed of matter. Matter is that which physics studies. Many philosophers are materialists and they argue that in the final analysis, a human being is composed of nothing but matter. Matter is studied in the physical sciences, and the physical sciences seem to be on the way to a complete explanation of all material phenomena. So, if we are purely material beings, and if physical science is the study of matter, and if the physical sciences are on the way to a complete explanation of all material phenomena, then one might suppose that a human act of free will must in principle be a scientifically explicable phenomenon.

Some philosophers conclude from this that agent causation theory doesn't fit in with the modern scientific view of a human being, a view that sees a human being as a material object whose every aspect can in principle be explained scientifically.

Here is one final problem for agent causation theory. Recall that materialism is the view that everything in the universe is either a particle of matter or is composed of particles of matter. On a materialistic view, a human being is composed entirely of nothing but particles of matter. Consider the following question: According to the agency theory, actions ultimately trace back to and originate from an agent. But what sort of entity is this agent? That which we call the agent can't, it seems, be the person's brain or even a part of a brain, such as a single neuron or a molecule, for physical structures such as brains and neurons do not behave the way an agent-cause behaves. Physical things don't cause things the way agent-causes supposedly cause things. This is so because a brain or a part of a brain causes something only when it is first caused to do so by a prior cause, whereas an agent-cause brings about an event without being caused to do so by a prior event. So, if we accept the agency theory, it looks as if we ought to suppose that the part of a human being that makes choices and initiates actions is a nonphysical part of the human being. But if we suppose this, then we must reject materialism and accept the view known as *mind–body dualism* or "dualism" for short. Dualism is the view that a person is composed of two fundamental parts:

1. a physical part, which consists of a body;
2. a nonphysical part, which consists of a nonphysical mind or soul.

Mind–body dualism offers one way to make sense of the phenomenon of agent causation. If one accepts dualism, one can hypothesize that the part of a person that exercises the power to initiate an action—the power of agent causation—is the nonphysical or spiritual part of the person. The part that carries out the action— that puts the will into effect—is the physical body. One might argue that if agent causation theory is true, then dualism is true. What is wrong with this? As we will see in the next section, many philosophers find dualism to be an unacceptable theory of the mind.

• PROBLEMS •

1. Do we possess free will? We have examined four philosophical theories: hard determinism, soft determinism, quantum indeterminism, and agent causation theory. Which of the four theories makes the most sense of *your* experience? Present an argument for your answer.

2. Explain the difference between logical fatalism and hard determinism.

---

3. WHAT IS THE NATURE OF THE MIND?

•— DUALISM and MATERIALISM

Materialism is the view that everything in the universe is composed of matter. Matter is that which physics studies, namely, subatomic particles, fields, and the like.[20]

As far as the mind is concerned, materialism claims that the mind is no more than a physical phenomenon. Since our mental operations are obviously closely related to the operations of our physical brains, most materialists would identify the mind with the (physical) brain. On this view, your beliefs, feelings, hopes, and other thoughts and mental events are no more than wholly physical things and events occurring within your brain.

Opposing materialism is the view that your mind is not your brain or any other physical part of you. On this view, your mind is a nonphysical entity. If by 'spiritual' we mean 'nonphysical', then on this view the mind is a spiritual entity of some sort. Since our mental operations are obviously closely related to our physical brains, most who hold this view claim that the nonphysical mind—although a separate entity from the brain—is linked in some way to the brain.

A theory is *monistic* if it supposes that only one kind of thing exists. According to materialism, matter is the only kind of thing that exists. Therefore, materialism is a monistic theory. Any theory is *dualistic* if it supposes that two kinds of things exist. The claim that the mind is a nonphysical entity distinct from the physical brain implies that there exist two kinds of things: material things and nonmaterial things.

---

[20] For a modern scientific account of the nature of matter, see Mulvey [1981]. Also highly recommended is Barrow [1988]; Close, Martin, and Sutton [1987]; Trefil [1980]; and Weinberg [1977].

This view is therefore a form of dualism. To distinguish it from other forms of dualism, the view is usually called mind–body dualism.

Why would anyone suppose mind–body dualism is true? Before we examine some of the arguments for dualism, you will need a bit of conceptual background.

The Concept of Identity   Suppose Joe Doakes has been charged with robbing a bank. The prosecutor's claim is:

Joe Doakes is the bank robber.

Another way to put this is:

Joe Doakes is the same person as the person who is the bank robber.

Let us express this by saying:

Joe Doakes *is identical with* the bank robber.

In general, if we say that x is identical with y, we mean that x is the same entity as y.

Now, suppose it has been established, from an analysis of a video recording of the robbery, that the bank robber was exactly six feet tall. However, suppose Doakes is measured and is precisely five feet tall. If the person who robbed the bank is exactly six feet tall, and Doakes is exactly five feet tall, is there any possibility that Doakes is the bank robber? Obviously not. Why? If Doakes has a feature or property that the bank robber doesn't have, then Doakes and the bank robber must be two distinct individuals. In this case, the bank robber has the property of being six feet tall, and Doakes has the property of being five feet tall. It follows that Doakes must not be identical with the bank robber, that is, Doakes and the bank robber must be two different individuals.

Leibniz's Principle   The general principle in this case is a principle that has come to be called *Leibniz's Principle*, since he formulated a noted version of it. That principle may be put this way:

> *If x is identical with y, then any property possessed by x is possessed by y, and any property possessed by y is possessed by x.*

Another way to express Leibniz's Principle is:

> *If x has a property that y does not have or y has a property that x does not have, then x is not identical with y.*

Notice that these two versions of Leibniz's Principle are logically equivalent.

For example, consider the singer who, in 1963, turned down an appearance on the Ed Sullivan Show because the producer wouldn't let him sing "The John Birch Society Blues." Before he became a folk singer, Bob Dylan's name was Robert Zimmerman. Within this context, Robert Zimmerman is identical with Bob Dylan. That is, the person named 'Robert Zimmerman' is the same person as the person named 'Bob Dylan'. Consequently, any property possessed by Robert Zimmerman must be a property possessed by Bob Dylan, and vice versa. Thus, just as it is true that Robert Zimmerman was born in Minnesota, so it is true that Bob Dylan was born in Minnesota; just as it is true that Bob Dylan performed in Greenwich Village in 1962, so it is true that Robert Zimmerman performed in Greenwich Village in 1962, and so on.

A General Format for Arguments for Dualism   With this material on identity as background, we can now turn to some arguments for dualism. Arguments for dualism often fit the following general format:

1.   The argument begins with the claim that our mental states possess a property that is not possessed by our (physical) brain states.

2.   An appeal is next made to Leibniz's Principle. That is, it is claimed that if x has a property that y does not have, then x must not be identical with y, that is, x and y must be two separate and distinct entities.

3.   It follows, the argument continues, that our mental states must be distinct from our brain states, that is, our mental states are not the same as our brain states.

4.   But our mental states are states of our minds and our brain states are states of our brains. Therefore, the argument concludes, our minds are not identical with our (physical) brains, that is, our minds and our brains are distinct and separate entities.

This general format may be converted into a specific argument for dualism by specifying, at the first premise, a property that is supposedly possessed by our mental states but that is not possessed by our physical brain states.

Philosophers supporting dualism have argued that there are a number of properties possessed by mental states and mental entities that are not possessed by physical states and physical entities. For instance, it has been argued that certain types of mental states possess—essentially—the property of *intentionality* or "aboutness," while physical states and physical entities are inherently nonintentional. It has been argued that mental states are nonspatial, that is, do not occupy a volume of space and do not possess a location in physical space, while physical states and physical objects are inherently spatial. It has also been argued that we directly experience our mental states but do not directly experience our brain states. Of course, philosophers supporting materialism have criticized the various arguments for dualism. The issue belongs to the field of philosophy known as the *philosophy of mind*.

A Modal Argument for Dualism   Alvin Plantinga, a contemporary philosopher, has offered a modal argument for dualism that may be summarized in the following terms:[21]

1. It is logically possible that you exist and that your physical body—the body that you currently have and that we will label 'B'—doesn't exist.

For instance, through a process of cell-by-cell replacement, a person could theoretically be given a replacement body completely distinct from the body the person now has. So, in terms of possible worlds:

2. There's a possible world w in which you exist but B doesn't exist.
3. So, you have the modal property of existing-at-w, but B does not have that property.
4. If some thing x has a property and some thing y does not have that property, then x cannot be the same thing as y, that is, x and y must be different things.
5. Therefore, you are not identical with your body. You and your body are different entities.

If this argument is sound, then you are a compound composed of two distinct parts: a nonphysical component and a physical body. Within the context of the major religions of the world, that which the dualist claims is a nonphysical component of a person is called the "soul."

The second premise of this argument seems to presuppose that in some cases one and the same individual may exist in more than one possible world. Some philosophers argue that it makes no sense to suppose individuals exist in more than one possible world. They argue that the idea that an individual in one world is identical with an individual in another possible world is contradictory. If these philosophers are right, then the modal argument for dualism is unacceptable. We do not have the space to enter into this very difficult issue here. The issue is taken up briefly in Appendix 2 at the end of the book and the interested reader may consult the references given there for further reading.

The Mental Image Argument   Close your eyes and imagine a stop sign. What color is the sign? If you imagined an ordinary stop sign, it is of course red. Now, as you form the image in your mind of the red sign, there are certain things occurring in your brain. If a neuroscientist were to look inside your brain while you form in your mind an image of a stop sign, would she see a red spot shaped like a stop sign somewhere in or on your brain? Certainly not. When you form the image in your mind of a stop sign, no brain cell or stop sign-shaped region of your brain turns red. The color of your brain remains the color it always is.

[21] See Plantinga [1974], Chapter 3.

But the mental image of the stop sign is red. In your mind you can see definitely that it is red. You can experience its redness directly. This gives rise to the following argument for dualism:

1. When I form the image in my mind of a red stop sign, my mental image is red. But no part of my physical brain turns red when I form the image in my mind of a stop sign.

2. So, my mental state has a property—the property of redness—that my brain state lacks.

3. If x has a property that y lacks, then x must not be identical with y.

4. Therefore, my mental state is not identical with my brain state, that is, my mental state and my brain state are two distinct entities.

5. Since my mental state is a state of my mind and my brain state is a state of my brain, it follows that my mind is not identical with my brain, that is, my mind and my brain are different entities.

Incidentally, even if a red stop sign-shaped image does appear on your brain when you form a mental image of a stop sign, that physical spot on your brain can't literally be your mental image of a stop sign. You can see the red image in your mind, but you don't have eyeballs inside your head that allow you to look at a spot on your own brain!

Objections   If we suppose, as the dualist claims, that the mind and the brain are distinct entities, we must suppose that they at least *interact* with each other. That is, if dualism is true, we must suppose that the physical body sends signals to the brain and then to the mind, and the mind sends signals and commands through the brain to the body. For example, if you touch a hot object, your nerves inform your mind very quickly. And when you wake up in the morning, you sometimes have to give your body a stern command before it gets moving. Thus, dualism is sometimes called "interactionism" or "interactionist dualism," since it hypothesises that the mind and the body interact.

However, this raises a very difficult question for anyone who accepts dualism. If dualism is correct, then the mind is nonphysical, which is to say that the mind is not composed of physical particles, has no location in physical space, and has neither mass, shape, nor momentum. However, the brain and central nervous system constitute a physical object with all the standard physical properties. According to mind–body dualism, signals pass from mind to body and from body to mind in such a way that mind and body affect each other. But how could a massless, nonphysical entity such as a nonphysical mind have any kind of effect on a physical object such as a brain cell or a state of the brain? There is something very puzzling in the claim that a nonphysical mind makes causal contact with a physical brain. How do the mind and the brain interact?[22]

---

[22] My discussion relies at many points on the presentation in Dennett [1991].

There are no easy answers to this question. Physics teaches us that the state of a physical object such as a brain cell will not change in any way unless a physical object or physical force affects it. In general, in order for the position or momentum of a physical object to change, an expenditure of energy is required. Mind–body dualism hypothesizes that signals pass from mind to body and from body to mind in such a way that the mind affects the body and the body affects the mind. What kind of signals might these be? The signals that pass from mind to body can't be physical signals, and they cannot involve an expenditure of energy, for according to dualism, the mind is a wholly nonphysical entity. (Matter and energy both count as physical in nature.) But according to current physical theory, the state of a physical object such as a brain cell won't change unless the brain cell is acted upon by a physical object or physical force. So, mind–body dualism seems to require a form of interaction that violates the laws of physics. Additionally, we have no scientific theory that might explain dualistic interaction. Many philosophers conclude from this that dualism is in conflict with modern science.

Furthermore, if dualism is correct, the mind is a nonphysical entity. But science cannot, by its nature, detect or explain something that is nonphysical. Thus, if dualism is correct, the mind is ultimately a scientifically inexplicable entity. It follows that if we accept mind–body dualism, then we must give up any hope of one day explaining the mind in terms of the laws of physics. But one of the aims of cognitive science is the development of a scientific theory of the mind—a theory that explains how the mind works purely in terms of the laws of physics. The acceptance of mind–body dualism would mean rejecting that goal. This is another reason why some philosophers refuse to accept dualism.[23]

In reply, dualists raise the following questions. Why suppose that the mind must be a scientifically explicable entity? We have no good reason to suppose that science can explain everything. If the arguments for dualism are any good, the human mind is a nonphysical entity ultimately beyond the reach of physical science. And why suppose that all forms of interaction must be scientifically explicable? Perhaps the world contains forms of interaction that are not understood by physical science. Perhaps the mind–body interaction postulated by the dualist is beyond the reach of—rather than in violation of—the laws of physics. If the arguments for dualism are sound, then there is more to reality than meets the scientific eye.

[23] See Dennett [1991] Chapter 2.

MODAL LOGIC: METHODS OF PROOF

1. INTRODUCTION

R ecall from Chapter 6 that a modal argument is an argument whose validity or invalidity depends on the arrangement of its modal operators. *Modal logic* may be defined as the branch of logic concerned with modal arguments. Many of the most intriguing arguments in the history of philosophy are modal in nature. Unfortunately, as we saw in Chapter 6, the methods of truth-functional logic presented in Chapters 2 through 4 simply don't apply to modal arguments. In order to understand and evaluate modal arguments, we must employ a system of modal logic.

A system of modal logic consists of two parts: (i) a formal language for abbreviating sentences containing modal operators; and (ii) a set of principles for the evaluation of modal arguments. A number of systems of modal logic have been developed, and philosophers do not generally agree on which system best represents the logic of possibility and necessity. This chapter will introduce formal proof techniques for two systems of modal logic. After working with these two systems, the reader may then decide on her or his own which system makes the most sense of modality.

Formal methods of proof for truth-functional arguments were presented in Chapter 3 in the form of natural deduction rules and in Chapter 4 in the form of truth-tree rules. Remember that truth-trees and natural deduction are merely different procedures for accomplishing the same thing, somewhat as binary addition and decimal addition are different methods for adding numbers. This chapter will expand those two methods of formal proof so that they cover modal arguments as well. Consequently, truth-tree rules and natural deduction rules will be presented for each of the two systems of modal logic presented below. The rules for the truth-functional trees of Chapter 4 constitute the prerequisite for this chapter's unit on modal truth-trees, and the rules for truth-functional natural deduction presented in Chapter 3 are the prerequisite for this chapter's unit on modal natural deduction.

Our aim, then, is to develop rules of inference and truth-tree rules for arguments containing modal operators. But where do we begin? What should the rules be? The tree method and natural deduction system for modal logic are not going to be created here out of thin air. We are first going to formulate a set of principles concerning the modal operators that will guide us as we specify the various inference and tree rules for modal logic.

First, it might be helpful to recall two assumptions that will be presupposed throughout our discussion. First, if a sentence expresses a truth (or a falsity) we will simplify things by saying that the sentence itself is true (or false). So, if you hold that the bearer of truth is an abstract entity that is expressed by a sentence but that is distinct from a sentence token, then you may interpret an attribution of truth (or falsity) to a sentence as merely a shorthand way of attributing truth (or falsity) to the abstract entity that the sentence *expresses*. Second, recall that we are using the term 'proposition' to refer just to whatever it is that is the bearer of truth. So, each may interpret the term 'proposition' in the light of a favored theory of the truth-bearer. If you hold that a sentence token, rather than some abstract entity, is the bearer of truth, then you will want to interpret the term 'proposition' as 'sentence token'.

●— FIVE PRINCIPLES OF MODALITY

If a sentence **P** is a tautology, then the final column of **P**'s truth-table contains all T's. Since each row of a truth-table represents a set of possible worlds, and since collectively the rows of a truth-table represent all possible worlds, it follows that **P** is true in all worlds. Consequently, every tautology is a necessary truth. Thus:

> **Principle 1:** If **P** is a tautology, then **P** is necessarily true.

If a sentence **P** is necessarily true, then **P** is true in every world. But the actual world is one of the many possible worlds. So, if **P** is necessarily true, then **P** is true in the actual world. That is, every necessary truth is actually true. Thus:

> **Principle 2:** If **P** is necessarily true, then **P** is actually true as well.

Suppose that a sentence **P** is necessarily true and that **P** implies a sentence **Q**. Since **P** is necessarily true, **P** is true in every world. Since **P** implies **Q**, there is no world in which **P** is true and **Q** false. It follows that **Q** must be true in every world as well. For if there were to be even one world in which **Q** is false, that would be a world in which **P** is true and **Q** is false, which would mean, contrary to our assumption, that **P** does not imply **Q**. Thus:

> **Principle 3:** If **P** is necessarily true, and **P** implies **Q**, then **Q** is also necessarily true.

That is, whatever is implied by a necessary truth is itself also necessarily true.

It would seem that if a sentence **P** is true in a world—call it world 1—then in every world it's true that **P** is true in world 1. Suppose **P** is true in all worlds. It would seem that from the perspective of any world, it's true that **P** is true in every world.

So, if **P** is true in every world, then in every world it must be true that **P** is true in every world. That is, if it is necessarily true that **P**, then it is necessarily true that **P** is necessarily true. Thus:

> **Principle 4:** If **P** is necessarily true, then **P** is necessarily necessarily true.

This is a controversial principle. You will see later why some philosophers reject this principle. Principle 4 amounts to the claim that the necessity of a proposition is itself also a matter of necessity. That is, it is not a contingent matter if **P** is a necessary truth. Rather, if **P** is necessarily true, it is necessary that **P** is necessarily true.

Suppose that a sentence **P** is possibly true. That is, suppose **P** is true in at least one possible world. It would seem that from the perspective of any world, it's true that there is a world in which **P** is true. But then it seems plausible to suppose that in every possible world, it's true that there is a possible world in which **P** is true. So, if **P** is possibly true, then it's necessarily true that **P** is possibly true. Thus:

> **Principle 5:** If **P** is possibly true, then **P** is necessarily possibly true.

This principle is also controversial. According to principle 5, possibility is also a matter of necessity. Later on, we will see a reason why some philosophers reject this principle.

These five principles of modality may be expressed more formally as follows:

$$\text{Principle 1:}^1 \quad \vDash P \supset \Box P$$
$$\text{Principle 2:} \quad \Box P \supset P$$
$$\text{Principle 3:} \quad [\Box P \,\&\, (P \rightarrow Q)] \supset \Box Q$$
$$\text{Principle 4:} \quad \Box P \supset \Box\Box P$$
$$\text{Principle 5:} \quad \Diamond P \supset \Box\Diamond P$$

●— ALTERNATIVE SYSTEMS OF MODAL LOGIC

Each system of modal logic presented in this chapter will use as a formal language the language ML. In order to develop a system of modal logic, we must add a set of principles that can be used to evaluate modal arguments. Some philosophers argue that all five of the above modal principles are necessarily true. Consequently, the principles they specify—in the form of tree rules or natural deduction rules—are logical consequences of principles 1 through 5. The system of modal logic that is based on all five of the modal principles is the system S5.

Some philosophers doubt the truth of the fifth modal principle and accept only the first four principles. Accordingly, they specify tree or natural deduction rules that are designed to reflect only principles 1 through 4. The system of modal logic that is based on the first four modal principles is called the system S4. The systems

---

1   $\vDash P$ abbreviates 'P is a tautology.'

S5 and S4 were first formulated and named by C. I. Lewis, a philosopher who helped develop the beginnings of modern modal logic.

In addition, there are philosophers who accept only the first three modal principles. Consequently, the system of modal logic they favor, the system T, is based only on principles 1 through 3 above. Quite a number of alternative systems of formal modal logic have recently been developed, each with a different set of natural deduction and tree rules and each based on a different set of modal principles. Each system constitutes a unique way of representing formally what we believe about the modal operators and modal reasoning.

The system S5 is stronger than the system S4 in the sense that any argument that can be proven valid using S4 can also be proven valid using S5, but not vice versa. So, many arguments that can be proven valid in S5 cannot be proven valid in S4. S4 is likewise stronger than T in that any argument that can be proven valid using T can also be proven valid with S4, but not vice versa. Another way to put this is to say that S5 "contains" S4 and S4 "contains" T. So, T is the most fundamental system of the three, while S4 and S5 are expanded systems of modal logic.

Many philosophers now believe it is the system S5 that best formalizes the truth about logical necessity and possibility. However, after comparing S5 and S4, it will be left to the reader to decide which system makes the most sense of the phenomenon of modality.

2. S5 MODAL TRUTH-TREES

The tree rules for the system S5 presented here are based upon those formulated by Daniel Bonevac.[2] Trees constructed with these rules may be called 'S5 trees'.

● RULES FOR THE BOX

Suppose we begin by making the assumption that the truth-value of $\Box P$ is true. Our assumption, in English, is simply that it is true that P is a necessary truth. On a truth-tree, this assumption is represented by placing $\Box P$ on the true side of a tree:

$$\Box P \ \Big|$$

Now, what validly follows from this assumption? Clearly, it follows that P is true, for $\Box P$ implies P. So, we place P underneath $\Box P$ on the true side of the tree:

$$\checkmark \ \Box P \ \Big|$$
$$P$$

---

[2] Daniel Bonevac, *Deduction* (Mayfield, Palo Alto; 1987).

The inference from □**P** to **P** reflects principle 2, the principle that if **P** is necessarily true, then **P** is true. Recall from Chapter 4 that the check mark next to our assumption serves as a dispatch sign to signal that we have drawn an inference from our assumption. The rule in this case will be called *Box Left*:

Box Left

*If □**P** is situated on the* true *side of a tree, then you may carry down **P** by placing it under □**P** on the* true *side of the tree. More formally:*

$$✓ \; □P$$
$$P$$

The following examples each involve an application of Box Left:

$$✓ \; □A$$
$$A$$

$$✓ \; □(A∨B)$$
$$(A∨B)$$

$$✓ \; □[A\&(B∨C)]$$
$$[A\&(B∨C)]$$

Suppose we begin with the assumption that the truth-value of □**P** is false. In English, our assumption is simply that **P** is not a necessary truth. On the truth-tree, the assumption appears as:

$$□P$$

What validly follows? It follows that in some world or other, **P** is itself false. However, based solely on our assumption, we have no way of knowing which world is the world in which **P** is false. Since we will often work with information about previously specified worlds, we must be careful not to infer here that some already specified world is the world in which **P** is false. So, we must have a convention that allows us to "move" to a new world and specify that a given proposition is true or false in that new world. We will indicate a "move" to a new, unspecified world using a device called a *world shift line*. The rule in this case is as follows:

Box Right

*If □**P** is situated on the* false *side of a tree, then you may introduce a new world by drawing a* world shift line *and you may then carry **P** down into that new world as long as **P** remains on the* false *side of the tree. More formally:*

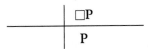

The horizontal line across the middle of the tree is a world shift line. It introduces a new and previously unspecified possible world. By placing **P** under the shift line, we are specifying that the world in which **P** is false—the world we have carried **P** down into—is a new and unspecified world. The reasoning is that if □**P** is false, then we know there must be at least some worlds in which **P** is false. The shift line introduces one of those worlds. Here are some applications of Box Right:

□A ✓		□(AvB) ✓		□[A&(BvC)] ✓
A		(AvB)		[A&(BvC)]

●— RULES FOR THE DIAMOND

Suppose next we assume that ◇**P** is false. This assumption, in English, is the claim that it's not the case that there is at least one world in which **P** is true. On the truth-tree, the assumption is represented as:

What validly follows from our assumption? Clearly, **P** must be false in every world. Consequently, no world shift line is needed when we infer from this that **P** is false, and we can bring **P** down on the false side underneath ◇**P**. Thus:

Diamond Right

> If ◇**P** is situated on the false side of a tree, then you may carry down **P** by placing it under ◇**P** on the false side of the tree. More formally:

The following examples each involve an application of Diamond Right:

◇(PvQ) ✓		◇H ✓		◇(S&G) ✓
(PvQ)		H		(S&G)

If we assume that ◇**P** is true, we write:

◇**P**

The assumption here is that **P** is possibly true. Does it follow from this that **P** is actually true? Surely not. However, if ◇**P** is true, it follows that **P** is true in some world or other. Since our assumption does not tell us which world this is, we must shift to a new, undesignated world before we infer that **P** is true. Thus:

Diamond Left

> If ◇**P** is situated on the true side of a tree, then you may draw a world shift line and carry down **P** into a new world as long as **P** remains on the true side of the tree. More formally:

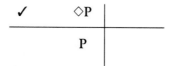

●— RULES FOR THE FISHHOOK

Before we specify a rule for the fishhook, it will be helpful to explore the relation between → and ⊃. If a sentence **P→Q** is true, then **P** implies **Q**. It follows that in no possible world is it the case that **P** is true and **Q** false. However, **P⊃Q** is false just in case **P** is true and **Q** is false. Therefore, if **P→Q** is true, then in every world **P⊃Q** is true. Therefore,

$$(\mathbf{P{\rightarrow}Q}) \rightarrow \Box\,(\mathbf{P{\supset}Q})$$

Suppose now that a sentence **P→Q** is assumed true:

$$\mathbf{P{\rightarrow}Q}$$

What follows? In every world, **P⊃Q** is true. We saw in Chapter 4 that **P⊃Q** is true just in case either **P** is false or **Q** is true. So, in this case we simply split the tree as we did for the horseshoe in Chapter 4:

Hook Left

> If **P→Q** is situated on the true side of a tree, then you may infer that either **P** is false or **Q** is true. More formally:

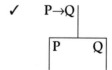

For example:

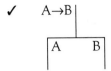

Suppose that a sentence **P→Q** is false. The assumption here is that it is false that **P** implies **Q**. On the tree, we write:

$$P{\rightarrow}Q$$

Given this, it does not follow that **P** is true and **Q** is false. So, what follows if we assume that **P→Q** is false? It follows only that in some world or other, **P** is true and **Q** false. Thus, the rule for this case is:

Hook Right

> If **P→Q** is situated on the false side of a tree, then you may draw a world shift line *and carry down into a new world the consequence that* **P** is true and **Q** is false. More formally:

$$
\begin{array}{c|c}
 & P{\rightarrow}Q \quad \checkmark \\
\hline
P & Q
\end{array}
$$

Here are some examples:

$$
\begin{array}{c|c}
 & A{\rightarrow}B \; \checkmark \\
\hline
A & B
\end{array}
\qquad
\begin{array}{c|c}
 & (H\&S){\rightarrow}(R{\vee}B) \; \checkmark \\
\hline
(H\&S) & (R{\vee}B)
\end{array}
$$

●— RULES FOR THE DOUBLE HOOK

Suppose a sentence **P↔Q** is true. Two sentences **P** and **Q** are equivalent just in case their truth-values match in every possible world. Consequently, if **P** is equivalent to **Q**, then in any and every world, either both are true or both are false. Thus:

Double Hook Left

> If **P↔Q** is situated on the true side of a tree, then you may infer that either **P** and **Q** are both true or they are both false. More formally:

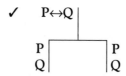

For example:

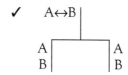

In professional boxing circles, this rule is sometimes referred to as the "Muhammad Ali rule."

If a sentence P↔Q is assumed false, then the assumption is that **P** and **Q** are not equivalent. If **P** and **Q** are not equivalent, then there must be some world or other in which the truth-values of the two differ. There are only two ways that **P** and **Q** might differ in truth-value: Either (1) **P** is true, **Q** is false, or (2) **P** is false, **Q** is true. Thus, if **P** and **Q** are not equivalent, then in some world or other, either **P** is true while **Q** is false or **P** is false while **Q** is true. Thus:

Double Hook Right

*If **P↔Q** is situated on the false side of a tree, then you may draw a world shift line and infer that either **P** is true and **Q** is false or **P** is false and **Q** is true. More formally:*

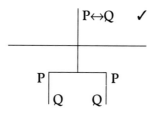

Notice that in this case, we shift to a brand new world before we bring down our consequence. Here is an example:

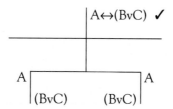

●— CLOSING A MODAL TREE

A branch of a modal tree closes just in case the same formula appears live on both sides within one world. For example, the following tree closes:

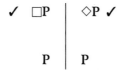

In this example, the branch represents a contradiction, for it would be contradictory for a proposition to be true and yet also false in one world.

However, if a branch has the same formula live on both sides but the two formulas are separated by a world shift line, then the branch does not close. For example:

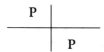

This branch has not generated a contradiction, for there is nothing contradictory about a proposition being true in one world and false in another.

Finally, remember that a tree closes just in case every branch closes.

●— PUTTING THE RULES TO WORK: EXAMPLES OF S5 TREES

We can now put the tree method to work by proving some necessary truths. In the next few examples, let P abbreviate your favorite proposition-expressing sentence. It won't matter which proposition-expressing sentence you pick, for what we will say about P will apply equally to any proposition-expressing sentence. According to the second modal principle, if any sentence is necessarily true, it follows that the sentence is true. So, applying the second modal principle to the case of P, if P is necessarily true, it follows that P is true. Thus, a particular case or application of the second modal principle is:

$$□P⊃P$$

Next, we will assume this false by placing it on the false side of a tree:

$$|\ □P⊃P$$

Our assumption is that it is false that if P is necessarily true then P is true. Now, since the main connective is a horseshoe, we begin by applying Horseshoe Right:

$$□P\ |\ \begin{array}{c} □P⊃P\ ✓ \\ P \end{array}$$

Next, we apply Box Left:

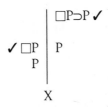

The tree closes, proving that our assumption implies a contradiction. Since only that which is contradictory can ever imply a contradiction, it follows that our assumption must itself be contradictory. So our assumption must be false. Since our assumption was that □P⊃P is false, it follows that it would be impossible for □P⊃P to be false, which proves that □P⊃P is necessarily true.

● THEOREMS

If a formula can be proven true using only the procedures of logic, without investigating the physical world, the formula is said to be *logically true*. If a formula can be proven true using only the procedures of the system S5, it is said to be *true in S5* and it is also said to be a *theorem in S5* (or a theorem *of* S5). The above tree thus proved that □P⊃P is a theorem in S5.

Similarly, if a formula can be proven false using only the procedures of logic, the formula is said to be *logically false*. If a formula can be proven false using only the procedures of S5, it is said to be *false in S5*.

According to the Law of Excluded Middle, in every world, every sentence **P** is either true or false. A particular case or application of this principle, using the sentence P again, is:

□(Pv~P)

In order to prove this formula true in S5, let us begin by assuming it false:

| □(Pv~P)

The first rule to apply is Box Right:

$$\frac{\quad}{\quad} \ \left| \ \begin{array}{l} □(Pv~P) \ ✓ \\ \\ Pv~P \end{array}\right.$$

Next, we apply Wedge Right:

$$\frac{\quad}{\quad} \ \left| \ \begin{array}{l} □(Pv~P) \ ✓ \\ \\ Pv~P \quad ✓ \\ P \\ ~P \end{array}\right.$$

Finally, when we apply Tilde Right, the tree closes:

$$\square(Pv\sim P) \checkmark$$

$$
\begin{array}{c}
Pv\sim P \quad \checkmark \\
P \\
\sim P \quad \checkmark \\
P \\
X
\end{array}
$$

The closed tree proves that our assumption is contradictory. It follows that $\square(Pv\sim P)$ can't possibly be false. The sentence $\square(Pv\sim P)$ is therefore logically true. It is also a theorem of S5.

In Chapter 5, it was argued that there's no possible world in which an explicit contradiction—a statement of the form **P&~P**—is true. In other words, it was argued that:

$$\sim\lozenge(\mathbf{P\&\sim P})$$

Let's assume, in the case of P, that P&~P is possibly true. That is, let us assume that it is false that:

$$\sim\lozenge(P\&\sim P)$$

The tree follows:

$$\sim\lozenge(P\&\sim P) \checkmark$$

$$
\begin{array}{c}
\checkmark \lozenge(P\&\sim P) \\
\checkmark P\&\sim P \\
P \\
\checkmark \sim P \quad | \quad P \\
X
\end{array}
$$

We first applied Tilde Right and carried $\lozenge(P\&\sim P)$ to the other side of the tree. Then we applied Diamond Left and carried P&~P across the world shift. From there, the tree closed after an application of Ampersand Left and then Tilde Left. This demonstrates that in the case of P, it is not possible that the conjunction of P with its negation is true. The sentence $\sim \lozenge(P\&\sim P)$ is therefore a theorem of S5.

It certainly seems to be a necessary truth that if the sentence P is true, then P is also possibly true, that is, $P\supset\lozenge P$. If we assume this false, the tree easily closes:

$$P\supset\lozenge P \checkmark$$

$$
\begin{array}{c}
P \quad | \quad \lozenge P \checkmark \\
P \\
X
\end{array}
$$

Thus, there is no possibility that P⊃◇P is false. P⊃◇P is therefore logically true and a theorem of S5.

If P implies Q, where Q is someone else's favorite proposition-expressing sentence, then P⊃Q must be true. In ML, this claim is abbreviated (P→Q)⊃(P⊃Q). On the true side of the following tree, we assume that P implies Q and on the false side we assume that P⊃Q is false:

$$P{\rightarrow}Q \quad \Big| \quad P{\supset}Q$$

An application of Horseshoe Right gives us:

$$P{\rightarrow}Q \quad \Bigg| \quad \begin{array}{c} P{\supset}Q \checkmark \\ Q \end{array}$$
$$P$$

The tree closes when we next apply Hook Left:

$$\checkmark \quad P{\rightarrow}Q \,\big|\, P{\supset}Q \ \checkmark$$
$$P \,\big|\, Q$$

$$\boxed{P} \qquad Q$$
$$X \qquad\quad X$$

This closed tree proves that it would be impossible for P→Q to be true while P⊃Q is false, which proves that P→Q implies P⊃Q.

Consider the following symbolized argument:

1. P⊃□Q
2. □P
3. □Q

Does it seem valid to you? According to the first premise, if P is true, then Q is necessarily true. According to the second premise, P is necessarily true. Does it follow that Q must be necessarily true? On the following tree, we assume the premises are true and the conclusion is false:

$$\begin{array}{c} P{\supset}\square Q \\ \square P \end{array} \quad \Bigg| \quad \square Q$$

The tree closes, proving the assumption contradictory and the argument valid:

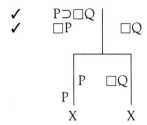

If an argument can be proven valid using only the procedures of S5, the argument is said to be *valid in S5*. Consequently, this argument is valid in S5. Later, more will be said on the notion of validity in S5.

● THE TRANSFER and RETRIEVAL RULES

The rules introduced so far are sufficient for most of the things you might want to prove within modal logic. However, certain types of arguments and formulas cannot be evaluated using just those rules. Two more rules must be added to our system. Our first additional rule presupposes the concept of a *modally closed* formula. A sentence **P** of ML is said to be modally closed just in case every sentence letter within **P** appears within the scope of a modal operator. Thus, each of the following sentences is modally closed:

$\Box A$ $\Box(A\&B)$ $\Diamond Hv\Box G$ $D{\rightarrow}R$ $A{\leftrightarrow}H$

And each of the following is not modally closed:

AvB $\Box AvB$ $P{\supset}Q$ $B{\supset}\Box A$

A very important consequence of the five principles of modality is that every modally closed sentence of ML represents a *noncontingent* proposition. So, if a formula is modally closed, the formula is either true in every world or it is false in every world, that is, it is either necessarily true or it is necessarily false. Therefore, if a modally closed formula is assumed true, it follows from this assumption that it is necessarily true, and if a modally closed formula is assumed false, it follows from this assumption that the formula is necessarily false. The general principle may be called the *Modal Closure Principle*:

> Any proposition expressed by a modally closed sentence is a noncontingent proposition.

The logical consequences of the modal closure principle may be incorporated into our set of tree rules in the form of a new rule called the *Transfer Rule*:

The Transfer Rule

If a formula is modally closed, it may be carried down (transferred) into any world whatever, provided that the formula remains on the same side of the tree throughout the transfer.

If this rule doesn't seem quite right to you, consider the following. If we place a modally closed formula on the true side of a tree, we are assuming the formula has the truth-value true. But if a modally closed formula is true in even one world, it is true in all worlds. Therefore, given the initial assumption, the formula must be true in all worlds and we are justified in carrying it down—on the true side—into any world. If a modally closed formula is placed on the false side of a tree, we are assuming that it is false. But if a modally closed formula is false in even one world, it is false in all worlds. Therefore, given the initial assumption, the formula must be false in all worlds, and we are justified in carrying it down—on the false side—into any world.

Trees Requiring the Transfer Rule According to the third modal principle, if P is a necessary truth, and P implies Q, then Q must be a necessary truth as well. A necessary truth only implies another necessary truth. In other words, the following argument is valid:

$$\Box P$$

$$P \rightarrow Q$$

$$\Box Q$$

In the following tree, we will assume that the premises are true and that the conclusion is false:

$$
\begin{array}{c|c}
\Box P & \\
P \rightarrow Q & \Box Q
\end{array}
$$

Our first step is an application of Box Right to $\Box Q$:

$$
\begin{array}{c|c}
\Box P & \\
P \rightarrow Q & \Box Q \checkmark \\
\hline
 & Q
\end{array}
$$

This required drawing in a world shift line.

Next, we twice apply the Transfer Rule and carry $\Box P$ and then $P \rightarrow Q$ across the world shift line:

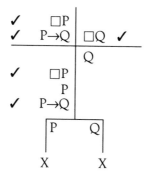

Finally, we apply Box Left and then Hook Left:

Since the tree closes, our assumption implies a contradiction. Consequently, the assumption that it is possible for the premises of the argument to be true and the conclusion false is a contradictory assumption, which implies that it is not possible for the premises to be true and the conclusion false. The argument must therefore be valid in S5. If we hadn't used the Transfer Rule on this problem, the tree would not have closed.

According to the fourth modal principle, if a proposition is necessarily true, then it is necessary that it is necessarily true. In the case of the sentence P, this is to say that if P is necessarily true, then it is necessary that P is necessarily true. In ML, this is symbolized:

$$\Box P \supset \Box\Box P$$

On the following tree, we begin by assuming this is false:

$$\mid \quad \Box P \supset \Box\Box P$$

Next, we apply Horseshoe Right:

$$\Box P \supset \Box\Box P \;\checkmark$$
$$\Box P \quad \mid \quad \Box\Box P$$

If we apply Box Right to □□P on the right, the tree extends to:

Next we apply Box Right to □P on the right and also transfer □P on the left across the shift:

$$
\begin{array}{c|c}
 & \Box P \supset \Box\Box P \checkmark \\
\checkmark\ \Box P & \Box\Box P \checkmark \\
\hline
 & \Box P \checkmark \\
\hline
\Box P & P
\end{array}
$$

Finally, by an application of Box Left to □P, the tree closes:

$$
\begin{array}{c|c}
 & \Box P \supset \Box\Box P \checkmark \\
\checkmark\ \Box P & \Box\Box P \checkmark \\
\hline
 & \Box P \checkmark \\
\hline
\checkmark\ \Box P & P \\
P &
\end{array}
$$

$$X$$

Thus, □P⊃□□P is a theorem of S5.

According to the fifth principle of modality, if a proposition is possibly true, then it is necessarily true that it is possibly true. Applied to the case of P, this is to say that if P is possibly true, then it is necessarily true that P is possibly true. In ML, this claim is:

$$\Diamond P \supset \Box \Diamond P$$

The next tree proves this logically true. As usual, we begin by assuming the sentence false:

$$\Diamond P \supset \Box \Diamond P$$

By applying Horseshoe Right, we get:

$$
\begin{array}{c|c}
 & \Diamond P \supset \Box \Diamond P \checkmark \\
\hline
\Diamond P & \Box \Diamond P
\end{array}
$$

We have to make a world shift as we apply Box Right:

$$
\begin{array}{c|c}
 & \Diamond P \supset \Box \Diamond P \checkmark \\
\hline
\Diamond P & \Box \Diamond P \checkmark \\
\hline
 & \Diamond P
\end{array}
$$

The Diamond Left Rule justifies the following extension of the tree:

$$
\begin{array}{c|c}
 & \Diamond P \supset \Box \Diamond P \checkmark \\
\hline
\checkmark\ \Diamond P & \Box \Diamond P \checkmark \\
\hline
 & \Diamond P \\
\hline
P &
\end{array}
$$

Finally, on the right, when we transfer $\Diamond P$ across the shift and apply Diamond Right, the tree closes:

$$
\begin{array}{c|c}
 & \Diamond P \supset \Box \Diamond P \checkmark \\
\hline
\checkmark\ \Diamond P & \Box \Diamond P \checkmark \\
\hline
 & \Diamond P \checkmark \\
\hline
P & \\
 & \Diamond P \checkmark \\
 & P \\
\hline
 & X
\end{array}
$$

This proves that $\Diamond P \supset \Box \Diamond P$ is a theorem of S5.

The Retrieval Rule In some cases, after you have drawn a series of inferences and extended a tree several stages, you will need to use a formula that is not modally closed and that was left behind as you moved down the tree into new worlds. Since the formula is not modally closed, the rules do not permit you to carry it down into any of the new worlds you have moved to. And the rules do not permit you to travel up a tree and make inferences that take you back to that formula. Yet you might need to draw an inference from that formula. We therefore need a rule for this type of case.

Here is an example: Within S5, it's obviously true that P⊃□◇P. For if P is true, then P is possibly true. And according to S5's fifth principle, if P is possibly true, then P must be necessarily possibly true. However, if we try to prove P⊃□◇P using the rules we have so far, the tree does not close:

$$
\begin{array}{c|c}
 & \text{P⊃□◇P} \checkmark \\
\hline
\text{P} & \text{□◇P} \checkmark \\
\hline
 & \text{◇P} \checkmark \\
 & \text{P}
\end{array}
$$

Given the rules specified so far, we have no way to carry the P on the left across the world shift. A new rule is needed.

The Retrieval Rule

> At any point on a tree, you may write down an undispatched formula from a previous stage of a tree provided that: (a) you first draw a world shift line and open up a new world; and (b) after the shift, the formula remains on the same side of the tree.

Since the formula is brought down into a new world, we aren't making any illegitimate assumptions as to which world it is true or false in. The Retrieval Rule lets us, in effect, carry our chain of reasoning back to a world left behind.

To return to the case of P, let us use the Retrieval Rule to prove that if P is true, then it is necessarily true that P is possibly true. That is, let us prove that P⊃□◇P is logically true:

$$
\begin{array}{c|c}
 & \text{P⊃□◇P} \checkmark \\
\text{P} & \text{□◇P} \checkmark \\
\hline
 & \text{◇P} \checkmark \\
\hline
\text{P} & \\
 & \text{◇P} \checkmark \\
 & \text{P} \\
 & \\
\text{X} &
\end{array}
$$

We began by applying Horseshoe Right. We then drew a world shift and applied Box Right to □◇P. Then we used the Retrieval Rule to bring P down the true side of the tree across the second world shift line. We also transferred ◇P across that line. In so doing, we in effect went back to the world in which P is assumed true—the world at the top of the tree—and we in effect carried ◇P back to that world using the Transfer Rule. A contradiction was then reached. If the Retrieval Rule had not been available, this tree could not be closed and we could not prove P⊃□◇P true.

Finally, here are a few additional trees for your examination. The first four trees each prove an argument valid. Consider the argument:

$$P \rightarrow Q$$
$$\underline{P}/Q$$

The premises appear above the line and the conclusion is written after the slanted slash mark. According to this argument, if P implies Q, and P is true, then Q is true. To prove this valid, we place the premises on the true side and the conclusion on the false side:

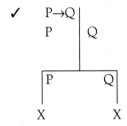

The tree closes, proving the argument valid in S5.
 Next, consider:

$$P \rightarrow Q$$
$$\underline{\sim Q}/\sim P$$

According to the premises, P implies Q and Q is false. The conclusion is that P is false. The tree closes in three moves:

The argument is therefore valid in S5. So, if P implies Q, and Q is false, then P is false.
 The next argument doesn't seem valid at first glance:

$$P \rightarrow Q$$
$$\underline{PvQ}/Q$$

According to this argument, if P implies Q, and PvQ is true, then Q must be true.

The tree for this argument follows:

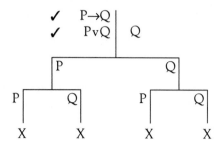

On this tree, the first split is due to an application of Hook Left to P→Q. The second split is due to an application of Wedge Left to PvQ. The tree closes, proving the argument valid in S5.

Consider next the argument:

$$PvQ$$
$$Q$$
$$\sim\Diamond(P\&Q)/\sim P$$

The tree follows:

```
                PvQ |
                 Q  |
  ✓   ~ ◇ (P&Q) | ~P   ✓
                 P  |
                    |    ◇ (P&Q) ✓
                    |    P&Q      ✓
                 |P     |Q
                 X       X
```

The following tree proves that □P→□(PvQ) is necessarily true:

```
              |  □P→□(PvQ)✓
  ✓   □P      |  □(PvQ) ✓
  ✓   □P      |  PvQ ✓
        P     |
              |
              |    P
              |    Q
         X
```

On this tree, Hook Right produced the first world shift. The second shift occurred when Box Right was applied. □P was then transferred across that shift. Finally, applications of Box Left and Wedge Right closed the tree. Thus, □P→□(PvQ) is a theorem of S5.

It is easy to show that ◊Pv◊~P is also a theorem. The following tree was generated by applying Wedge Right, Diamond Right twice, and Tilde Right.

$$
\begin{array}{c|l}
 & \text{◊Pv◊~P ✓} \\
 & \text{◊P ✓} \\
 & \text{◊~P ✓} \\
 & \text{P} \\
\text{P} & \text{~P ✓} \\
 & \text{X}
\end{array}
$$

The following tree demonstrates that (P→Q)⊃(P⊃Q) is necessarily true:

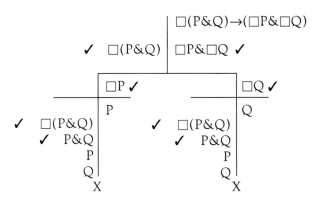

Finally, the following tree proves that □(P&Q)→(□P&□Q) is a theorem:

We first applied Hook Right. An application of Ampersand Right split the tree. Box Right caused the two world shifts. Finally, □(P&Q) was transferred across the shifts and broken up into its components.

A Summary of the Tree Tests A truth-tree constructed according to the rules of this section is an "S5 tree." We may state a set of S5 tree tests in the form of a concise set of principles. In the following, 'P' and 'Q' are metavariables ranging over sentences of ML:

1. **P** is a *theorem* in S5 if and only if it is the case that if **P** is placed on the false side of an S5 tree, the resulting completed tree closes.

2. **P** is *false in S5* if and only if it is the case that if **P** is placed on the true side of an S5 tree, the resulting completed tree closes.

3. **P** is *contingent in S5* if and only if it is the case that if **P** is placed on the false side of an S5 tree the resulting completed tree is open and it is the case that if **P** is placed on the true side of an S5 tree, the resulting completed tree is open.

4. An argument is *valid in S5* if and only if it is the case that if the premises are placed on the true side of an S5 tree and the conclusion is placed on the false side, the resulting completed tree closes.

5. **P** and **Q** are *inconsistent in S5* if and only if it is the case that if the two are placed on the true side of an S5 tree, the resulting completed tree closes.

6. **P** and **Q** are *consistent in S5* if and only if it is the case that if the two are placed on the true side of an S5 tree, the resulting completed tree is open.

7. **P** *implies* **Q** *in S5* if and only if it is the case that if **P** is placed on the true side of an S5 tree and **Q** is placed on the false side, the resulting completed tree closes.

8. **P** and **Q** are *equivalent in S5* if and only if it is the case that if the two are placed on opposite sides of an S5 tree, the resulting tree closes.

3. TRUTH-TREES WITHIN S4

Recall the difference between S5 and S4: S5 is based on all five of the five modal principles set out at the beginning of this chapter, and S4 is based only on the first four of those principles. If we make two changes in the tree rules for S5, then we produce the tree rules for S4. The two changes that must be made as we move to S4 are the following:

1. The retrieval rule must be dropped, for it does not belong to S4.

2. The transfer rule must be altered.

Trees constructed with this new set of rules will be called S4 trees.

●— THE S4 TRANSFER RULE

The system S5 accepts and the system S4 rejects the fifth principle of modality, that is, the principle that if **P** is possibly true, then **P** is necessarily possibly true. However, S4 and S5 both accept the fourth principle of modality, namely, the principle that if **P** is necessarily true, then it is true in every world that **P** is necessarily true. In S5, if a proposition is necessarily true, then in every world it is true that it is necessarily true, and if a proposition is possibly true, then in every world it is true that it is possibly true. Since S4 treats only necessary truths as necessarily true in every world, it follows that only necessary truths can be transferred across world shifts on an S4 tree. The *S4 Transfer Rule* is the logical consequence of this:

The S4 Transfer Rule

A formula of either the form $\Box P$, $P \rightarrow Q$, *or* $P \leftrightarrow Q$ *on the true side of a tree, may be transferred into any world as long as it remains on the true side of the tree throughout the shift.*

A formula of the form $\Diamond P$—*situated on the false side of a tree*—*may be transferred into any world as long as it remains on the false side of the tree throughout its shift.*

The reason $\Diamond P$ on the false side transfers across world shifts is that an assumption that $\Diamond P$ is false is an assumption that $\Box \sim P$ is true, and necessary truths transfer across world shifts in S4. Similarly, $P \rightarrow Q$ transfers on the true side in S4 because $P \rightarrow Q$ is equivalent to $\Box(P \supset Q)$ and $P \leftrightarrow Q$ on the true side transfers in S4 because it is equivalent to $\Box(P \equiv Q)$.

An S4 Tree As noted above, S4 rejects the fifth principle of modality. According to that principle, if a proposition is possibly true, then it is necessary that the proposition is possibly true. Again, let P abbreviate your favorite proposition-expressing sentence. Applied to the case of P, the fifth modal principle indicates that:

$$\Diamond P \supset \Box \Diamond P$$

must be true. Let us try to prove that $\Diamond P \supset \Box \Diamond P$ is true by constructing an S4 tree. We will begin by assuming the sentence is false:

$$\Big| \quad \Diamond P \supset \Box \Diamond P$$

If the tree closes, the formula is proven true. Let us first apply the Horseshoe Right Rule:

$$\Diamond P \quad \Big| \quad \begin{array}{c} \Diamond P \supset \Box \Diamond P \checkmark \\ \Box \Diamond P \end{array}$$

Next, let us apply Box Right:

$$\Diamond P \quad \Big| \quad \begin{array}{c} \Diamond P \supset \Box \Diamond P \checkmark \\ \underline{\Box \Diamond P \checkmark} \\ \Diamond P \end{array}$$

In S4, a modally closed sentence such as $\Diamond P$ on the left will not transfer across world shifts. Therefore, we have reached a dead end. This S4 tree cannot be closed.

◇P⊃□◇P cannot be proven on an S4 tree and is therefore not a theorem of S4.

Here is an S4 tree that proves □P⊃P necessarily true. We begin with an application of Horseshoe Right and then apply Box Left:

$$
\begin{array}{c|c}
 & \text{□P⊃P ✓} \\
\text{✓ □P} & \text{P} \\
\text{P} & \\
\end{array}
$$

X

Thus, □P⊃P counts as a theorem of S4.

4. S5 NATURAL DEDUCTION

An S5 natural deduction system is a natural deduction system that (a) incorporates all of the truth-functional rules, (b) includes rules for handling the modal operators, and (c) is designed to reflect all five of the principles of modality presented in section 1 above. In this section, we will use an S5 natural deduction system to evaluate modal arguments. The system we will use includes all of the truth-functional rules and techniques studied in Chapter 3, and in addition will include a set of specifically modal rules and procedures. Consequently, this section presupposes a working knowledge of the truth-functional natural deduction rules and techniques covered in Chapter 3.

Some arguments containing modal operators can be proven valid using only the truth-functional rules of Chapter 3. Consider, for example, the following:

1. □P∨□Q
2. ~□P /□Q
3. □Q DS 1,2

1. □P⊃□Q
2. ~□Q /~□P
3. ~□P MT 1,2

1. ◇P⊃◇Q
2. ◇Q⊃◇R / ◇P⊃◇R
3. ◇P⊃◇R HS 1,2

Of course, the reason these arguments can be proven valid using only truth-functional rules is that although the three arguments above contain modal operators, the main operators of each of the premises of those arguments are all truth-functional connectives, and the truth-functional rules used in those three proofs apply only to main operators.

However, when it comes to arguments in which the main operator of each premise is a modal operator, additional rules are needed. If we use only truth-functional rules, we cannot properly evaluate such arguments. The following valid argument, for instance, can't be proven valid if only truth-functional rules are used:

$$P \rightarrow Q$$
$$\Box P / \Box Q$$

In order to handle modal arguments such as this one, we need rules that allow us to draw inferences from sentences whose main connectives are modal operators.

● THE FIRST SEVEN S5 INFERENCE RULES

Let us begin with some fairly obvious rules first. The following rules of inference should be fairly easy to work with. Each reflects one or more of the five principles of modality.

Recall the second of the five principles of modality: If **P** is necessarily true, then **P** is true. That is,

$$\Box P \supset P$$

The corresponding inference rule in this case is pretty obvious:

Rule M1

> *From a statement* $\Box$**P** , *one may infer the statement* **P** . *More formally:*

$$\frac{\Box P}{P}$$

Here are a number of applications of M1:

$\Box A$	$\Box B$	$\Box(A \& B)$	$\Box \sim (A v B)$
A	B	(A&B)	~(AvB)

The following additional rules should seem fairly obvious:

— **M2**	**M3**	**M4**	**M5** → ●
P	P→Q	P→Q	P→Q
◇P	P	~Q	Q→R
	Q	~P	P→R

— **M6**		**M7** → ●
P↔Q P↔Q		P→Q
P or Q		Q→P
Q P		P↔Q

In M4, the tilde may be read as 'the opposite of'.

●— PROOFS USING THE FIRST SEVEN RULES

Here are two proofs that use these new rules:

1. A&B
2. ◇(A&B)→□E
3. □E→◇S/◇S
4. ◇(A&B)   M2,1
5. □E   M3, 2,4
6. ◇S   M3, 3,5

1. □(E≡S)
2. ◇B→~(E≡S)
3. ~◇B→□G /G
4. E≡S   M1, 1
5. ~◇B   M4, 2,4
6. □G   M3, 3,5
7. G   M1, 6

●— ANSELM'S PRINCIPLE

In order to understand the next inference rule, you will need to know what a *modally closed formula* is. A sentence **P** of ML is said to be modally closed just in case every sentence letter within **P** appears within the scope of a modal operator. Thus, each of the following sentences is modally closed:

□A   □(A&B)   ◇Hv□G   D→R   A↔H

And each of the following is not modally closed:

AvB   □AvB   P⊃Q   B⊃□A

An important consequence of the five principles of modality is that every modally closed sentence of ML represents a noncontingent proposition. If a formula is modally closed, the formula is either true in every world or it is false in every world, that is, it is either necessarily true or it is necessarily false. If a modally closed formula is assumed true, it follows from this assumption that the formula is necessarily true, and if a modally closed formula is assumed false, it follows from this assumption that the formula is necessarily false. The general principle may be called the *Modal Closure Principle*:

> *Any proposition expressed by a modally closed sentence is a noncontingent proposition.*

The following rule of inference should now seem valid:

— <u>M8</u> ⟶•

$$P{\rightarrow}Q$$
$$\underline{\diamond P}$$
$$\Box Q \quad \text{(Q must be modally closed)}$$

According to M8, from a line of the form **P→Q** and another of the form ◇**P**, we may infer □**Q** if **Q** is a modally closed formula.

If this rule of inference doesn't seem valid, think of it this way. Suppose **P** implies **Q**. Then there is no possibility that **P** is true and **Q** false. Suppose further that there is a world in which **P** is true. Then, since **P** implies **Q**, **Q** must be true in that world also. However, if **Q** is modally closed, then **Q** is either true in every world or true in no worlds. Since we know that **Q** is true in at least one world, it follows that **Q** must be true in all worlds.

St. Anselm's ontological argument for God's existence was discussed in the Philosophical Interlude. When Anselm's argument is symbolized, one of the inferences within the argument is an instance of M8. Since Anselm's famous argument employs this modal inference rule, M8 may be called "Anselm's Principle".

●— THE NECESSITATION RULE

The previous eight rules should be pretty easy to work with. However, the next four rules are a bit more complicated. According to the third of the five principles of modality, whatever follows from a necessary truth is itself a necessary truth. That is:

$$\Box(P{\supset}Q){\supset}(\Box P{\supset}\Box Q)$$

Another way to put this is:

$$[\Box P \& \Box(P{\supset}Q)]{\supset}\Box Q$$

Suppose that we begin with one or more modally closed formulas and, using only valid rules of inference, derive a formula **P**. What conclusion may be drawn concerning **P**? First, if the modally closed formulas we began with are true, then they are necessarily true, for a modally closed formula is true in all worlds if it is true in even one world. But the third modal principle assures us that whatever follows from a necessary truth is itself also necessarily true. Therefore, if **P** follows from one or more modally closed formulas, then if those modally closed formulas are true, then they are necessarily true and **P** is also necessarily true as well. The conclusion we may draw from this is that **P** is necessarily true. In this case, we may therefore prefix a box to **P**.

The rule this reasoning is leading up to requires the reiteration of formulas. A formula that appears as a whole line of a proof is *reiterated* if it is copied from an earlier line and entered as a later line in the proof. Suppose that at some point in a proof you indent, reiterate one or more modally closed lines, and derive from these reiterated lines a sentence **P**. Such a sequence of indented lines shall be called an S5

Necessitation Subproof. Since **P** has been derived from only modally closed formulas, **P** must be necessarily true if the modally closed formulas are true. It follows that we are justified in prefixing a box to **P** in such a case. The corresponding rule is as follows:

S5 Necessitation Rule

At any point in a proof, you may indent and construct an S5 necessitation subproof in which every line is either justified by the S5 reiteration rule or follows from previous lines of the subproof by a valid rule of inference. You may then end the indentation, draw a line around the indented lines, write down any line derived within the subproof, and prefix a box to that line. (Write as justification 'S5 Nec' and the line numbers of the subproof.)

S5 Reiteration Rule

You may reiterate any line provided that the entire line consists of just one modally closed formula and that the formula does not lie within the scope of a discharged assumption or a terminated necessitation subproof. (Write as justification 'S5 Reit' and the line number of the reiterated formula.)

This rule sounds complicated, but it is actually quite easy to apply once you get used to it. The S5 necessitation rule allows us to derive statements from modally closed premises and then assert that the statements thus derived are, given the truth of those premises, necessarily true. Since anything that follows from a necessary truth is itself necessarily true, and since a modally closed formula is necessarily true if it is true at all, this is a valid rule of inference.

Here is another way to put this. An S5 necessitation subproof serves to demonstrate that a particular formula follows validly from only modally closed premises. So, once a formula has been derived within an S5 subproof, we have shown that it follows from modally closed formulas. Since modally closed formulas are necessarily true if true at all, and since only necessary truths follow from necessary truths, any formula derived within an S5 subproof must be necessarily true if the premises it is derived from are true. We are therefore justified in prefixing a box to any formula so derived.

Here is a simple example to begin with:

1. □(P⊃Q)
2. □(Q⊃R)/□(P⊃R)

3. □(P⊃Q)   S5 Reit 1
4. □(Q⊃R)   S5 Reit 2
5. P⊃Q   M1, 3
6. Q⊃R   M1, 4
7. P⊃R   HS 5, 6

8. □(P⊃R) S5 Nec. 3–7

Notice that after completing an S5 necessitation subproof, we box the indented lines off as in a conditional or indirect proof.

The following proof contains an S5 necessitation subproof:

1. □(PvQ)
2. □(~P&~R)
3. □(~QvS)/□(SvR)

4. □(PvQ)   S5 Reit 1
5. □(~P&~R)   S5 Reit 2
6. □(~QvS)   S5 Reit 3
7. PvQ   M1, 4
8. ~P&~R   M1, 5
9. ~QvS   M1, 6
10. ~P   Simp 8
11. Q   DS 7, 10
12. S   DS 9, 11
13. SvR   Add 12

14. □(SvR) S5 Nec 4–13

THE TAUTOLOGY NECESSITATION RULE

According to the first principle of modality, every tautology is necessarily true. Thus, if we can prove that a sentence **P** is a tautology, we are justified in prefixing a box to **P**. The following rule incorporates this reasoning.

Tautology Necessitation Rule

If a statement **P** is proven tautological, we may infer from this the statement □**P**. (Write as justification 'TN' and the line the tautology appears on.)

Recall from Chapter 3 that we prove a statement tautological by proving the statement with a premise-free conditional or indirect proof sequence.

Here are two very simple examples of Tautology Necessitation:

(1) _____/□(P⊃P)

1. | P | AP

2. P⊃P   CP 1
3. □(P⊃P)   TN 2

(2) _____/□[P⊃(PvQ)]

1. | P   AP
2. | PvQ   Add 1

3. P⊃(PvQ)   CP 1–2
4. □[P⊃(PvQ]   TN 3

●— MODAL CONDITIONAL PROOF

Suppose that somewhere within a proof we construct a conditional proof sequence in which we assume **P** and reach, at the bottom of the sequence, **Q**. For reasons given in Chapter 3, we are then justified in asserting, by the conditional proof rule, **P⊃Q**. Suppose, however, that in such a sequence, every line after the assumption is justified by the S5 reiteration rule or follows from a previous line in the conditional proof sequence by a valid rule. That is, if we derive anything within the sequence using one or more lines outside the sequence, the lines used are brought into the sequence by S5 reiteration. In this type of case, we are justified in concluding **P→Q**, rather than just **P⊃Q**. Thus, we have the Modal Conditional Proof Rule:

Modal Conditional Proof

> At any point in a proof, you may indent, write in an assumed premise **P**, and then derive a formula **Q**, provided that every line in the indented subproof after **P** is justified by S5 reiteration or follows from previous lines in the indented subproof by a valid rule of inference. From this, you may end the indentation, draw a line around the indented steps, and assert **P→Q**. (Write as justification 'MCP' and the line numbers in the subproof.)

The validity of this rule may be verified if the reader will think through the following three cases.

— Case 1 —●

Suppose in a modal conditional proof the assumed premise **P** is necessarily false. Since a necessary falsehood implies any formula, then no matter which formula **Q** we derive at the bottom of the sequence, **P→Q** follows validly.

— Case 2 —●

Suppose in a modal conditional proof the assumed premise **P** is necessarily true. Since only modally closed formulas are reiterated into the conditional proof sequence, since modally closed formulas are necessarily true if true at all, and since only necessary truths follow from necessary truths, the line **Q** reached at the bottom of the sequence must be necessarily true if the premises cited are true. Hence, since a necessary truth implies any necessary truth, the assumed premise, being necessarily true, implies any other line in the conditional proof sequence and so **P→Q** follows.

— Case 3 —●

If the assumed premise **P** is contingent, we must derive the last line **Q** from either **P** itself or from reiterated modally closed formulas. If we derive **Q** only from reiterated modally closed formulas, then **Q** is necessarily true if the reiterated premises are true. But since a contingent formula implies any necessary truth, **P** implies **Q** and therefore **P→Q** validly follows in such a case. If we assume **P** and, using no outside premises, derive **Q**, then **P⊃Q** follows by conditional proof, no matter what **P**'s modal status is. Furthermore, it follows that **P⊃Q** is a tautology, since it was proven with a premise-free proof. Therefore, by tautology necessitation, □(**P⊃Q**) follows. Since □(**P⊃Q**) is equivalent to **P→Q**, then **P→Q** validly follows in this case as well.

Here are some applications of this new rule. Recall from Chapter 6 one of the paradoxes of implication: If **P** is a necessary truth, then any proposition implies **P**. Consider the following modal conditional proof:

1.   □P/Q→P

2. | Q   A P
3. | □P   S5 Reit. 1
4. | P   M1, 3

5. Q→P   MCP 2–4

Here is another application of MCP:

1.   □Q/P→(P&Q)

2. | P   A P
3. | □Q   S5 Reit. 1
4. | Q   M1, 3
5. | P&Q   Conj. 2, 4

6.   P→(P&Q) MCP 2–5

MODAL REPLACEMENT RULES

Recall from Chapter 3 that if a proposition-expressing sentence within an argument is removed and replaced by a sentence expressing an equivalent proposition, the validity of the argument will not be affected. That is, if the argument happened to be valid before the replacement, it will remain valid after the replacement. Let us use the following replacement rules:

E1. **P→Q** // □(**P⊃Q**)
E2. **P↔Q** // □(**P≡Q**)
E3. □P // ~◇~P
E4. ◇P // ~□~P
E5. ~□P // ◇~P
E6. ~◇P // □~P

The last four replacement rules above, E3 through E6, may be summed up in this simple algorithm:

The Diamond Exchange Rule (DE)

If you want to convert a formula's box into a diamond or a formula's diamond into a box, then: (i) change the box to a diamond or the diamond to a box and (ii) take the opposite of each side of the box (or diamond). The resulting formula will be equivalent to the original formula and may therefore replace or be replaced by the original formula anywhere in a proof.

Here are several proofs employing some of the equivalence rules:

1.	~□~P	
2.	~□P	
3.	(◇P&◇~P)⊃□H/H	
4.	◇P	DE 1
5.	◇~P	DE 2
6.	◇P&◇~P	Conj 4,5
7.	□H	MP, 3,6
8.	H	M1, 7

1.	P→Q	
2.	Q→S	
3.	~□(P⊃S)v□R/ RvZ	
4.	P→S	M5, 1,2
5.	□(P⊃S)	E1, 4
6.	□R	DS, 3,5
7.	R	M1, 6
8.	RvZ	Add, 7

• REPLACEMENT RULES FOR THE REDUCTION OF MODALITIES

Consider the following four arguments:

1. If **P** is necessarily true, then in every world it is true that **P** is necessarily true. That is, □P→□□P. Furthermore, if in every world it's true that **P** is necessarily true, then **P** is necessarily true. That is, □□P→□P. One sentence is equivalent to another just in case the two imply each other. Since □P and □□P imply each other, it follows that:

□□P is equivalent to □P

2.   If there is at least one world in which **P** is true, then there is a world in which it's true that there is a world in which **P** is true. That is, ◇**P**→◇ ◇**P**. Furthermore, if there is a world in which it's true that there is a world in which **P** is true, then there is a world in which **P** is true. That is: ◇ ◇**P**→◇**P**. Since ◇ ◇**P** and ◇**P** imply each other, it follows that:

◇ ◇P is equivalent to ◇P

3.   It is either true in all worlds that **P** is a necessary truth, or it is true in no world that **P** is a necessary truth. So, if there is even one world in which it's true that **P** is necessarily true, then **P** is necessarily true. That is, ◇□**P**→□**P**. Furthermore, if **P** is necessarily true, then there is at least one world in which it's true that **P** is necessarily true. That is, □**P**→◇□**P**. Thus, since □**P** and ◇□**P** imply each other, it follows that:

◇□P is equivalent to □P

4.   If in every world it's true that there is at least one world in which **P** is true, then there is at least one world in which **P** is true. That is, □◇**P**→◇**P**. And if there is at least one world in which **P** is true, then in every world it's true that there is at least one world in which **P** is true. That is, ◇**P**→□◇**P**. Therefore, since □◇**P** and ◇**P** imply each other:

□◇P is equivalent to ◇P

Two monadic operators are *iterated* if the first applies directly to the second. Thus, □□**P** contains iterated operators but □(□**P**&**Q**) does not. It follows from the four arguments set out above that:

> If we remove the left member of a pair of iterated monadic modal operators—no matter what that operator is—the shortened formula represents a proposition equivalent to the one it formerly represented.

Let us say that a sentence shortened in this way is a *reduced* sentence. So if a sentence **P** reduces to a sentence **Q**, then **P** and **Q** express equivalent propositions.

A sentence of ML that contains no iterated operators will be called a *fully reduced* modal sentence. Thus, the following formulas are all fully reduced:

□P   ◇Q   □(◇P&◇Q)

And these formulas are not fully reduced:

□□P   ◇□P   ◇◇□Q

Within S5, any sequence of iterated monadic modal operators in a formula, no matter how long, can be reduced to the last member on the right simply by reducing its operators two at a time until only one operator is left. The resulting reduced formula will express a proposition equivalent to the proposition expressed before the reduction. For example:

☐◇☐P reduces to ☐P
◇◇◇P reduces to ◇P
◇◇☐☐P reduces to ☐P
☐☐◇(☐☐P&◇◇Q) reduces to ◇(☐P&◇Q)

This may all be summed up in the following replacement rule:

The S5 Reduction Rule

> *Any sequence of iterated monadic modal operators in a formula may be reduced to the last member on the right, and the resulting reduced formula may replace the original formula anywhere within a proof. (Write as justification 'S5 Red' and the number of the line the rule was applied to.)*

The following proof employs the S5 Reduction Rule:

1. ☐◇◇☐P⊃◇Q
2. ~◇~P/◇Q

3. ☐P   DE 2
4. ☐P⊃◇Q   S5 Red, 1
5. ◇Q   MP 3, 4

Validity in S5   The system of modal logic consisting of (1) the language ML and (2) the S5 natural deduction rules of this chapter shall be called the system S5D (for 'S5 deduction'). A *proof in S5D* is a sequence of formulas of ML that is such that every formula is either a premise, an assumption, or follows from previous formulas by an S5D rule, and in which every line that is not a premise is justified and every assumption has been discharged. An argument is *valid in S5D* if and only if it is possible to construct a proof in S5D whose premises are the premises of the argument and whose conclusion is the conclusion of the argument. Any argument that is valid in S5D may also be said to be *valid in S5* or *S5 valid*.

●─ PROVING FORMULAS LOGICALLY TRUE

If a formula can be proven true using a premise-free proof, it is said to be *logically true*. In Chapter 3, we proved formulas logically true using premise-free conditional and indirect proofs. Within modal logic, we will use the same basic procedures to prove various modal logical truths. The following premise-free proof establishes that P→☐◇P is logically true:

1.	P AP
2.	◇P M 2, 1

3.	◇P S5 Reit 2

4.	□◇P S5 Nec 3

5. P→□◇P   MCP 1–4

The following proves that P→(PvQ) is logically true:

1.	P AP
2.	PvQ Add 1

3. P⊃(PvQ)   CP 1–2
4. □[P⊃(PvQ)] TN, 3
5. P→(PvQ)   E 1, 4

Formulas such as these are said to be logically true because they can be proven true using just the techniques of logical theory without investigating the physical world, without conducting experiments, without using any procedures outside of logical theory.

If it is possible to construct a premise-free S5D proof whose last line is the sentence **P**, then **P** is *logically true in S5D*. A sentence that is logically true in S5D is also said to be a *theorem in S5D*. Any sentence that is a theorem in S5D may also be said to be a *theorem in S5*.

Finally, here are a number of proofs for your examination. In each case, try to understand the justifications given for each line. This will help familiarize you with the proof procedures of the system S5D.

1.

1.	□(P≡Q)	
2.	□P/Q	
3.	P↔Q	E 2 1
4.	P	M1, 2
5.	Q	M6, 3, 4

2.

1.	□(P⊃Q)	
2.	□(Q⊃P)	
3.	(P↔Q) ⊃ □R/R	
4.	P → Q	E1 1
5.	Q → P	E1 2
6.	P↔Q	M7 4, 5
7.	□R	MP 3,6
8.	R	M1, 7

3.

1.	◇S⊃□Q	
2.	□~S ⊃ ◇P	
3.	□~P/Q	
4.	~◇P	DE 3
5.	~□~S	MT 2, 4
6.	◇S	DE 5
7.	□Q	MP 1, 6
8.	Q	M1, 7

4.

1.	◇(P & Q) ⊃ □~(RvS)	
2.	◇(R v S)/ □~(P & Q)	
3.	~□~(R v S)	DE 2
4.	~◇(P & Q)	MT 1, 3
5.	□~(P & Q)	DE 4

5.
1. □P ⊃ □Q
2. ◇~Q/ ◇~P
3. ~□Q   DE, 2
4. ~□P   MT 1, 3
5. ◇~P   DE, 4

6.
1. □(P ⊃ Q)
2. ~◇Q/ ~P
3. P → Q   E1, 1
4. □~Q   DE 2
5. ~Q   M1, 4
6. ~P   M4, 3, 5

7.
1. □□□Q ⊃ ◇◇S
2. ◇S ⊃ □P
3. (□Q ⊃ □P) ⊃ ◇R
4. ~◇R v □A/A
5. □Q ⊃ ◇S   S5 Red 1
6. □Q ⊃□P   HS 2, 5
7. ◇R   MP 3, 6
8. □A   DS 4, 7
9. A   M1, 8

8.
1. □◇□P
2. ◇~P v □Q
3. Q ⊃S/ S
4. □P   S5 Red 1
5. ~◇~P   DE 4
6. □Q   DS 2, 5
7. Q   M1, 6
8. S   MP 3, 7

9.
1. ◇□◇P v □Q
2. ◇~Q
3. ◇P ⊃ □P/P
4. ~□Q   DE 2
5. ◇□◇ P   DS 1, 4
6. ◇P   S5 Red 5
7. □P   MP 3, 6
8. P   M1, 7

10.
1. □◇◇□P
2. Q ⊃ ~P
3. ~□Q ⊃ □S/ S
4. □P   S5 Red 1
5. P   M1, 4
6. ~Q   MT 2, 5
7. ◇~Q   M2, 6
8. ~□Q   DE 7
9. □S   MP 3, 8
10. S   M1, 9

11.
1. □□P → ◇Q
2. ◇□P
3. ~◇Q v □R
4. R → □S/ S
5. □P → ◇Q   S5 Red 1
6. □P   S5 Red 2
7. ◇Q   M 3, 5, 6
8. □R   DS 3, 7
9. R   M1, 8
10. □S   M3, 4, 9
11. S   M1, 10

12.
1. ◇◇◇◇□P
2. P → Q
3. Q → R/ R
4. □P   S5 Red 1
5. P   M1, 4
6. P → R   M5, 2, 3
7. R   M3, 5, 6

13.
1. □◇◇P
2. □S → □~P
3. ◇~S → □G/ G
4. ◇P   S5 Red 1
5. ~□~P   DE 4
6. ~□S   M4, 2, 5
7. ◇~S   DE 6
8. □G   M3, 3, 7
9. G   M1, 8

14.

1. $\Box(P \equiv Q)$
2. $(P \leftrightarrow Q) \rightarrow (Q \leftrightarrow R)$
3. $\Box(Q \equiv R) \supset \Box\Box S$
4. $\sim\Box A \supset \sim S/\ A$
5. $P \leftrightarrow Q$ E2 1
6. $Q \leftrightarrow R$ M3, 2, 5
7. $\Box(Q \equiv R)$ E2 6
8. $\Box\Box S$ MP3, 7
9. $\Box S$ M1, 8
10. S   M1, 9
11. $\Box A$ MT 4, 10
12. A   M1, 11

15.

1. $\Box\Box\Diamond P$
2. $\Diamond P \rightarrow \sim\Box Q$
3. $\Box R \rightarrow \Box Q$
4. $\Diamond\sim R \rightarrow \Diamond\sim S$
5. $\Diamond\sim G \supset \Box S/\ G$
6. $\Diamond P$ S5 Red 1
7. $\sim\Box Q$ M3, 2, 6
8. $\sim\Box R$ M4, 3, 7
9. $\Diamond\sim R$ DE 8
10. $\Diamond\sim S$ M3, 4, 9
11. $\sim\Box S$ DE 10
12. $\sim\Diamond\sim G$ MT 5, 11
13. $\Box G$ DE 12
14. G   M1, 13

5. NATURAL DEDUCTION WITHIN S4

S4 NATURAL DEDUCTION

While the system S5 presupposes all five principles of modality, S4 presupposes only the first four. In order to specify rules of deduction for S4, we must alter our rules so that they reflect principles 1 through 4 but not principle 5. This requires the following fairly simple changes.

First, we must alter the necessitation rule so that it reflects the fourth but not the fifth principle of modality. The fourth principle specifies that necessity is itself a matter of necessity, that is, $\Box P \supset \Box\Box P$. The fifth principle states that possibility is itself also a matter of necessity, that is, $\Diamond P \supset \Box\Diamond P$. In S5, we are permitted to prefix a box to anything that follows from any modally closed formula. The fifth principle of modality is dropped when we switch to S4. In S4, a formula may be reiterated only if the main connective is a box or a fishhook. Consequently, the *S4 Necessitation Rule* takes the following form:

The S4 Necessitation Rule

At any point in a proof, you may indent and construct a subproof in which every line is either justified by the S4 reiteration rule or follows from previous lines of the subproof by a valid rule of inference. You may then end the indentation, draw a line around the indented lines, write down any line derived within such a subproof, and prefix a box to that line. (Write as justification 'S4 Nec' and the line numbers of the subproof.)

S4 Reiteration Rule

> *A formula may be reiterated if and only if the formula's main connective is either a box or a fishhook and provided that the formula does not lie within the scope of a discharged assumption or a terminated necessitation subproof. (Write as justification 'S4 Reit'.)*

Three further changes must be made as we shift from S5 to S4. First, rule M8 must be dropped, for that rule presupposes the fifth principle of modality. Second, when we do modal conditional proofs in S4, we must follow the S4 Reiteration Rule instead of the S5 Reiteration Rule when we reiterate. Third, S4 has its own replacement rules for the reduction of modalities. (These rules are specified below.)

The Tautology Necessitation Rule carries over into S4, since the TN rule reflects the first modal principle, a principle accepted within both S4 and S5. And the replacement rules E1 through E6 also carry over into S4.

The system of modal logic consisting of (1) the language ML and (2) the S4 natural deduction rules of this chapter shall be called the system S4D (for 'S4 deduction'). A proof in S4D is a sequence of formulas of ML which is such that every formula is either a premise, an assumption, or follows from previous formulas by an S4D rule of inference, and in which every line that is not a premise is justified and every assumption has been discharged. An argument is *valid in S4D* if it is possible to construct a proof in S4D whose premises are the premises of the argument and whose conclusion is the conclusion of the argument. Any argument that is valid in S4D may be said to be *valid in S4* or *S4 valid*.

If it is possible to construct a premise-free S4D proof whose last line is the sentence **P**, then **P** is *logically true in S4D*. Sentences that are logically true in S4 are also called *theorems of S4*.

The following is a proof in S4D.

$$
\begin{array}{lll}
1. & \quad \Box P & AP \\
2. & \quad\quad \Box P & S4 \text{ Reit } 1 \\
3. & \quad \Box\Box P & S4\ \text{Nec } 2 \\
4. & \Box P \supset \Box\Box P & CP\ 1\text{–}3 \\
\end{array}
$$

This establishes that $\Box P \supset \Box\Box P$ is a theorem of S4.

If you will reflect upon the relationship between the S4 and S5 rules, you will see that the following holds true: Any argument that can be proven valid in S4 can also be proven valid in S5 and any theorem of S4 is a theorem of S5. However, it's not the case that any argument that can be proven valid in S5 can be proven valid in S4, for an infinity of arguments are provable in S5 that are not provable in S4. And not every theorem of S5 is a theorem of S4. Thus, we say that S5 "contains" S4 or that S5 is "stronger" than S4.

— REDUCTION OF MODALITIES IN S4

Because it rejects the fifth principle of modality, the system S4 contains only two re-duction rules:

$$R1 \quad \Box\Box P//\Box P$$
$$R2 \quad \Diamond\Diamond P//\Diamond P$$

Let us call the conjunction of these two rules 'S4 Red'. So, in S4, any sequence of two iterated modal operators may be reduced to a sequence containing just one of the operators, provided that the two iterated operators are the same, that is, both are boxes or both are diamonds.

Here is a simple application of S4 reduction:

1. $\Box\Box P \supset Q$
2. $\Box P / \Diamond Q$
3. $\Box P \supset Q$ S4 Red, 1
4. Q MP 2, 3
5. $\Diamond Q$ M2, 4

6. PHILOSOPHICAL DIFFERENCES BETWEEN VARIOUS SYSTEMS OF MODAL LOGIC

As noted earlier, the five principles of modality presented at the start of this chapter are not accepted by all philosophers. All five principles constitute the basis of the system S5, and one who accepts all five is logically committed to the S5 system of modal logic. However, some philosophers accept only the first four principles and reject the fifth. Since the first four principles form the basis of the system S4, these philosophers are logically committed only to the system S4. If only the first three modal principles are accepted, one is committed to accept only the system T.

The main area of disagreement surrounds the fourth and fifth principles of modality. Let us begin with an examination of principle 4.

Principle 4, you will recall, may be expressed as

$$4. \quad \Box P \rightarrow \Box\Box P$$

This says that if some proposition is necessarily true, then it is necessary that it is necessarily true. That is, if **P** is true in all worlds, then in every world it's true that **P** is true in all worlds. What is necessarily true is not a contingent or accidental mat-ter that varies from world to world; rather, what is necessary is itself a matter of ne-cessity that does not vary from world to world.

Why would someone reject this principle? Consider the following line of reasoning:

Logical necessity is not an objective feature of an independent reality, that is, of a reality existing independently of our minds. Rather, logical necessity is merely a product of our human thinking that we project onto reality. However our human thinking is itself determined by contingent features of the world, features such as (a) the course of evolution; and (b) the societal conditions that shape our thoughts. Therefore, we do not necessarily think as we do, and even though something is—according to the way we think—necessarily true, it nevertheless might not have been had we developed a different way of thinking about modality. So, what is necessarily true is not also necessarily necessarily true. The fourth modal principle is therefore false.

Call the conclusion of this argument the *Relativity of Necessity* thesis, or 'RN' for short. If RN is true, then the fourth principle of modality must be rejected. But is RN true? Not if the following opposing view is true:

The five principles of modality are *ratiocinative* truths. That is, through an inner process of pure reasoning, we can know with certainty that these five principles can't possibly be false, much as we can know, through reasoning alone, that certain truths of mathematics can't possibly be false.

This opposing view may be called the *Absoluteness of Necessity* thesis, or 'AN' for short. There are arguments for and against each of these opposing views. The student wishing to explore the issue further may consult the Further Reading section at the back of the book.

Consider next the fifth principle of modality, namely:

$$\Diamond P \rightarrow \Box \Diamond P$$

Why would anyone reject this principle? It certainly seems to be a necessary truth. If some proposition **P** is true in one or more worlds, then it seems that in every world it's true that in one or more worlds **P** is true. Many philosophers claim this to be a ratiocinative truth.

In order to understand why someone would reject this principle, we must first examine the concept of physical possibility. Consider two worlds, named 'W1' and 'W2'. Suppose that W1 is structured by a set of physical laws of nature L1. Suppose W2 is structured by a totally different set of physical laws of nature L2 in such a way that many events in W2 are physically impossible in W1 and many events in W1 are physically impossible in W2. Surely there are pairs of worlds such as this.

Now, suppose someone were to reject the concept of logical possibility on the grounds that the concept itself doesn't make sense. Suppose further that this person accepts the concept of physical possibility, perhaps on the grounds that this concept is an integral part of the scientific view of the world. Then as far as physical possibility goes, it is not the case that if something is possible in one world then it is possible in all worlds. It would follow that principle 5 is false, for principle 5 holds that if something is possible in one world, it is possible in all worlds.

Call this view the *Relativity of Possibility* thesis, or 'RP' for short. One who accepts this view should reject principle 5 and, to be consistent, should reject system S5 as well. So, according to RP, it's not the case that what is possible in one world is possible in all worlds.

In contrast, according to the fifth principle of modality, what is possible in one world is automatically also possible in every world. Call this view the *Absoluteness of Possibility* thesis, or 'AP' for short.

Which view is correct? Unfortunately, an examination of the arguments for and against each of these views would take us far beyond the scope of an introductory symbolic logic text. The student who wishes to pursue the matter further may turn to the Further Reading section at the back of the book.

EXERCISES

Exercise 7.1

Use either truth-trees or natural deduction to establish the validity of the following arguments in S5.

1. 1. $\Diamond$(H & S) $\supset$ G
    2. ~R
    3. R v (H & S)/G

2. 1. R $\rightarrow$ M
    2. H & R/M v G

3. 1. S $\leftrightarrow$ Q
    2. H $\supset$ Q
    3. F & H/S

4. 1. ~H
    2. S $\rightarrow$ H
    3. $\Diamond$~S $\supset$ R/R v $\Diamond$M

5. 1. H $\rightarrow$ G
    2. G $\rightarrow$ (M & N)
    3. [H $\rightarrow$ (M & N)] $\supset$ S/S

6. 1. ~R
    2. R v (A $\supset$ B)
    3. (A $\supset$ B) $\supset$ (P $\rightarrow$ Q)
    4. Q $\rightarrow$ P/P $\leftrightarrow$ Q

7. 1. A $\rightarrow$ $\Diamond$S
    2. A/$\Box$$\Diamond$S

8. 1. $\Box$R
    2. R $\supset$ G
    3. ~G v M/$\Diamond$M

9. 1. $\Box$A $\rightarrow$ $\Diamond$B
    2. H $\supset$ ~$\Diamond$B
    3. ~H $\supset$ G
    4. ~G/~$\Box$A

10. 1. F $\rightarrow$ ($\Box$A & $\Box$B)
    2. G $\supset$ F
    3. H & G/$\Box$($\Box$A & $\Box$B)

11. 1. G $\rightarrow$ $\Diamond$M
    2. H & (G & R)
    3. $\Box$$\Diamond$M $\supset$ A/$\Diamond$A

12. 1. M v (H & W)
    2. M $\rightarrow$ ~G
    3. $\Box$(G & S)

4. $\diamond W \supset \Box B / A \supset B$

13. 1. $\Box (A \supset B)$

    2. $H \lor A$

    3. $\sim(F \lor H)/B$

14. 1. $\Box (A \supset B)$

    2. $B \supset G$

    3. $\sim G \lor S$

    4. $\sim(S \lor R)/\diamond\sim A$

15. 1. $\Box (A \lor B)$

    2. $B \supset H$

    3. $\sim R \& \sim H/A$

16. 1. $A \leftrightarrow G$

    2. $\sim G$

    3. $\sim A \supset \Box E/E$

17. 1. $\Box (A \lor B)$

    2. $\Box \sim B$

    3. $\Box A \supset \Box E/E$

18. 1. $\Box (H \supset S)$

    2. $\Box (S \supset R)$

    3. $\Box (H \supset R) \supset G/\diamond G$

19. 1. $\Box (\sim R \supset S)$

    2. $\Box (\sim S \lor G)$

    3. $\Box [(R \lor G) \supset H]$

    4. $\sim \Box H \lor S/S$

20. 1. $\Box (A \supset B)$

    2. $\Box (\sim B \lor \sim A)$

    3. $G \supset \sim \Box \sim A/\sim G$

21. 1. $\Box A \lor \Box B/\Box (A \lor B)$

22. 1. $H \lor S/\diamond H \lor \diamond S$

23. 1. $\Box (P \& Q)$

    2. $\Box (Q \supset S)/H \to S$

24. 1. $\Box (H \supset S)$

    2. $\Box (S \supset P)$

    3. $\Box [\diamond (H \supset P) \supset G]/M \to G$

25. 1. $\Box (Z \equiv M)$

    2. $\Box Z$

    3. $\Box [M \supset (S \& A)]/Q \to (A \lor R)$

26. 1. $P \to Q$

    2. $\Box P / \Box Q$

27. 1. $\Box A / \sim A \supset \sim B$

28. 1. $\Box \sim A / \sim \Box A$

29. 1. $\Box A / \diamond \Box A$

30. 1. $\sim \diamond A / \diamond \sim A$

31. 1. $\Box A / \diamond \diamond \Box A$

32. 1. $\sim \diamond (A \& \sim B) / \Box (A \supset B)$

33. 1. $\Box (A \supset B)$

    2. $\Box (B \supset \sim C) / \sim \diamond (A \& C)$

34. 1. $A \supset \Box B$

    2. $\sim B / \diamond \sim A$

35. 1. $\Box (A \supset B)$

    2. $\diamond \sim B / \diamond \sim A$

36. 1. $\diamond (A \& B) / \diamond A$

37. 1. $\Box (A \supset B)$

    2. $\diamond A / \diamond B$

38. 1. $\Box A$

    2. $\sim \Box B / \sim \Box (A \supset B)$

39. 1. $\Box (A \supset B) / \Box A \supset \Box B$

40. 1. $\diamond A \lor \diamond B / \diamond (A \lor B)$

41. 1. $\diamond A \to B / A \to B$

42. 1. $A \to \Box B / A \to B$

43. 1. $\diamond A \to \Box B / A \to B$

44. 1. $A \to B / \Box A \to B$

45. 1. A → B/ A → ◇B   3. Q → R / □R

46. 1. A → B/ □A → ◇B   56. 1. □□□A/ A

47. 1. □A/ □□A   57. 1. ◇□A/ □A

48. 1. □□A/ □A   58. 1. □□◇A/ ◇A

49. 1. □□□A/ □A   59. 1. ◇◇A/ ◇A

50. 1. □(A & B) / □A & □B   60. 1. ◇A/ □□◇A

51. 1. □A & □B / □(A & B)   61. 1. ◇A/ ◇◇A

52. 1. ◇(A & B) / ◇A & ◇B   62. 1. ◇A/ ◇◇◇A

53. 1. ◇(A v B) / ◇A v ◇B   63. 1. □□A/ ◇□A

54. 1. □[(P & Q) & R] /□P & (□Q & □R)   64. 1. ◇□A/ □□A

55. 1. P → Q   65. 1. □◇A/ ◇◇A

     2. □P   66. 1. ◇◇A/ □◇A

Exercise 7.2

Using the S4 system, produce either truth-trees or natural deduction proofs for as many of the arguments in Exercise 7.1 as you can. Not all can be proven valid in S4.

Exercise 7.3

Use either truth-trees or natural deduction to prove that each of the following is a theorem of S5.

1. _____/ □[~P⊃(P⊃Q)]

2. _____/ □[(PvQ)⊃(~Q⊃P)]

3. _____/ □[Q⊃(P⊃Q)]

4. _____/ □{[(P⊃Q)&P]⊃Q}

5. _____/ □[(P⊃Q)⊃(~Q⊃~P)]

6. _____/ □{[(P⊃Q)&~Q]⊃~P}

7. _____/ □{[(PvQ)&~P]⊃Q}

8. _____/ □{[(P⊃Q)&(Q⊃R)]⊃(P⊃R)}

9. _____/ □{{[(P⊃Q)&(R⊃S)]&(PvR)}⊃(QvS)}

10. _____/ □{[(PvQ)&(~QvS)]⊃(PvS)}

11. _____/ □{[(P&Q)&(P⊃R)]⊃R}

12. _____/ □[P⊃(P⊃P)]

13. _____/ □[(P⊃Q)⊃~(P&~Q)]

14. _____/ □[(PvQ)⊃(~Q⊃P)]

15. _____/ P → P

16. _____/ □□P → □P

17. _____/ (P → Q) → (□P ⊃ □Q)

18. _____/ □P → □(P v Q)

19. _____/ P → (P v Q)

20. _____/ ~□P → ◇~P

21. _____/ □[(P & Q) ⊃ (P v Q)]

22. _____/ ◇~P → ~□P

23. _____/ □~P → ~◇ P

24. _____/ ◇P v ◇~P

25. _____/ ~◇P → □~P

26. _____/ □(P ≡ Q) → (P ↔ Q)

27. _____/ (P → Q) → □(P ⊃ Q)

28. _____/ (P → Q) → □(P → Q)

29. _____/ (P → Q) → ~◇(P & ~Q)

30. _____/ (P → Q) → (~Q → ~P)

31. _____/ (P & ~Q) → ~(P → Q)

32. _____/ ~◇(P & ~P)

33. _____/ (P ↔ Q) → □(P ≡ Q)

34. _____/ ◇(P ↔ Q) → (P ↔ Q)

35. _____/ □(P ⊃ Q) → (P → Q)

36. _____/ ~(P → Q) → ◇(P & ~Q)

37. _____/ ◇(P → Q) → (P → Q)

Exercise 7.4

Which of the formulas in Exercise 7.3 are theorems of S4? Using the S4 system, produce either truth-trees or natural deduction proofs for as many of the formulas in Exercise 7.3 as you can. Not all are theorems of S4.

Exercise 7.5

Symbolize the following arguments and then use S5 trees or S5 natural deduction to prove each valid.

1. It's not possible that Spock and Kirk both beam down to the planet. If Spock doesn't beam down to the planet, then the mission will not be accomplished. If Kirk doesn't beam down to the planet, the mission will not be accomplished. Therefore, it's not possible that the mission be accomplished.

2. If Art belongs to the Peace and Freedom Party, then Art is a dangerous revolutionary and a threat to the entire political system. It's not possible that Art is a dangerous revolutionary. Therefore, it's not possible that Art belongs to the Peace and Freedom Party.

3. Bob either belongs to the SDS or to the PLP. If Bob is a staunch Republican, then it is impossible that Bob belongs to the PLP. Bob does not belong to the SDS. Bob is not a staunch Republican.

4. It's not possible that Glenn and Larry occupy the same place at the same time. If Glenn and Larry both deliver the closing speech, then Glenn and Larry will occupy the same place at the same time. If Glenn and Larry do not both deliver the closing speech, then Melissa and Val will be disappointed and so will Alan and Charlie, not to mention Ron, Mel, and Patty. Therefore, Patty will be disappointed.

5. If Bob and Val are late, then Randy and Bill will wait. If Janet can't wait, then it's not possible that Randy will wait. Bob and Val will be late. So, Janet can wait.

QUANTIFICATIONAL LOGIC I: THE LANGUAGE QL

1. SINGULAR and GENERAL SENTENCES

I n Chapters 2, 3, and 4 we developed a formal language, TL, and techniques to prove arguments truth-functionally valid. In Chapters 6 and 7, we developed a formal language, ML, and techniques to prove modal arguments valid. It is now time to introduce techniques to apply to a third type of argument.

Consider the following piece of reasoning:

— **Argument 1** —•

1. All elephants are large.
2. Some elephants are old.
3. So, some large things are old.

If you reflect upon this argument for a moment, you will see that it is valid. There is simply no possibility of true premises with a false conclusion. However, suppose we wish to prove this argument valid, using either the truth-functional or modal proof techniques learned so far? How should we proceed? The argument contains no truth-functional operators and no modal operators. So we can't symbolize it using the ampersand, wedge, horseshoe, box, diamond, or any of the other logical operators we have studied. Therefore, the only way we can abbreviate the argument, in either TL or ML, is by assigning a single sentence constant to each premise and to the conclusion:

$$\frac{\begin{array}{c} P \\ Q \end{array}}{R}$$

Although argument 1 itself is clearly valid, this abbreviation of the argument is an instance of a form that is invalid within TL and within ML. So, if we translate argument 1 into either TL or ML in order to develop a proof, the argument instantiates an invalid form. Consequently, using only the techniques of truth-functional and modal logic, argument 1 cannot be proven valid, even though we can see, intuitively, that it is indeed valid.

Why does argument 1 instantiate an invalid form when we try to symbolize it in TL or ML? The answer is essentially this. The logical features of the argument that make it valid are not represented within either TL or ML. Therefore, the techniques of truth-functional and modal logic do not apply to the very aspects of the argument we want them to apply to, namely, the aspects accounting for the argument's validity. This is why the procedures of truth-functional and modal logic cannot be used to prove argument 1 valid.

Argument 1 is an example of a type of argument we haven't studied yet—the quantificational argument. This is an argument built out of quantifiers and the units these attach to. In order to explore the nature of this type of argument, we must enter a third branch of logic, *quantificational logic*.

First, we will identify the various parts of a quantificational argument. As we go, we will also specify logical symbols that abbreviate those parts. Then we will put all this together in the form of a third formal logical language, which will be named 'QL' for 'quantificational logic'. In the next chapter, a semantics for QL along with tree rules and deduction rules for quantificational arguments will be presented, thus rounding out our study of a third branch of logic.

●— SINGULAR SENTENCES

Terminology The first step in the development of quantificational logic is to characterize *singular sentences*. First, a *singular term* is an expression that refers or purports to refer to one specific thing. Two types of singular terms will concern us: (i) proper names, and (ii) definite descriptions. Proper names such as 'Mario Savio', 'John F. Kennedy' and 'the Space Needle' are used within a given context to refer by name to one single thing. Definite descriptions such as 'The third husband of Zsa Zsa Gabor', 'The first person to walk on the moon', or 'The teacher of Plato' are used within a given context to refer by unique description to one single thing. Essentially, a singular term is used to single out a specific thing so that we may then say something about it.[1]

Terms such as 'curved', 'tall', 'astronaut', and 'blue' are *general terms*. These terms are not typically used to refer, either by name or definite description, to one single thing. Rather, terms such as these represent properties (qualities or characteristics)

---

[1] Some singular terms do in fact denote or refer to one specific thing; others don't. Singular terms that actually refer to one specific thing are called *referring* singular terms; singular terms that do not actually refer to a specific thing are *nonreferring* singular terms. Examples of referring terms are 'the Sears Tower', 'the President of the United States', and 'the Moon'. Examples of nonreferring terms are 'Frosty the Snowman', 'the tallest unicorn', 'the swamp thing', and 'the Grinch who stole Christmas'.

that a number of things might have. The properties represented by the four general terms above are the properties of being curved, of being tall, of being an astronaut, and of being blue. For example, the flight path of a Frisbee has the property of being curved; Wilt Chamberlain has the property of being tall; Neil Armstrong has the property of being an astronaut; and the sky over Seattle (occasionally) has the property of being blue. If a number of things have the same property, then we may say they "share" that property. Thus, many fire engines, cars, coats, apples, and sunsets share the property of being red. That is, they are all red. Many persons share the property of being an adult. That is, a large number of persons are adults.

Terms such as 'curved', 'tall', 'astronaut', and 'blue' are called 'general terms' because instead of singling out a particular thing for discussion (as singular terms do), these terms may be used to characterize a general category of things. For example, the general term 'astronaut' characterizes the general category of individuals having the property of being an astronaut. This general category includes all those who are astronauts. The general term 'blue' may be used to characterize the general category of things having the property of being blue. This category includes all those things that are blue.

In sum, a singular term is an expression that refers or purports to refer to one specific thing while a general term represents a property that a number of things might possess. A general term, as we have seen, also characterizes a general category of things, namely, that general group of things having the property represented by the general term.

With this terminology in hand, we are now ready to explain just what a singular sentence is. The simplest type of singular sentence we will consider contains just two parts:

a. a *singular subject expression* in which one specific thing is singled out for discussion by a singular term;

b. a predicate expression that contains a general term and that attributes a property to the specific thing designated by the subject expression.

Here are several examples of singular sentences:

1. John F. Kennedy commanded a PT boat during World War II.
2. Mario Savio was a leader of the Berkeley Free Speech Movement.
3. The Archbishop of Canterbury wrote several important works of philosophy.
4. The galaxy M31 can be seen with the unaided eye.
5. The tallest building in Chicago was built in 1972.

The subject expressions in these five sentences are, in order:

1. John F. Kennedy
2. Mario Savio
3. The Archbishop of Canterbury

4. The galaxy M31

5. The tallest building in Chicago

Note that the subject clauses of 1, 2, and 4 are proper names, and the two other subject clauses are definite descriptions.

The predicate expressions in the five sentences are, in order:

1. _____ commanded a PT boat during World War II.
2. _____ was a leader of the Berkeley Free Speech Movement.
3. _____ wrote several important works of philosophy.
4. _____ can be seen with the unaided eye.
5. _____ was built in 1972.

The blank placed at the beginning of each predicate expression represents the place where the singular term of the subject expression goes.

If you study again the five singular sentences, you will notice that in each of those sentences, the subject expression singles out one thing for discussion and the predicate expression says something about the thing that has been singled out. Each of the predicate expressions says something about the subject by attributing to it a property—the property represented by the predicate's general term. Thus, sentence (1) says something about the subject, John F. Kennedy, by attributing or ascribing to him the property of having commanded a PT boat during World War II; sentence (4) says something about the subject, M31, by attributing to it the property of being visible to the unaided eye; sentence (5) says something about its subject, the tallest building in Chicago, by attributing to it the property of having been built in 1972. On this way of understanding a singular sentence, then, a sentence such as 'George Harrison is a musician' attributes a property (in this case the property of being a musician) to the individual designated by the subject clause (in this case the individual named George Harrison).

Thus, singular subject expressions purport to refer to or designate things for discussion and predicate expressions specify properties things may have. We use singular subject expressions to single things out for discussion, and we use predicate expressions to say something about those things, that is, to attribute properties to those things. (Some predicate expressions represent relations between things; this type of predicate will be examined in a later section of this chapter.)

The next step in the process of introducing quantificational logic is to specify the symbolic elements that will abbreviate singular sentences. Let us begin with the predicate expressions, which are repeated here:

1. _____ commanded a PT boat during World War II.
2. _____ was a leader of the Berkeley Free Speech Movement.
3. _____ wrote several important works of philosophy.
4. _____ can be seen with the unaided eye.
5. _____ was built in 1972.

These expressions are called *open sentences* because they have an open space in them where a subject expression may be placed. Since they lack a subject expression, open sentences are incomplete sentences. The absence of a subject expression also means that open sentences do not make assertions about anything and so are neither true nor false. Open sentences, in other words, do not have truth-values. However, once an appropriate subject expression is inserted into the blank in an open sentence, the result is a complete proposition-expressing sentence that has one of the two truth-values.

We will use the letters 'x', 'y', and 'z', called *individual variables*, to represent the open spaces—the blanks—in these predicate expressions. Within predicate expressions, variables will serve as *placeholders* for singular subject expressions. The five predicate expressions may now be abbreviated:

1. x commanded a PT boat during World War II.
2. x was a leader in the Berkeley Free Speech Movement.
3. y wrote several important works of philosophy.
4. y can be seen with the unaided eye.
5. z was built in 1972.

Notice that in each open sentence, the variable simply marks the place where a singular subject expression may be added to produce a complete sentence. In effect, the variable serves as a placeholder for whatever the subject expression might be. In the case of each of these open sentences, it makes no difference whether we use as a variable 'x', or 'y', or 'z'—the choice of variable is completely arbitrary.

So far, we have used variables to represent the blanks in the predicate expressions. Next, we will abbreviate the remainder of those expressions using capital letters A, B . . . Z. These letters, when used to abbreviate predicates, will be called *predicate constants*. Thus, if we abbreviate 'commanded a PT boat during World War II' with 'C', 'was a leader in the Berkeley Free Speech Movement' with 'L', 'wrote several important works of philosophy' with 'W', 'can be seen with the unaided eye' with 'S', and 'was built in 1972' with 'B', then the predicate expressions may be completely abbreviated as:

1. Cx
2. Lx
3. Wx
4. Sx
5. Bx

It is the custom in quantificational logic to place the predicate constant to the left of the variable. Notice that in each case we chose a predicate constant that reminds us of the predicate it abbreviates.

We shall use the lower case letters 'a', 'b', 'c' . . . 't', called *individual constants*, to abbreviate singular terms. If we abbreviate 'John F. Kennedy' with 'j', 'Mario Savio' with 'm', 'The Archbishop of Canterbury' with 'a', 'The galaxy M31' with 'g', and 'the tallest building in Chicago' with 't', then the five subject terms become:

1. j
2. m
3. a
4. g
5. t

We now have all the elements necessary to symbolize singular sentences. If we go back to the open sentences—the predicate expressions—and replace the variables in those open sentences with the individual constants abbreviating the subject expressions, we get the complete abbreviations of our five singular sentences:

1. Cj
2. Lm
3. Wa
4. Sg
5. Bt

Again, notice that the predicate constants are placed to the left.

The formula 'Cj' may be understood as indicating that the individual represented by 'j' has the property represented by 'C'. That is, John F. Kennedy has the property of having commanded a PT boat during World War II, which is to say that John F. Kennedy commanded a PT boat during World War II. Incidentally, this property, represented in English by 'commanded a PT boat during World War II', and represented in our symbols by either 'C' or by the open sentence 'Cx', is shared by a number of individuals, namely, by those who commanded the little plywood patrol torpedo boats later featured in the 1960s television series *McHale's Navy*.

• EXERCISE 8.1 •

Symbolize the following:

1. Beth is happy.
2. George is quitting the team.
3. The oldest building in town will be demolished.
4. New York is a city that never sleeps.
5. The first person to step on the moon was an American astronaut.
6. Pat is a happy camper.

●— TRUTH-FUNCTIONAL COMPOUNDS OF SINGULAR SENTENCES

Truth-functional combinations of singular sentences—singular sentences joined by truth-functional operators—will also be counted as singular sentences. The abbreviation of such sentences should be fairly obvious. Here are some examples:

Ann is happy or Bob is sad.	Ha v Sb
If the Space Needle is open to the public, then Betty will visit it.	Os ⊃ Vb
Doug is impatient and his boss is waiting.	Id & Wb
It's not the case that Sue is walking.	~Ws
It's not true that both Chris and Pat are artists.	~(Ac & Ap)

●— SUBSTITUTION

When a variable in an open sentence is replaced by a singular term, the result is called a *substitution instance* of the open sentence. For example:

OPEN SENTENCE	SOME SUBSTITUTION INSTANCES:
1. Cx	Cj, Ck, Cl, Cr
2. Lx	La, Lb, Lc, Ld, Le
3. Wy	Wa, Wb, Wc, Wd, We
4. Sx	Sa, Sb, Sc, Sd, Se
5. Bx	Ba, Bb, Bc, Bd, Be

A specific individual item, be it a person, tree, rock, galaxy, molecule, or whatever else, *satisfies* an open sentence just in case replacement of the variable in the open sentence with a singular term or individual constant designating that item results in a substitution instance that expresses a true proposition. For example, if we let 'Wx' abbreviate 'x was President of the United States', then John F. Kennedy satisfies this open sentence, since the substitution instance Wj, where j abbreviates 'John F. Kennedy', expresses a truth. However, Al Capone does not satisfy this open sentence, since he was never President. Here are some further examples:

> Birmingham satisfies 'x is a city'.
> Africa does not satisfy 'x is a city'.
> The Earth satisfies 'x is spherical'.
> The Space Needle does not satisfy 'x is spherical'.
> Janis Ian satisfies 'x is a musician'.
> The Moon does not satisfy 'x is a musician'.

Notice that when we say an item satisfies an open sentence, we are saying, in effect, that the item has the property represented by the open sentence. In the above examples, for instance, we are merely noting that Birmingham has the property of being a city but Africa does not; the earth has the property of being spherical but the Space Needle does not; and Janis Ian has the property of being a musician but the Moon does not. Of course, we could attribute this same list of properties using the more ordinary sentences 'Birmingham is a city', 'Africa is not a city', 'The Earth is spherical', 'the Space Needle is not spherical', and so on.

If a specific item satisfies an open sentence, let us say the open sentence *applies* to the item. The set of things that an open sentence applies to is called the *extension* of the open sentence. Thus, the extension of 'x is red' includes apples, fire trucks, sunsets, Christmas presents, and anything else having the property of redness.

● GENERAL SENTENCES

Recall that a singular sentence attributes a property to one specifically identified individual item or thing. In contrast, a *general* sentence makes a claim not about a specific individual but about some or all members of a group of individuals. Two types of general sentences will concern us here. *Universal* general sentences make a claim about all of the members of a group. *Existential* general sentences make a claim about some members of a group, where 'some' is understood as meaning 'at least one' or (equivalently) 'one or more'. The subject phrase of a universal general sentence typically contains the quantity word 'all' or one of its cognate terms, such as 'every'. The subject phrase of an existential general sentence typically contains the quantity word 'some' or one of its cognate terms such as 'at least one'.

● UNIVERSAL QUANTIFICATION

Materialism is the view that absolutely everything is material. Something is material just in case it is a particle of matter or is composed of particles of matter. Particles of matter are the particles studied in physics, particles such as photons, protons, neutrons, electrons, quarks, and so on. Suppose we wish to symbolize the thesis of materialism, that is, the view that:

Everything is made of matter

One suggestion that first comes to mind is this. Let 'Mx' abbreviate 'x is made of matter', let 'e' abbreviate 'everything', and then put the two together to get:

Me

However, this won't do. The letter 'e' is an individual constant. Recall that individual constants may be used only to abbreviate singular terms, terms which purport to refer by name or definite description to one specifically identified item. Since 'everything' is most definitely not a singular term, we can't use 'e' to abbreviate it. Of course, we could use 'e' to abbreviate a singular term such as 'Ernest's bowling ball'. In this case, 'Me' would abbreviate 'Ernest's bowling ball is made of matter'. So, since 'e' can't be used for the term 'everything', we are back to where we started.

How shall we abbreviate the materialist thesis? We could try:

Mx

However, this is just the open sentence 'x is made of matter'. The materialist thesis means more than this. 'Everything is made of matter' says that all things satisfy the open sentence 'Mx', that is, all things have the property of being made of matter.

In order to symbolize the thesis of materialism, we will proceed by a series of paraphrases. First, let us paraphrase the materialist thesis

Everything is made of matter

into

Every thing is such that it is made of matter.

Next, let us paraphrase this by replacing the word 'thing' with the variable 'x':

Every x is such that it is made of matter

Since the pronoun 'it' relates back to the variable 'x' in the subject clause, let us also replace 'it' with 'x':

Every x is such that x is made of matter

Next, the predicate expression 'x is made of matter' may be abbreviated 'Mx' and enclosed within parentheses. (We have already seen the rationale for this abbreviation.) Finally, the quantifier word 'every' will be abbreviated using the symbol $(\forall x)$. This symbol is called a *universal quantifier*.

Putting this all together, the materialist's thesis 'everything is made of matter' may be abbreviated:

$$(\forall x)(Mx)$$

which may be read as:

For all x, x is M.

This formula may be understood in a number of equivalent ways:

1. Every x is such that it is material.
2. For every x, it is the case that x is material.
3. Every x satisfies 'x is material'.
4. Every x satisfies 'Mx'.
5. For all x: x is material.

6. For all x: Mx.
7. All x's are material.
8. Everything is material.

We will also use, as individual variables, 'y', and 'z'. So 'Everything is material' might just as well be symbolized as

$$(\forall y) \, (My)$$

or

$$(\forall z) \, (Mz)$$

The symbols $(\forall x)$, $(\forall y)$, and $(\forall z)$ are all universal quantifiers.

Universe of Discourse   In logic, a variable is said to *range* over a *domain*. The domain of the variable is the set of things the variable can take as values. The values of a variable are just the things represented by the singular terms that can replace the variable. The domain of a variable is also called the *universe of discourse* for the sentence containing the variable. Thus, in the case of $(\forall x)(Mx)$, the domain of the variable—the sentence's universe of discourse—is everything in the entire universe. When we specify the universe of discourse for a sentence, we specify what the sentence is making its claim about. When the domain of a variable is everything in the universe, the domain is said to be "unrestricted" or "universal." Throughout the rest of our discussion, unless it is specified otherwise, assume the universe of discourse is unrestricted. In other words, if we do not explicitly restrict the universe of discourse, suppose that the variables range over everything in the entire universe.

Back to the Quantifier   Suppose next that we need to symbolize a sentence such as

All dogs have fleas

One suggestion that comes to mind at first is this. Let 'Fx' abbreviate 'x has fleas' and let 'd' abbreviate 'all dogs':

Fd

However, this won't work. The individual constant 'd' may only be used to abbreviate a singular term—either a proper name or a definite description. Remember that a singular term singles out one specific thing for discussion. Since the expression 'all dogs' is neither the proper name of one specific thing nor a definite description of one specific thing either, we cannot abbreviate 'all dogs' using the individual constant 'd'. Of course, if a specific dog named 'Dave' had fleas, we could abbreviate 'Dave has fleas' with 'Fd'; but we can't abbreviate 'all dogs have fleas' this way.

We shall again work through a series of paraphrases to get to the proper symbolization. We may first paraphrase

> All dogs have fleas

into

> Every dog is such that it has fleas.

Next, this is equivalent to:

> Every thing such that it is a dog is also such that it has fleas.

This in turn may be rewritten as

> Every x such that x is a dog is such that x has fleas.

That is,

> Every x such that Dx is such that Fx

where 'Dx' abbreviates the open sentence 'x is a dog' and 'Fx' abbreviates 'x has fleas'.

We are not finished symbolizing the sentence under consideration. However, before we look at the next paraphrase, it will be helpful if we perform a brief thought experiment. Afterward, the next paraphrase will make more sense. Suppose you are a contestant on one of those afternoon television game shows. Behind the curtain is a hidden object selected completely at random. You must answer the following question:

> Does it have fleas?

If you win, you will receive a lifetime subscription to a symbolic logic magazine. (Every month you will get new logic problems to solve!) Let us assume that it is true that all dogs have fleas and suppose that you are allowed to ask one question. Wouldn't you want to know whether the object is a dog? This could be valuable information, for no matter what object is behind the curtain, if the object behind the curtain is a dog, then it has fleas. After reflecting on this, consider the next step in this series of paraphrases. We earlier paraphrased

> Every thing that is a dog is such that it has fleas

as

> Every x such that Dx is such that Fx

Now, we will say instead

> For any and every thing you might pick, if it is a dog, then it has fleas

That is,

> For every x: if Dx then Fx

In other words, for anything you might pick, if it happens to be a dog, then it has fleas. That is one way to indicate that all dogs have fleas.

Finally, this may be abbreviated:

$$(\forall x)(Dx \supset Fx)$$

This expression may be understood in a number of equivalent ways:

1. For every x, if x is a dog, then x has fleas.
2. For any x, if x is a dog, then x has fleas.
3. For every x, if x satisfies 'x is a dog' then x satisfies 'x has fleas'.
4. For every x, if x satisfies Dx then x satisfies Fx.

●— EXISTENTIAL QUANTIFICATION

Materialism is the thesis that all things are material. The denial of materialism is the claim that not all things are material. If not all things are material, then there exists at least one thing that is not material, that is, there exists at least one nonmaterial thing. So, the denial of materialism is the view that

> At least one thing is nonmaterial

In quantified logic, the word 'some' is given the same meaning as the phrase 'at least one'. Consequently, the denial of materialism may also be expressed as the view that

> Some things are nonmaterial.

How shall we symbolize this thesis?

One suggestion that comes to mind is to let 's' abbreviate 'some' or 'at least one' and let 'M' abbreviate 'is material':

$$\sim Ms$$

However, this won't work. The letter 's' is an individual constant. Therefore, it can only be used to abbreviate a singular term. Since 'some' and 'at least one' are not singular terms—they do not single out one specifically identified item for discussion—they cannot be represented by an individual constant such as 's'.

So we are back to our original question: How shall we abbreviate 'at least one thing is nonmaterial'? Again, we shall proceed through a series of paraphrases. The denial of materialism, namely:

> At least one thing is nonmaterial

may be paraphrased as

> At least one thing is such that it is nonmaterial.

Substituting a variable for the word 'thing', we can rewrite this as:

> At least one x is such that it is nonmaterial.

If we replace the pronoun 'it' with the variable it relates to, we get:

> At least one x is such that x is nonmaterial.

Now, we already know how to abbreviate 'x is material'. This is symbolized 'Mx'. We'll abbreviate the quantifier phrase 'at least one' with the symbol '(∃x)', called an *existential quantifier*. Putting this together, we get:

$$(\exists x)(\sim Mx)$$

which may be read:

> There is at least one x such that ~Mx

or more simply,

> There is an x such that ~Mx.

This formula may be understood in a number of alternative but equivalent ways.

1. There exists at least one x such that x satisfies 'x is nonmaterial'.
2. There exists at least one x such that 'x is nonmaterial' applies to x.
3. For at least one x: x is nonmaterial.
4. Some x are such that they satisfy 'x is nonmaterial'.
5. Something is nonmaterial.

Consider next the sentence

> Some cars are noisy.

How shall we symbolize this? First, let us paraphrase this as:

There is at least one noisy car.

This is equivalent to

There exists at least one thing such that it is a car and it is noisy.

That is,

There exists at least one x such that x is a car and x is noisy.

And this goes into symbols easily as:

$$(\exists x)(Cx \;\&\; Nx)$$

We now have identified two types of general subject clauses:

    a. Those containing the quantifier 'all' or one of its cognate terms.
    b. Those containing the quantifier 'some' or one of its cognate terms.

Sentences whose subject phrase is of the first type may be called *universal quantifications*. Sentences whose subject phrase is of the second type may be called *existential quantifications*.

Notice, in the symbolizations above, that we captured the sense of the 'are' in 'all are' by using a horseshoe and the sense of the 'are' in 'some are' by using an ampersand.

2. THE LANGUAGE QL

We have been speaking in somewhat informal terms so far. However, it is time to specify a formal language for the logical theory we are exploring. Let us call the language 'QL' for 'Quantificational Logic'. This section sets out the syntax for QL. The semantics for QL will be developed in the next chapter.

VOCABULARY OF QL

Sentence constants: A ... Z A_1, B_1, .. Z_1 .. with or without subscripts.
n-place predicate constants: A', B', ... Z', A_1', B_1', .. with or without subscripts and each with a prime mark.
Individual constants: a, b, c, .. w, a_1, b_1, c_1... w_1... with or without subscripts.
Individual variables: x, y, z
Sentence operators: ~ v ⊃ & ≡
Quantifier symbols: ∀, ∃
Grouping indicators: ( ) [ ] { }

Notice that the prime marks distinguish the predicate constants from the sentence constants. When symbolizing predicates, we will normally omit the primes on the predicate constants and simply use the capital letter alone as an abbreviation of the predicate constant. This shouldn't lead to confusion. In any context of use, a sentence constant stands alone and a predicate constant is paired with one or more individual constants, and this makes it easy to distinguish the two types of constant.

Formation Rules of QL   In the following, '**c**' is used as a metalinguistic variable ranging over individual constants of QL and '**v**' is used as a metalinguistic variable ranging over variables of QL. '**P**' and '**Q**' serve as metalinguistic variables ranging over wffs of QL.

Q1. Any sentence constant is a wff.

Q2. Any n-place predicate followed by n individual constants is a wff.

Q3. If **P** is a wff, ~**P** is a wff.

Q4. If **P**, **Q** are wffs, then **(P&Q)**, **(PvQ)**, **(P⊃Q)**, **(P≡Q)** are wffs.

Q5. If **P** is a wff containing a constant **c**, and **v** is a variable that does not appear in **P**, then (∀**v**)[**P** with **c/v**] and (∃**v**)[**P** with **c/v**] are wffs, where [**P** with **c/v**] abbreviates Sentence **P** with one or more occurrences of **c** replaced uniformly by **v** and the whole surrounded by a pair of grouping indicators.

Any formula that can be constructed by a finite number of applications of these rules is a sentence or well-formed formula of QL. Nothing else is a sentence or well-formed formula of QL.

If you compare the syntax for QL with the syntax for TL, you will notice that any wff of TL will count as a wff of QL, although not every wff of QL will count as a wff of TL.

Recall the difference between an object language and a metalanguage. In the present context, English is the metalanguage and QL is the object language. Also, recall the difference between using an expression and mentioning it. In order to simplify things a bit, when the context makes it clear that an expression is being mentioned rather than used, we will sometimes omit the quotation marks.

3. SYMBOLIZING WITHIN QL

•— CATEGORICAL SENTENCES

A general sentence, which asserts that all or some members of a group of things have or lack a specified property, is called a *categorical* sentence. Aristotle was the first logician to study the logic associated with categorical sentences. Since categorical sentences are still of importance in modern logic, it will be instructive to examine the symbolizations of such sentences.

A categorical sentence is *universal* if it asserts something about all the members of a group, and it is *existential* if it asserts something about some of the members of a group. In addition, a categorical sentence is *affirmative* if it asserts that things have a certain property and it is *negative* if it denies that things have a certain property. This gives us four different types of categorical or general sentences: universal affirmative, universal negative, existential affirmative, and existential negative.[2]

Universal Affirmative   The *standard form* for a universal affirmative is the following:

> All _____ are _____ .

In an actual universal affirmative sentence, the first blank is filled in with an expression referring to a group of things and the second blank is filled in with an expression attributing a property to the things belonging to that group. One of the sentences we examined earlier, 'All dogs have fleas', is an example of a universal affirmative. If we paraphrase 'All dogs have fleas' as 'All dogs are flea-infested', we can see that it fits the standard form of a universal affirmative sentence. Here are a number of additional examples of this type of categorical sentence:

1. All whales are mammals.
2. All maples are trees.
3. All mosquitos are annoying.
4. All persons are mortal.

We have already seen how to symbolize sentences such as these. However, since a bit of repetition can be helpful when learning a new language, let us work through the steps involved in the symbolization of (1). First,

> All whales are mammals

is equivalent to

> Every thing that is a whale is such that it is a mammal.

---

[2] For ease of reference, the four categorical forms are traditionally labeled A, E, I, and O as follows:

> A: Universal affirmative
> E: Universal negative
> I: Existential affirmative
> O: Existential negative

This may be paraphrased

>Every x such that x is a whale is such that x is a mammal.

In other words,

>For every x, if x is a whale, then x is a mammal.

If we now abbreviate 'x is a whale' with 'Wx', and 'x is a mammal' with 'Mx', then (1) goes into symbols as

$$(\forall x)\ (Wx \supset Mx)$$

This simply says: For any x, if x is a whale, then x is a mammal. That is how we shall express, within logic, the claim that all whales are mammals.

Of course, when symbolizing universal affirmatives, there is no need to actually work through the above four steps. Once one understands the rationale behind the symbolization process, universal affirmatives can be translated into symbols in one quick step.

The three other universal affirmative sentences, (2)–(4), are symbolized thus:

2.	All maples are trees	$(\forall x)(Mx \supset Tx)$
3.	All mosquitos are annoying	$(\forall x)(Mx \supset Ax)$
4.	All persons are mortal	$(\forall x)(Px \supset Mx)$

A word of caution is in order. Some persons, when they are learning to symbolize universal affirmative sentences, try using an ampersand in place of a horseshoe. For example, when symbolizing

>All maples are trees

they try:

$$(\forall x)(Mx \& Tx)$$

This is a very inaccurate symbolization, and it is important that you understand why. To assert $(\forall x)(Mx \& Tx)$ is to assert that

>For every x, x is a maple *and* x is a tree.

In other words, $(\forall x)(Mx \& Tx)$ asserts that each and every thing in the entire universe is both a maple and a tree. That is to say, you are a maple tree, the chair you are sitting on is a maple tree, the moon is a maple tree, each and every atom in this room is a maple tree, and so on. Generally then, although there are some exceptions, the main connective of the open sentence attached to the right of the universal quantifier expression should not be an ampersand.

There are a number of alternative but equivalent ways to express a universal affirmative sentence. For example, instead of saying

All whales are mammals.

we could just as well say any of the following:

1. Whales are mammals.
2. Any whale is a mammal.
3. A whale is a mammal.

These are all symbolized as:

$$(\forall x)(Wx \supset Mx)$$

Universal Negative   A universal negative sentence is used to deny that any of the members of a group have a certain property. Since to deny that any of the members of a group have a certain property is just to assert that none of the members of the group have the property, the standard form for a universal negative may be represented as:

No _____ are _____ .

As in the case of a universal affirmative, the first blank is to be filled in with an expression referring to a group of things and the second blank gets an expression representing the property the things in the group purportedly do not have.

Examples of this type of sentence include:

i. No birds are reptiles.
ii. No rocks are alive.
iii. No horses have wings.

Let us examine the symbolization of the first example. Notice first that:

No birds are reptiles

is equivalent to

All birds are nonreptiles

That is, 'no bird has the property of being reptilian' is equivalent to 'All birds have the property of being nonreptilian'. So, we could also represent the standard form of a universal negative as:

All _____ are not _____ .

Let us therefore paraphrase the first example as:

For every x, if x is a bird, then x is *not* reptilian.

In symbols, this is simply

$$(\forall x)(Bx \supset \sim Rx)$$

Notice that universal negatives are symbolized exactly as we symbolized universal affirmatives except for the addition of a suitably placed negation operator.

Existential Affirmative   An existential affirmative sentence is used to assert that some (at least one) of the members of a group of things have a certain property or characteristic. The standard form of such a sentence is:

Some _____ are _____ .

where the first blank is filled in with an expression referring to a group of things and the second blank contains an expression attributing a property to some of the members of the group. One of the sentences we examined earlier, 'Some cars are noisy', is an example of an existential affirmative. Other examples include:

   i. Some trees are pretty.
  ii. Some horses are brown.
 iii. Some light bulbs are burned out.

In order to translate the third example into symbols, we can again think through a series of paraphrases. First, 'Some light bulbs are burned out' is equivalent to

At least one thing is such that it is a light bulb and it is burned out.

That is,

At least one x is such that x is a light bulb and x is burned out.

Letting 'Lx' abbreviate 'x is a light bulb' and letting 'Bx' abbreviate 'x is burned out', this translates into symbols as:

$$(\exists x)(Lx \& Bx)$$

The other two existential affirmatives above, (i) and (ii), go into symbols as follows:

Some trees are pretty. $(\exists x)(Tx \& Px)$
Some horses are brown. $(\exists x)(Hx \& Bx)$

Another word of caution is in order. Some persons, after using a horseshoe in the symbolization of a universal quantification, attempt to use a horseshoe in the symbolization of an existential affirmative. For instance, they try to symbolize 'Some trees are maples' with (∃x)(Tx⊃Mx). It is important that you understand that this is not an accurate symbolization. Consider why. The formula (∃x)(Tx⊃Mx) asserts that:

a. There is at least one x such that *if* x is a tree then x is a maple.

If you will reflect on the meaning of this expression, you will notice that this is not equivalent to 'some trees are maples'. Furthermore, it is not even clear what sentence (a) is claiming. That is, it is just not clear what it means to say, of an undesignated thing, that if it is a maple then it is a tree. Sentence (a) seems to lack a truth-value altogether. So, although there are exceptions, the main connective of an existential sentence is typically not a horseshoe.

Existential Negatives An existential negative sentence is used to claim that some of the members of a group of things do not have a certain property. The standard form of such a sentence is:

Some _____ are not _____ .

For example:

1. Some students are not poor.
2. Some basketball players are not rich.
3. Some particles are not charged.
4. Some trees are not maples.

These existential negatives are symbolized exactly as we symbolized existential affirmatives, except for the addition of an appropriately placed negation operator:

1.	Some students are not poor.	(∃x)(Sx&~Px)
2.	Some basketball players are not rich.	(∃x)(Bx&~Rx)
3.	Some particles are not charged.	(∃x)(Px&~Cx)
4.	Some trees are not maples.	(∃x)(Tx&~Mx)

EXERCISE 8.2

Symbolize the following.

1. All aardvarks are cute.

2. Pat is a happy camper but Chris is not a happy camper.

3. No real hamburger contains mayonnaise.

4. If Joe sings, then Sue will get physically ill.

5. A whale is a thing of beauty.

6. If everybody is happy, then Chris will be happy.

7. Some books are long and dull.

8. Either everything is material or all things are nonmaterial.

9. Not all teachers are infallible.

10. Anything that harms nobody is not against the law.

11. If everyone claps, then someone will object.

12. Hamsters do not eat rocks.

13. Some plays are philosophical and some are not.

14. There are flying dragons.

15. No rocks are edible.

16. At least one nondetectable strain of the AIDS virus exists.

17. None but the brave are free.

18. There is a green car in the garage.

19. Nothing that is harmful is allowed by law.

20. Every contestant will be given a prize.

21. Something is rotten in the refrigerator.

22. Many people do not have enough to eat.

23. A human being is a featherless biped.

24. A hot dog with peanut butter on it is a tasty treat.

25. A penny saved is a penny earned.

26. An alpha particle is the nucleus of a helium atom.

27. Not every book is a good book.

28. Whoever eats a diet of nothing but hamburgers has a high cholesterol level.

29. If anything is physical, then something has mass.

SWITCHING QUANTIFIERS

Sentences expressed with a universal quantifier can be translated into sentences expressed with an existential quantifier, and vice versa. For example, notice that

Everything is material.

is equivalent to

>Not even one thing is not material.

Thus,

$$(\forall x)(Mx)$$

may be expressed equivalently as

$$\sim(\exists x)(\sim Mx)$$

And notice that:

>Something is spiritual.

Is equivalent to:

>It is not the case that everything is not spiritual.

Thus:

$$(\exists x)(Sx)$$

may be expressed equivalently as:

$$\sim(\forall x)(\sim Sx)$$

Similarly, $\sim(\exists x)(Sx)$ is equivalent to $(\forall x)(\sim Sx)$ and $\sim(\forall x)(Sx)$ is equivalent to $(\exists x)(\sim Sx)$ as well.[3]

●— IT DOESN'T EXIST

Sentences denying the existence of something are easily symbolized. For example, a sentence such as

>Cadborosauruses do not exist.

---

[3] A, E, I, and O sentences may be symbolized using either universal or existential quantifiers:

>A: $(\forall x)(Hx \supset Bx)$ or $\sim(\exists x)(Hx \& \sim Bx)$

>E: $(\forall x)(Hx \supset \sim Bx)$ or $\sim(\exists x)(Hx \& Bx)$

>I : $(\exists x)(Hx \& Bx)$ or $\sim(\forall x)(Hx \supset \sim Bx)$

>O: $(\exists x)(Hx \& \sim Bx)$ or $\sim(\forall x)(Hx \supset Bx)$

Is equivalent to

There are no cadborosauruses

Which is symbolized as

$$\sim(\exists x)(Cx)$$

If we switch quantifiers, this may also be expressed as

$$(\forall x)(\sim Cx)$$

which abbreviates 'Each thing is not a cadborosaurus'. (If a real Loch Ness "monster" is ever discovered, the scientific name for the beast will apparently be cadborosaurus.)

●— MORE COMPLEX SENTENCES

Consider next:

Some dogs are either noisy or obnoxious

This is similar in form to an existential affirmative sentence. First, let us paraphrase it as

There is at least one x such that x is a dog and x is either noisy or x is obnoxious.

This goes into symbols easily as:

$$(\exists x)[Dx \& (Nx \vee Ox)]$$

where 'Dx' abbreviates 'x is a dog', 'Nx' abbreviates 'x is noisy', and 'Ox' abbreviates 'x is obnoxious'.

Consider next the sentence

Any player who uses drugs will be suspended.

This is similar in form to a universal affirmative sentence. First, we may paraphrase this as:

Every x such that x is a player and x uses drugs is such that x will be suspended.

This is equivalent to

For every x, if x is a player and x uses drugs, then x will be suspended.

Employing obvious abbreviations, this becomes:

$$(\forall x)[(Px\&Ux) \supset Sx]$$

So a relative clause such as 'who uses drugs' translates as a predicate constant and variable conjunction.

Notice that adjectives typically translate into symbols as monadic predicates:

All old dogs growl.	$(\forall x)[(Ox\&Dx)\supset Gx]$
All friendly old cats purr.	$(\forall x)[(Fx\&(Ox\&Cx)) \supset Px]$
Some fish are blue.	$(\exists x)(Fx\&Bx)$
Some tall, sleepy giraffes are cute.	$(\exists x)[(Tx\&(Sx\&Gx))\&Cx]$

Notice that we choose predicate constants that remind us of the English predicates they abbreviate.

●— THE ONLY WAY TO GO

It is sometimes difficult to figure out the proper symbolization of a sentence containing "only." Suppose we wish to abbreviate

> **i.** Only politicians are honest.

This is equivalent to

> The only persons who are honest are politicians.

Another way to put this is:

> **ii.** All honest persons are politicians.

Notice that (ii) is not equivalent to

> **iii.** All politicians are honest.

Thus, to claim that *only* politicians are honest is *not* to claim that all politicians are honest. Paraphrased, (ii) becomes

> For any x, if x is honest, then x is a politician.

In symbols, this is:

$$(\forall x)(Hx\supset Px)$$

Notice that 'All honest persons are politicians' is equivalent to 'All nonpoliticians are not honest'. Thus, $(\forall x)(Hx\supset Px)$ is equivalent to $(\forall x)(\sim Px\supset \sim Hx)$.

Here is a more complicated case:

> Only musicians with union cards were hired.

That is,

> All who were hired were musicians who had union cards.

We may paraphrase this as:

> For any x, if x was hired, then x was a musician and x had a union card.

In symbols:

$$(\forall x)[Hx \supset (Mx \& Ux)]$$

4. DYADIC PREDICATION

We earlier divided singular sentences into two components: subject phrases and predicate phrases. We then used open sentences to symbolize predicate phrases. For instance, the predicate phrase '_____ is happy' was symbolized 'Hx'. Recall that if the blank in a predicate phrase is filled in with a singular term, the result is a completed sentence and if the variable in 'Hx' is replaced with an individual constant, the result, say 'Ha', is a substitution instance of the open sentence and represents a complete sentence.

All of the predicate phrases we have examined so far have been *monadic*, or one-place predicate phrases. A monadic predicate phrase contains just one blank and is used to attribute a property to just one thing. In contrast, a *dyadic*, or two-place, predicate expression contains two blanks. Here are some examples of dyadic predicate expressions:

> _____ knows _____
> _____ is older than _____
> _____ is the employer of _____
> _____ is taller than _____

In order to turn a dyadic predicate phrase into a complete proposition expressing sentence, two singular terms must be inserted, one in each blank. If we go back to the four dyadic predicates above and insert, 'Pat' in the left blank and 'Chris' in the right blank of each we get:

> Pat knows Chris
> Pat is older than Chris

Pat is the employer of Chris

Pat is taller than Chris

A dyadic predicate phrase represents a two-place property or, as it is more often called, a *relation* between two things. We use a dyadic predicate to assert the existence of a relation between two individual things.

The dyadic predicate phrases listed above will be symbolized just as we symbolized the monadic predicates, except for the addition of a second variable representing the second blank. Thus:

'_____ knows _____' is abbreviated as 'Kxy'

'_____ is older than _____' is abbreviated as 'Oxy'

'_____ is the employer of _____' is abbreviated as 'Exy'

'_____ is taller than _____' is abbreviated as 'Txy'

In each case, we used the variable x in place of the left blank and the variable y in place of the right blank.

If we abbreviate 'Pat' with 'p' and 'Chris' with 'c', we get the following four substitution instances of these open sentences:

Kpc

Opc

Epc

Tpc

Notice that we place the predicate constant to the left and the two singular terms to the right. These four formulas abbreviate, in order:

Pat knows Chris

Pat is older than Chris

Pat is the employer of Chris

Pat is taller than Chris

Here are a few additional examples:

Joanie likes Chachi

easily translates as

Ljc

and

Fred is taller than Sam

becomes

>Tfs

Likewise,

>Jean traveled to Arizona

becomes

>Tja

● COMBINING QUANTIFIERS WITH TWO-PLACE PREDICATES

Let us consider a sentence that combines a quantifier with a two-place predicate. For example:

>Jan knows everybody.

First, notice that this is equivalent to

>For all x such that x is a person, Jan knows x.

This may be paraphrased as:

>For all x, if x is a person, then Jan knows x.

If we now abbreviate 'Jan' with 'j', 'Jan knows x' with 'Kjx', and 'x is a person' with 'Px', then our sentence may be symbolized as:

$$(\forall x)(Px \supset Kjx)$$

Consider next the sentence

>Fred knows somebody.

We may rewrite this as

>There exists at least one x such that x is a person and Fred knows x.

In symbols this is:

$$(\exists x)(Px \& Kfx)$$

Compare this with the sentence

>Somebody knows Fred.

This sentence is equivalent to:

There exists at least one x such that x is a person and x knows Fred.

In symbols this becomes

$$(\exists x)(Px \& Kxf)$$

Notice the difference in meaning between Kfx and Kxf in the two QL sentences above.

The following symbolizations should now make sense to you:

Everybody knows Fred.	$(\forall x)(Px \supset Kxf)$
Not everybody knows Fred.	$\sim(\forall x)(Px \supset Kxf)$
Fred doesn't know anybody.	$\sim(\exists x)(Px \& Kfx)$
Fred doesn't know everybody.	$\sim(\forall x)(Px \supset Kfx)$

Here are some further examples. Let 'Txy' abbreviate 'x is taller than y'. Then:

All brontosauruses are taller than Pat.	$(\forall x)(Bx \supset Txp)$
Pat is taller than all brontosauruses.	$(\forall x)(Bx \supset Tpx)$
Not all cats are taller than Sam.	$\sim(\forall x)(Cx \supset Txs)$

We can also form compounds of such sentences. The following is a compound of two quantified sentences:

All giraffes are yellow and Bob is taller than any giraffe.

$$(\forall x)(Gx \supset Yx) \& (\forall x)(Gx \supset Tbx)$$

●— ANY and EVERY

'Any' and 'every' function differently in certain kinds of complex sentences. In the following example, the two words function in unison:

Pat likes *every* hamburger.	$(\forall x)[Hx \supset Lpx]$
Pat likes *any* hamburger.	$(\forall x)[Hx \supset Lpx]$

But here they function in opposing ways:

Pat does not like *every* hamburger.	$\sim(\forall x)[Hx \supset Lpx]$
Pat does not like *any* hamburger.	$(\forall x)[Hx \supset \sim Lpx]$

●— RELATIVE CLAUSES and REFLECTIVE SENTENCES

Relative Clauses   Relative clauses typically begin with 'that,' 'who,' 'whom,' 'which,' 'when,' or 'where' and are best represented with the aid of a dyadic predicate. Here are some examples:

> Some city that I used to live in is an old city.

Paraphrased, this is

> There is an x such that x is a city and I used to live in x and x is old.

Symbolized, this becomes:

$$(\exists x)[(Cx \,\&\, Lix)\,\&\, Ox]$$

where 'Lix' represents 'I used to live in x'. And:

> Susan is a doctor who once treated Michael Jackson.

symbolizes as:

> Ds & Tsm

Here are further examples:

Ann is a photographer who has a heart.	Pa & Ha
Bob is a person who likes to travel.	Pb & Lb
New York is a city that Frank Sinatra sings about.	Cn & Sfn
George, whom Susan knows, is late.	Lg & Ksg
Washington, which is a state, has fewer people than the city of Los Angeles.	Sw & Fwl

Reflexive Sentences   Here are examples of reflexive sentences and their proper symbolizations. To symbolize

> Susan is proud of herself

we may write

> Pss

with the understanding that 'Pxy' abbreviates 'x is proud of y' and 's' abbreviates 'Susan'.

To symbolize

> Pat gave himself a present

we write

$$(\exists x)[Px\& Gpxp]$$

where 'Px' abbreviates 'x is a present' and 'Gpxp' abbreviates 'Pat gave x to Pat'.

Some reflexive sentences are more general in nature. Consider the following examples and their symbolizations:

Somebody knows himself.	$(\exists x)(Px\&Kxx)$
Everybody knows herself.	$(\forall x)(Px\supset Kxx)$

where Kxx abbreviates 'x knows x'.

5. MULTIPLE QUANTIFICATION

We have been presupposing an unrestricted universe of discourse in the symbolizations so far specified. That is, we have been assuming that our variables range over all things in the universe. However, in this section, it will simplify things greatly if we stipulate a restricted domain or universe of discourse. In order to see how a restricted domain simplifies things, compare the following two symbolizations of the same English sentence. First, assuming an unrestricted universe of discourse, the sentence

All persons have rights

goes into symbols as

$$(\forall x)(Px\supset Rx)$$

However, if we now stipulate that the variable ranges only over persons, then 'All persons have rights' may be symbolized as

$$(\forall x)(Rx)$$

with the understanding that it is only persons that are under discussion. Of course, $(\forall x)(Rx)$ by itself—with an unrestricted domain—simply says that all things have rights. But since the universe of discourse is restricted to persons and nothing but persons, the 'things' referred to are all and only persons, and so $(\forall x)(Rx)$ asserts that all persons have rights.

Let us stipulate, then, that the variables in the following examples range over persons and nothing but persons. Consider now the sentence

Someone knows someone.

In order to symbolize this, let us paraphrase it into an appropriate form. First, 'someone knows someone' is equivalent to:

For some x and for some y, x knows y.

In symbols, this is abbreviated:

$$(\exists x)(\exists y)(Kxy)$$

where 'Kxy' abbreviates 'x knows y'. (Remember that we are assuming the variables range only over persons.)

Next, consider:

Everyone knows everyone.

Another way to put this is:

For every x, and for every y, x knows y.

In QL, this becomes:

$$(\forall x)(\forall y)(Kxy)$$

• **EXERCISE 8.3** •

Symbolize in QL:

1. Gerbils and hamsters are cute pets.

2. There are no abominable snowpersons.

3. Only an unreasonable person would favor initiative 94.

4. All those who attended were philosophy majors who belonged to the club.

5. Every suspect who assaults an officer will be booked in jail.

6. Only men are eligible for the draft.

7. Not everyone respects himself or herself.

8. Everybody knows Ed but nobody knows Dick.

9. If Pat knows anybody, Pat knows Rita.

10. The Beatles were all from Liverpool.

•— QUANTIFIER ORDER

If a sentence contains two universal quantifiers in a row or two existential quantifiers in a row, we can switch the order of the adjacent quantifiers without altering the meaning of the sentence. For example, 'Someone knows someone' may be symbolized by either:

$$(\exists x)(\exists y)(Kxy)$$

or:

$$(\exists y)(\exists x)(Kxy)$$

And 'Everyone knows everyone' may be symbolized by either:

$$(\forall y)(\forall x)(Kxy)$$

or:

$$(\forall x)(\forall y)(Kxy)$$

In general, if a sentence contains two universal quantifiers—each with a different variable attached—or two existential quantifiers with different variables, then the order of the adjacent quantifiers makes no difference.

However, when two quantifiers appear next to each other in a sentence, and one is universal and the other is existential, the order of the two quantifiers affects the meaning of the sentence. Changing the order of the quantifiers changes the meaning of the sentence. In order to see this, consider the following examples:

Somebody knows everybody.	$(\exists x)(\forall y)(Kxy)$
Somebody is known by everybody.	$(\exists y)(\forall x)(Kxy)$
Everybody knows somebody or other.	$(\forall x)(\exists y)(Kxy)$
Everybody is known by somebody or other.	$(\forall y)(\exists x)(Kxy)$

In most cases, when an existential quantifier appears first, followed by a universal quantifier, the sentence asserts that one thing stands in a relation to all things. When a universal quantifier appears first, followed by an existential quantifier, the sentence asserts that for each and every thing, some entity stands in a relation to that thing.

Let's pause and note the difference between two of these sentences.

Everybody knows somebody or other

is abbreviated as

$$(\forall x)(\exists y)(Kxy)$$

and

> Somebody is known by everybody

goes into symbols as

$$(\exists y)(\forall x)(Kxy)$$

Notice that these two sentences are not equivalent. The first tells us that each person knows somebody or other. Perhaps, given this, we each know a different person. The second tells us that there exists some one person who is the object of everyone's knowledge—we each know this one person.

Here is another example of the difference that quantifier order makes in the case of mixed quantifier sentences. According to the principle of universal causation (hereafter abbreviated 'UC'), everything has a cause. This is taken to mean just that each thing has some cause or other. Thus, although each thing has a cause, the cause of one thing might be different from the cause of another thing. If we let Cyx abbreviate 'y causes x' or 'y is the cause of x', then UC may be symbolized as

$$(\forall x)(\exists y)(Cyx)$$

This reads: For every x, there exists a y such that y is the cause of x. In other words, for every thing x, there is some thing or other y such that y is x's cause. Note that it does not follow from this that there exists some one thing (such as God) that is the sole cause of everything.

Suppose now that we switch the order of the two quantifiers in the formula above. This gives us:

$$(\exists y)(\forall x)(Cyx)$$

This reads: There exists a y such that for every x in the universe, y is the cause of x. In other words, there exists some one thing y that is itself the cause of everything. In this case, the sentence asserts the existence of a creator of the universe. Thus, $(\forall x)(\exists y)(Cyx)$ and $(\exists y)(\forall x)(Cyx)$ mean two different things.

• QUESTION •

> Does $(\forall x)(\exists y)(Cyx)$ imply $(\exists y)(\forall x)(Cyx)$? Give an argument for your answer.

In general then, when universal and existential quantifiers appear next to each other in a formula, the order of the two quantifiers affects the meaning of the sentence.

One more example of switched quantifier order will be instructive. Consider:

$$\text{a.} \quad (\forall x)(\exists y)(Cxy)$$

This says that for every x, there is a y such that x causes y. In other words, for anything whatsoever, there exists something or other that it causes. Each thing causes something. However, it doesn't follow from this that there exists some one thing that is itself caused by everything.

Now let us rearrange the order of the quantifiers in (a):

b. $(\exists y)(\forall x)(Cxy)$

This says something very different. According to (b), there is some y such that every x is the cause of y. In other words, there is some one thing that is such that it is caused by everything. Thus, (a) and (b) do not mean the same thing. Again, when dealing with "mixed" quantifier sentences, the order of the quantifiers matters.

● QUANTIFIER SCOPE

We now have six quantifiers to work with: $(\exists x)$, $(\forall x)$, $(\exists y)$, $(\forall y)$, $(\exists z)$, and $(\forall z)$. In a sentence containing just one quantifier, the scope of the quantifier is the quantifier itself plus the expression enclosed in parentheses to the quantifier's immediate right. In the following, the scope of each quantifier is underlined:

$\underline{(\forall x)(Fx \supset Gx)}$ $\underline{(\forall x)(Fx)} \supset Sa$ $\underline{(\forall x)(Fx)} \supset \underline{(\exists x)(Hx)}$

$\underline{(\exists y)(Fy \lor Gy)}$ $\underline{(\exists y)(Hy)} \lor Gb$ $\underline{(\forall x)(Hx)} \lor Gx$

If a quantifier appears immediately to the right of another quantifier, the scope of the first quantifier is that quantifier itself plus the scope of the second quantifier. In the following sentence, the second quantifier lies within the scope of the first quantifier:

$(\forall x)(\exists y)(Fxy)$

Notice that as far as scope is concerned, a quantifier functions exactly as if it were a tilde. In other words, the scope of a quantifier is just what the scope would be if the quantifier were to be removed and replaced with a tilde.

● FREE VARIABLES and OPEN SENTENCES

A quantifier is said to *bind* the variable it contains plus any occurrences of that variable that lie within the quantifier's scope. Thus, in the following sentence, the quantifier binds all three variables:

$(\forall x)(Fx \supset Gx)$

If a variable is not bound by a quantifier, then it is called a *free* variable. In each of the following, the variable 'x' is bound by the universal quantifier and the variable 'y' is a free variable.

$$(\forall x)(Fx) \supset Gy \qquad (\forall x)(Fx \supset Gx) \vee Sy$$

In the following, the third occurrence of the variable 'x' is a free occurrence of that variable:

$$(\forall x)(Fx) \supset Gx$$

Any sentence containing one or more free variables is an *open* sentence. Open sentences, expressions such as 'Fx' or 'Hxy', contain no subject terms and so do not express claims about things. It therefore makes no sense to ask, concerning an open sentence, 'Is it true or false?' Consequently, open sentences do not have truth-values.

If a sentence is not open, it is said to be *closed*. In order for an open sentence to be turned into a closed sentence, one or the other of two things must be done: (a) the variables must be replaced by constants; or (b) each variable must be bound by an appropriate quantifier, that is, by a quantifier containing the same variable. Unlike open sentences, closed sentences represent claims that are either true or false, and so every closed sentence has a truth-value.

Consider the sentence $(\forall x)(Hx \supset Fy)$. The parentheses to the right of the quantifier indicate that the scope of that quantifier covers the whole sentence. This puts 'Fy' within the scope of '$(\forall x)$'. However, the variable 'y' within 'Fy' is not bound by '$(\forall x)$', since the variable 'y' does not occur within the quantifier '$(\forall x)$'. In the sentence $(\forall x)(Hx \supset Fy)$, then, 'y' occurs as a free variable, even though it is within the scope of the quantifier.

Consider next the sentence $(\forall x)(\forall y)(Kxy)$. The second quantifier sits within the scope of the first quantifier. However, the variable 'y' in Kxy is bound only by the second quantifier, and the variable x is bound only by the first quantifier.

●— SOME ADDITIONAL TERMINOLOGY

In general, a predicate phrase is an *n-place* predicate phrase if it contains n blanks and is symbolized with n variables. For example, a predicate phrase with six blanks is a six-place predicate phrase. If a predicate phrase has two or more blanks, it is called a polyadic ('many place') predicate phrase. Thus, we first covered monadic predicate phrases and then polyadic phrases. When we first combined quantifiers with monadic predicate phrases we were studying *monadic quantification*, and when we went on to combine quantifiers with polyadic predicates we were studying *polyadic quantification*.

————————● EXERCISE 8.4 ●————————

Symbolize the following in QL.

1. If all things are material, then Descartes was wrong.

2. If everybody knows somebody, then somebody knows everybody.

3. Only vegetarians were admitted to the dinner.

4. Nobody except citizens may vote.

5. Everybody likes someone.

6. There's a person whom everyone knows.

7. Some people do not know anybody.

8. Wimpy bought himself a hamburger.

9. If John Candy says one word, people will laugh.

10. Electrons and positrons are leptons.

11. If Pat owes Jan money, then Jan owes Chris money.

12. If someone helps someone, then God is pleased.

13. There are songwriters and there are poets, and there are persons who are both.

14. Everything is temporary and perishable.

15. Everybody likes Bob's grandmother.

16. A mind hooked on drugs is a mind that is wasted.

17. Somebody gave something to Jane.

18. Everybody who knows Pat attended the ceremony.

19. Nobody who loves someone can be all bad.

20. Someone's dog barked at someone.

21. Someone barked at his or her dog.

22. Everybody knows Chris.

23. Only material objects reflect light.

24. Jan is a roadie who once worked for the Buffalo Springfield.

25. All elephants are larger than Nathan's pet mouse.

26. Lorraine likes any horse.

27. Somebody knows Katie.

28. Johnny is a friend of Elliot's.

29. Pigs will eat anything.

30. Many stars are larger than the sun.

31. Everything good is loved by God.

32. A cat or dog is a trustworthy friend.

33. Nicholas is an actor who once starred in "Swamp Thing."

34. If all things are perishable, then life has no meaning.

35. The only people who were hired were people who put pepper in their coffee.

36. If an irresistible force meets an immovable object, something has got to give.

37. There exists a barber who shaves only those who don't shave themselves. (Does the barber shave himself?)

38. A lawyer who represents himself or herself has a fool for a client.

39. Anyone who loves no one is to be pitied.

40. Pat is taller than any aardvark and is shorter than any elephant.

6. IDENTITY and DEFINITE DESCRIPTIONS

IDENTITY

The verb 'to be' can be used in three very different ways. First, it can be used to assert the existence of something, as in the following sentence:

There is a divine being.

Here, the word 'is', a form of the verb 'to be', is being used to assert the existence of a deity. If we let Dx abbreviate 'x is a divine being', this sentence is abbreviated:

$$(\exists x)(Dx)$$

Call this the *existential* use of the verb 'to be'.

Second, the verb 'to be' may be used to assert that a predicate expression applies to an object, as in the following sentence:

Fred is happy.

In symbols, this is:

$$Hf$$

Call this the *predicative* use of the verb 'to be'.

In this section, we shall not be concerned with either of these two uses of 'is'. Rather, we shall focus our attention on a third use of this verb. Consider the following sentence:

Robert Zimmerman is Bob Dylan

Here 'is' is being used to assert that the individual named Robert Zimmerman is the very same individual as the one named Bob Dylan. In other words, Robert Zimmerman, it is claimed, is identical with Bob Dylan. Call this the *identificational* use of the verb 'to be'.

An interesting example of the 'is' of identity concerns the heavenly object known to ancient astronomers as the Evening Star. In ancient times, the term

> the Morning Star

had one meaning, and the term

> the Evening Star

carried another meaning. It was thought that the two names referred to two different stars. Astronomical observations eventually proved that the two terms refer to the same individual object, namely, the planet Venus. That is, through observation, it was discovered that

> The Morning Star is the Evening Star

expresses a true proposition.

Here are additional examples of the identificational use of 'is':

> Taiwan is Formosa.
>
> Muhammad Ali is Cassius Clay.
>
> Olympia is the capital of Washington state.
>
> Richard Starkey is Ringo Starr.

It is important that you do not confuse the concept of identity with the concept of similarity. In everyday speech, we might point to two hamburgers and say that they are "identical", meaning just that the two are very similar. However, this is not the concept of identity used in logic, for to say that x and y are identical in the logical sense is to say that x and y are not two different things but are one and the same thing. Thus, two different hamburgers, no matter how similar, could never be identical as far as the logical concept of identity is concerned.

The Identity Sign Let us now introduce a symbol for the 'is' of identity. We will abbreviate

> _____ is identical with _____

as

$$x = y$$

and we will abbreviate

> _____ is not identical with _____

as

$$(x \neq y)$$

or

$$\sim(x = y)$$

In order to incorporate the identity sign into QL, we must return to QL's syntax and add:

> **Q6.** If **c** and **d** are individual constants, then (**c=d**) and (**c≠d**) are wffs.

In addition, we must add '=' and '≠' to QL's vocabulary.

Symbolizing with the Identity Sign   Using this notation, we can now abbreviate sentences that could not previously have been accurately symbolized. In the following, assume the domain is restricted to persons. Consider:

> There are at least two musicians.

Without the identity sign, one might try symbolizing this as

$$(\exists x)(\exists y)(Mx\&My)$$

That is, for some x and for some y, x is a musician and y is a musician. However, this won't do, for it is possible that in this case x and y stand in for the same person, in which case it is possible, given the symbolism under consideration, that only one musician exists. That is, if only one musician were to exist, $(\exists x)(\exists y)(Mx\&My)$ would nonetheless be true. So this attempted symbolization fails to capture the meaning of the sentence at hand. Similarly, suppose we know that:

> **a.**   Someone in the house is an electrician.

and

> **b.**   Someone in the house is a truck driver.

It does not follow that there are two persons in the house; perhaps the two occurrences of the variable word 'someone' refer to one and the same person.

The solution is to symbolize 'There are at least two musicians' as follows:

$$(\exists x)(\exists y)[(Mx\&My)\&(x \neq y)]$$

That is, there is an x and there is a y and x is a musician and y is a musician, and x is not the same individual as y. This guarantees that there are at least two musicians.

Next, consider

There are at least three musicians.

In order to avoid unnecessary complications, we shall leave out some of the parentheses in the long conjunctive clause:

$$(\exists x)(\exists y)(\exists z)[(Mx \& My \& Mz) \& (x{\neq}y) \& (y{\neq}z) \& (x{\neq}z)]$$

Since x is not identical with y, and y is not identical with z, and x is not identical with z, there must exist at least three musicians.

Suppose we wish to abbreviate:

There exists *at most one* Creator.

The symbolization is:

$$(\forall x)(\forall y)[(Cx \& Cy){\supset}(x{=}y)]$$

According to this, for any x and any y, if x is a creator and y is a creator, then x is identical with y. And to symbolize:

There exist *at most two* Creators.

we need:

$$(\forall x)(\forall y)(\forall z)\{[(Cx \& Cy \& Cz)]{\supset}[(x{=}y)v(x{=}z)v(y{=}z)]\}$$

This guarantees that at most two Creators exist.

Next, a slightly different construction:

There exists *exactly one* Creator.

This becomes:

$$(\exists x)(\forall y)[Cx \& (Cy{\supset}(y{=}x))]$$

And

There exist *exactly two* Creators.

abbreviates as:

$$(\exists x)(\exists y)(\forall z)\{[Cx \& Cy \& (x{\neq}y)] \& [Cz{\supset}(z{=}x \text{ v } z{=}y)]\}$$

We can also use the identity sign to abbreviate *exceptive* statements. Here are several examples:

Rita is happier than anyone else except Joe.

In QL, this may be rendered:

$$(\forall x)\{[Px \,\&\, (x \neq r) \,\&\, (x \neq j)] \supset Hrx\}$$

This says that for any x, if x is a person and x is neither Rita nor Joe, then Rita is happier than x. Next:

Everyone is happy except me and my monkey.

In QL, this is:

$$(\forall x)[Px \,\&\, (x \neq j) \,\&\, (x \neq m)] \supset Hx] \,\&\, (\sim Hm \,\&\, \sim Hj)$$

where j abbreviates 'my monkey' and m abbreviates 'me'. Note that we are simplifying things a bit by leaving out a pair of brackets within the triple conjunction. Here is another example:

Rodney Dangerfield is the only person who respects Rodney Dangerfield.

If the constant r designates Rodney Dangerfield and Rxy abbreviates x respects y, then this goes into QL as:

$$Rrr \,\&\, (\forall x)[(x \neq r) \supset \sim Rxr]$$

The identity sign also comes into play when we symbolize *superlative* statements. For instance:

Rita is the loveliest meter maid.

In QL, this may be put as:

$$(\forall x)[(Mx \,\&\, (x \neq r)) \supset Lrx] \,\&\, Mr$$

where Mx abbreviates 'x is a meter maid', Lxy abbreviates 'x is lovelier than y', and r designates Rita. This QL sentence says that for each x, if x is a meter maid and x is not Rita, then Rita is lovelier than x, and it adds that Rita is a meter maid.

EXERCISE 8.5

Symbolize the following.

1. Sue is the tallest on the team.

2. Betty is not the richest person in the world.

3. Rita has at least two jobs.

4. There is at most one President.

5. There is no greatest number.

6. Washington state has more than one Senator.

7. Only Chris knows Pat.

8. Something is greater than everything else.

9. Someone is older than everyone else.

10. There is only one deity.

11. There are exactly two deities.

12. God is greater than all other beings.

13. The Beatles had three drummers.

14. Ringo was the best drummer the Beatles ever had.

15. Only John Lennon was a member of both the Beatles and the Plastic Ono Band.

16. Pat loved everybody except himself.

17. All were hired except Chris.

18. Raegina is the oldest member of the team.

19. Everyone is happy except Pat and Chris.

20. Only two slide rules remain in existence.

21. Jupiter's moon, Io, is the only solar system body other than Earth known to have erupting volcanoes.

22. No two people see the same rainbow.

23. No more than three deities exist.

24. Millard is the only person more long-winded than Rob.

— DEFINITE DESCRIPTIONS

Recall that a singular term is an expression that purports to refer to or designate one specifically identified thing. Proper names—abbreviated by individual constants—constitute one type of singular term. Definite descriptions constitute another type of singular term. A definite description is an expression that purports to refer by description to one specified thing. Here are some examples:

1. The first TV I ever owned
2. The third house on the left

3. The inventor of bifocals

4. The twentieth President of the United States

Any definite description contains two parts. One part consists of a *descriptive phrase* describing a thing, and the second part asserts that this description refers descriptively to just one specific thing. In description (3) above, the expression 'inventor of bifocals' constitutes the descriptive phrase and the definite article 'the' tells us that the descriptive phrase describes just one thing.

Let us take description (3) and develop a way to symbolize it. First, (3) may be paraphrased as

    i.   The thing which the description 'the inventor of bifocals' describes

This is logically equivalent to:

    ii.   the x which is such that x invented bifocals

Consider next the descriptive phrase within (ii), namely, the words 'x invented bifocals'. This part may be abbreviated with the open sentence

    Ix

which produces:

    iii.   the x which is such that Ix

The final step in this process is to abbreviate the phrase 'the x which is such that'. The convention in logic is to represent this phrase as follows. The ninth letter of the Greek alphabet, the iota, is inverted and placed in front of the variable x, and the whole is enclosed within parentheses. The expression thus produced, '(ιx)', is called a *descriptive operator*. If we combine the descriptive operator with an open sentence, we generate a formula called a *descriptor*. The following descriptor fully abbreviates description (3):

$$(\iota x)(Ix)$$

There are a couple things to keep in mind as you work with this new symbolism. First, a descriptor simply abbreviates a definite description. As such, it is a type of singular term. Thus, a descriptor does not represent a complete sentence and so does not represent something that is either true or false. Second, it follows from this that the descriptive operator is not a quantifier, for when this operator is attached to an open sentence, the result does not express a sentence that is either true or false.

The descriptor is used as if it were an individual constant. For example, if we abbreviate

    x is a philosopher

with

> Px

and if we abbreviate

> Anselm

with

> a

we can abbreviate 'Anselm is a philosopher' as:

> Pa

Now, we can refer to Anselm by definite description instead of by name. Perhaps Anselm's most noted contribution to philosophy is the *Proslogium*, a book in which he first put forward the ontological argument for God's existence. If we paraphrase

> The author of the *Proslogium*

as

> The x such that x authored the *Proslogium*

then the descriptor is:

$$(\iota x)(Axp)$$

The sentence 'The author of the *Proslogium* is a philosopher' may now be abbreviated:

$$P(\iota x)(Axp)$$

Notice that the only change from 'Pa' to 'P(ιx)(Axp)', is that we replaced 'a' with '(ιx)(Axp)'.

Consider next the sentence:

> The most decorated soldier of World War II is Audie Murphy.

In order to abbreviate this we will need both a descriptor and an identity sign. First, we can abbreviate 'Audie Murphy' with 'a', and we can abbreviate 'the most decorated soldier of World War II' with (ιx)(Mx). Next, we place the identity sign between these two terms:

$$(\iota x)(Mx) = a$$

Suppose we wish to abbreviate:

God is the greatest of all beings.

Using 'Gxy' for 'x is greater than y' and 'g' for 'God', we get:

$$g=(\iota x)\{\forall y[(y\neq x) \supset Gxy]\}$$

Similarly, to abbreviate:

The greatest rock band of all time was the Beatles

we let 'Rx' abbreviate 'x is a rock band' and we abbreviate 'the Beatles' as 'b':

$$(\iota x)\{Rx\&(\forall y)[(Ry\&(y\neq x)) \supset Gxy]\}=b$$

• PROBLEM •

Write a rule adding the definite description symbolism to QL's syntax.

•— NONREFERRING DESCRIPTIONS

Definite descriptions generally function syntactically as if they were individual constants. A new problem arises when we consider nonreferring definite descriptions, that is, descriptions that do not describe any existing individual thing. Consider an example first discussed by Bertrand Russell:

a. The present King of France is bald.

This is abbreviated:

$$B(\iota x)(Kx)$$

where 'Kx' abbreviates 'x is presently King of France' and 'Bx' abbreviates 'x is bald'.

Does this express a truth or a falsehood? Since France has no King, $B(\iota x)(Kx)$ surely doesn't express a truth. However, suppose we say that the sentence expresses a falsehood. This seems to imply that France does have a King, but a King who is not bald. The problem is that it doesn't seem right to assign truth to the sentence, but it doesn't sound right to assign falsity to it, either.

One solution to this dilemma is to suppose that the sentence has no truth-value; it expresses neither truth nor falsity. As satisfactory as this solution sounds at first, the adoption of this solution raises a whole new set of problems. All of our definitions of validity, inconsistency, and so on have presupposed the Law of Excluded Middle. The definitions of the important logical terms have therefore all assumed that every sentence is either true or false. If we were to drop this assumption and allow that some sentences lack a truth-value, we would have to rewrite all of our logical definitions. The result would be a radical change in our logical principles, a far more complex set of definitions, and a less intuitive semantical theory.

Bertrand Russell developed a solution that is widely regarded as the best resolution of the difficulty. What are the truth-conditions of sentence (a)? In order for (a) to express the truth, Russell argued, three conditions must obtain:

1. At least one person is presently King of France.

2. At most one person is presently King of France.

3. Whoever is presently King of France is bald.

If we observe our world, we see that the first condition does not obtain. So, the truth-conditions of (a) do not obtain and the sentence must therefore be assigned the truth-value false.

Thus, on Russell's view, (a) has the same truth conditions as, and is therefore equivalent to, the following:

$$(\exists x)(Kx)\&(\exists x)(\forall y)(Ky \supset (x=y))\&(\forall x)(Kx \supset Bx)$$

where 'Kx' abbreviates 'x is presently King of France' and 'Bx' abbreviates 'x is bald'. Another way to put this is:

$$(\exists x)[Kx\&(\forall y)(Ky \supset (y=x))\&Bx]$$

In sum, according to Russell's solution, a sentence containing a definite description and purporting to attribute a property to a uniquely described thing is assigned the truth-value false when either (i) the description fails to refer to anything, or (ii) the thing referred to lacks the property attributed to it.

EXERCISE 8.6

Symbolize the following:

1. Bill Gates is the richest person in Washington state.

2. Everyone respects the head of the largest college in the country.

3. The strongest person in the world is also the tallest person in the world.

4. Susan's mother loves her.

5. Everyone wants to borrow money from the richest person in town.

6. Muhammad Ali is the greatest boxer of all.

7. The fool on the hill sees the sun.

8. The new janitor is a hard worker.

9. The President of the United States wears a toupee.

10. John Rawls is the author of A *Theory of Justice*.

●— PROPERTIES OF RELATIONS

The claim that Julie is older than Peter is abbreviated in QL as:

Ojp

where O represents the two-place relation of 'being older than'. Relations have properties of their own and important philosophical arguments often depend on which properties it is correct to suppose a particular relation actually has. We shall look at three especially important properties of relations.

Transitivity   A relation is *transitive* just in case if one thing bears the relation to a second thing and the second thing bears the relation to a third thing, then the first must bear the relation to the third. The relation of 'being older than' thus qualifies as transitive, for if one thing is older than a second and the second is older than a third, then the first must be older than the third.

In general, for any two-place relation **R, R** is transitive if and only if:

$$(\forall x)(\forall y)(\forall z)[(Rxy \& Ryz) \supset Rxz]$$

A relation qualifies as *intransitive* just in case if one thing bears the relation to a second and the second bears the relation to a third, then the first cannot possibly bear the relation to the third. The relation of 'being the grandmother of' thus qualifies as intransitive, for if one person is the grandmother of a second person, and the second is the grandmother of a third person, then the first person cannot possibly be the grandmother of the third.

In general, for any two-place relation **R, R** is intransitive just in case:

$$(\forall x)(\forall y)(\forall z)[(Rxy \& Ryz) \supset \sim Rxz]$$

Relations that are neither transitive nor intransitive are *nontransitive*. The relation 'is a friend of' is an example of a nontransitive relation, for if Betty is a friend of John and John is a friend of Beth, it does not follow that Betty and Beth are friends and it does not follow that they are not friends. Other examples of nontransitive relations are 'admires', 'is jealous of', 'loves', and 'gave a present to'.

Other examples of transitive relations are 'is the same weight as', 'owns more pencils than', 'has a larger bank account than', and 'received more votes than'. Other examples of intransitive relations are 'is exactly 10 years older than', 'ate exactly twice as many hamburgers as', and 'owns one more car than'.

Symmetry   A relation is a *symmetrical* relation just in case if one thing has the relation to a second thing, then the second thing has that relation to the first thing.

For example, the relation 'is the same age as' is symmetrical, for if one thing is the same age as a second thing, then the second thing is the same age as the first.

In general, for any two-place relation **R, R** is symmetrical if and only if:

$$(\forall x)(\forall y)(Rxy \supset Ryx)$$

Other examples of symmetric relations are 'is married to', 'lives in the same city as', 'is a cousin of', 'is the same height as', and 'is a friend of'.

A relation qualifies as *asymmetrical* just in case if one thing has that relation to a second then the second cannot possibly have that relation to the first. For example, 'is taller than' is asymmetrical, for if one person is taller than a second, then the second must not be taller than the first.

In general, a relation **R** is asymmetrical if and only if:

$$(\forall x)(\forall y)(Rxy \supset \sim Ryx)$$

Other asymmetrical relations include 'is older than', 'is father of', 'is mother of', and 'is east of'.

If a relation is neither symmetrical nor asymmetrical, then it is *nonsymmetrical*. Thus, 'knows the name of' is nonsymmetrical, for if one person knows the name of a second person, it doesn't follow that the second knows the name of the first and it also doesn't follow that the second does not know the name of the first. Other examples of nonsymmetric relations are 'is a sister of', 'painted a picture of', 'admires', and 'greeted'.

Reflexivity   A relation is said to be *reflexive* just in case if one thing has the relation to something or something has the relation to the first thing, then the first thing has that relation to itself. For example, the relation 'belongs to the same church as' is reflexive, for if Clyde belongs to the same church as Bonnie then Clyde belongs to the same church as himself.

In general, a relation **R** is reflexive just in case:

$$(\forall x)(\exists y)[(Rxy \lor Ryx) \supset Rxx]$$

Other examples of reflexive relations are 'weighs as much as', 'went to the same school as', 'is as tall as', and 'has the same parents as'.

A relation is *irreflexive* just in case nothing has the relation to itself. Thus, 'is older than' is irreflexive, for nothing could possibly be older than itself. In general, a relation **R** is irreflexive just in case:

$$(\forall x)(\sim Rxx)$$

Other examples of irreflexive relations are 'is a brother of', 'is older than', 'weighs more than', and 'owns more pencils than'.

If a relation is neither reflexive nor irreflexive, then it is *nonreflexive*. The relations 'gave a present to', 'looked at', 'sang to', and 'likes' are examples of nonreflexive relations.

EXERCISE 8.7

1. List additional examples of relations that are:

    a. transitive

    b. intransitive

    c. nontransitive

    d. symmetric

    e. asymmetric

    f. nonsymmetric

    g. reflexive

    h. irreflexive

    i. nonreflexive

2. Classify as transitive, symmetric, and so on: (Note: each relation will possess more than one of the properties covered in this section).

    a. is south of

    b. is the sister of

    c. is at least 10 pounds heavier than

    d. is proud of

    e. is a relative of

    f. is the boss of

    g. has more stamps than

    h. borrowed a book from

    i. is not taller than

    j. went to the circus with

    k. is acquainted with

    l. is identical with

QUANTIFICATIONAL LOGIC II: SEMANTICS AND METHODS OF PROOF

1. SEMANTICS FOR QL

Chapter 8 presented the syntax for the language QL. In this chapter, a semantical theory for QL is specified and methods of proof for quantificational arguments are developed.

The following three sections set out the semantics for QL.

•— INTERPRETATIONS

Suppose **P** is a formula of QL and suppose **P** contains one or more predicates, one or more variables, and one or more individual constants. An *interpretation* of **P** specifies three things:

  i. the universe of discourse or domain over which the variables of **P** range

  ii. which individual objects of that universe of discourse are designated by each of the constants occurring in **P**

  iii. which property or relation is designated by each predicate constant occurring in **P**

An interpretation of a quantified sentence, also called a *quantificational model,* is the analogue in quantified logic of a truth-table row or a truth-value assignment in truth-functional logic. An interpretation specifies what objects or things the formula is about and assigns meaning to the individual constants and to the predicate constants of the formula.

Here is an example of an interpretation of a QL sentence:

  Sentence: $(\forall x)(\exists y)(Gxy)$

  Interpretation: Let the domain be the rational numbers, and let Gxy abbreviate 'x is greater than y'.

On this interpretation the QL sentence expresses the claim that for every rational number x, there is a number less than x. Notice that this sentence expresses a true proposition on this interpretation. However, consider another interpretation of the same formula:

Domain: human beings

Gxy: x is the grandmother of y.

The sentence comes out false on this interpretation.

When you specify an interpretation for a sentence, you are specifying a set of objects (the interpretation's domain) and the properties those objects possess. An interpretation therefore describes a possible world.

We may also provide an interpretation of several sentences together. For example:

Sentences: $(\forall x)(Ax \vee Bx)$
         $(\exists x)(Ax)\&(\exists y)(By)$
         $(\exists x)(\exists y)(Gxy)$

Interpretation:
  Domain: the real numbers
  Ax: x is positive
  Bx: x is negative
  Gxy: x is the square root of y

On this interpretation, the three sentences read:

Each real number is either positive or negative.

There exists at least one positive number and at least one negative number.

Some number is the square root of another number.

All three sentences are true on this interpretation. However, consider another interpretation of the same set of sentences:

  Domain: subatomic particles
  Ax: x is a lepton
  Bx: x is a gauge boson
  Gxy: x is the antiparticle of y

All three sentences do not together come out true on this interpretation, since not all subatomic particles are either gauge bosons or leptons. (Other categories of subatomic particles include quarks, mesons, and baryons.)

We may also provide an interpretation for a set of quantified sentences that symbolizes an argument. For example:

$$(\forall x)(Bx \supset Gx)$$
$$\underline{(\forall x)(Gx \supset Px)}$$
$$(\forall x)(Bx \supset Px)$$

Domain: animals

Bx: x is a bird

Gx: x is graceful

Px: x is peaceful

On this interpretation, the argument is:

All birds are graceful.

All graceful things are peaceful.

All birds are peaceful.

●— SEMANTICS FOR QL

Now, using the notion of an interpretation, we may specify a set of semantical principles. In the following, 'P' and 'Q' are metavariables ranging over sentences of QL:

1. **P** is *quantificationally true* if and only if **P** is true on every interpretation.
2. **P** is *quantificationally false* if and only if **P** is false on every interpretation.
3. **P** is *quantificationally contingent* if and only if **P** is true on one or more interpretations and false on one or more interpretations.
4. An argument is *quantificationally valid* if and only if there is no interpretation making the premises true and the conclusion false. An argument is *quantificationally invalid* if and only if it is not quantificationally valid.
5. **P** and **Q** are *quantificationally inconsistent* if and only if there is no interpretation upon which both are true.
6. **P** and **Q** are *quantificationally consistent* if and only if there exists at least one interpretation making both true.
7. **P** *quantificationally implies* **Q** if and only if there is no interpretation upon which **P** is true and **Q** is false.
8. **P** and **Q** are *quantificationally equivalent* if and only if there is no interpretation making one true and the other false.

A Qualification One qualification must be presupposed. In the semantical clauses above, we must presuppose that every interpretation involves a nonempty universe of discourse. The case of an empty universe—a universe containing no individuals or objects—must be excluded or else absurd logical results will follow. For instance, assuming an empty universe of discourse, we must say that the following argument, which is clearly valid, has a true premise and a false conclusion:

$$\underline{(\forall x)(Fx)}$$
$$(\exists x)(Fx)$$

The reason this argument is invalid on an empty universe interpretation is the following: The premise is equivalent to ~(∃x)(~Fx), which is true on an empty universe interpretation. Yet the conclusion is false on that interpretation. Since this constitutes an interpretation making the premise true and the conclusion false, the argument is invalid, assuming we allow the possibility of an empty universe. Yet the argument seems intuitively valid. In order to avoid this paradox, we must stipulate that the semantic principles above speak only of interpretations involving membered universes—universes containing at least one individual or object.

● USING INTERPRETATIONS TO SHOW INVALIDITY

Recall that a quantificational argument is invalid if and only if there exists at least one interpretation on which the premises are true and the conclusion is false. That is, a quantified argument is invalid if and only if there exists one universe of discourse and at least one way of interpreting all predicate letters and constants occurring in the argument's sentences such that all the premises are made true and the conclusion is made false.

This suggests one method of establishing that a quantified argument is invalid: specify an interpretation that makes the premises true and the conclusion false. For example, consider the argument:

$$(\exists x)(Fx)$$
$$\underline{(\exists x)(Gx)}$$
$$(\exists x)(Fx \& Gx)$$

An interpretation that shows this to be invalid is:

> Domain: mammals
> Fx: x is an elephant
> Gx: x is a lion

On this interpretation, the argument is:

> There are elephants.
> There are lions.
> Something is both a lion and an elephant.

Consider next this argument:

$$(\forall x)(Fx \supset Gx)$$
$$\underline{(\forall x)(Fx \supset Hx)}$$
$$(\forall x)(Gx \supset Hx)$$

An interpretation showing this to be invalid is:

> Domain: automobiles
> Fx: x is a Model T

Gx: x is a Ford

Hx: x is black

On this interpretation, the argument reads:

All Model T's are Fords.
All Model T's are black.
All Fords are black.

For one more example, the following argument is invalid:

$$(\forall x)[(Ax \text{ v } Bx) \supset Fx]$$
$$(\exists x)(Ax \& Fx)$$
$$(\exists x)(Bx \& Fx)$$
$$(\forall x)(Fx \supset (Ax \text{v} Bx))$$

In order to see why, consider this interpretation:

Domain: subatomic particles

Ax: x is a proton

Bx: x is a neutron

Fx: x is a baryon

This interpretation makes the two premises true, for protons and neutrons are baryons. However, the conclusion comes out false on this interpretation, for the conclusion now says: anything that is a baryon is a proton or a neutron. But this is false, for protons and neutrons are not the only types of baryons in existence. Since we have found an interpretation making the premises true and the conclusion false, the argument is invalid.

──────────── • **EXERCISE 9.1** • ────────────

Specify an interpretation showing that each argument is invalid.

1. $(\forall x)(Ax \supset Bx) / (\forall x)(Bx \supset Ax)$

2. $(\forall x)(Ax \supset Bx)$

    $(\forall x)(Bx) / (\forall x)(Ax)$

3. $(\forall x)(Ax \supset Bx)$

    $(\forall x)(\sim Ax) / (\forall x)(\sim Bx)$

4. $(\forall x)(Ax \supset Bx)$

    $(\exists x)(Ax) / (\forall x)(Bx)$

5. $(\forall x)(Ax \lor Bx)$

    $\underline{(\forall x)(Ax)} \, / (\forall x)(Bx)$

6. $(\exists x)(Ax)$

    $\underline{(\exists x)(Bx)} \, / (\exists x)(Ax \& Bx)$

7. $(\exists x)(Ax \lor Bx)$

    $\underline{(\exists x)(Ax)} \, / (\exists x)(Bx)$

2. TRUTH-FUNCTIONAL EXPANSIONS

If a quantificational argument contains only monadic predicates, it is possible to translate the argument into a purely truth-functional form containing no quantifiers. Once this is accomplished, a truth-functional test for validity may be performed. Thus, for a limited class of quantified arguments, namely, monadic predicate arguments, truth-functional decision procedures exist. Let us proceed in stages.

Assume that the variable x ranges over a finite domain, one consisting just of the n objects $a_1, a_2, a_3, \ldots a_n$. Then, if we say that *every x is F*, where F is a monadic predicate, we are saying that each and every member of the domain is characterized by F, which is to say:

$$Fa_1 \& Fa_2 \& Fa_3 \ldots Fa_n$$

So, assuming a three-member domain, we may translate

$$(\forall x)(Fx)$$

into

$$Fa_1 \& Fa_2 \& Fa_3$$

Call this formula a *truth-functional expansion* of the quantified formula.

Next, suppose instead that at least one member of the same domain is F. This is to suppose that at least one of a_1, a_2, a_3 is characterized by F. In symbols, this is:

$$Fa_1, \lor Fa_2, \lor Fa_3$$

Consequently,

$$(\exists x)(Fx)$$

may be translated into the following truth-functional expansion:

$$Fa_1, \lor Fa_2, \lor Fa_3$$

Consider two more cases for a three-member domain:

Formula: Expansion:
$(\forall x)(Fx \supset Gx)$ $(Fa_1 \supset Ga_1)$ & $(Fa_2 \supset Ga_2)$ & $(Fa_3 \supset Ga_3)$
$(\exists x)(Sx \& Hx)$ $(Sa_1 \& Ha_1)$ v $(Sa_2 \& Ha_2)$ v $(Sa_3 \& Ha_3)$

A TRUTH-FUNCTIONAL TEST FOR MONADIC PREDICATE ARGUMENTS

A general principle has been proven for quantificational arguments, although we won't provide the proof here. That principle is this. Suppose a quantificational argument contains no polyadic predicates and n different monadic predicates. Given that the number of monadic predicate letters in the argument is n, if there exists no interpretation yielding true premises and a false conclusion for a domain of 2^n objects or individuals, then the argument is valid.

This result makes possible a decision procedure, a mechanical test, for any quantificational argument containing only monadic predicates. For any such argument, we simply go to the case of a 2^n universe or domain, where n stands for the number of monadic predicates in the argument, we construct the appropriate truth-functional expansions, and we mechanically test for validity using the methods of truth-functional logic.

However, it has also been proven that there exists no such mechanical procedure for quantificational arguments containing two-place or higher predicates. That is, no decision procedures exist for quantificational arguments containing polyadic predicates. This is not to say that a decision procedure for such arguments hasn't been discovered. Rather, it has been proven that no such procedure exists.

Let us test a monadic predicate argument for validity using the method of truth-functional expansions. Consider the following argument:

1. $(\exists x)(Sx \& Px)$
2. $(\exists x)(\sim Sx \& Px)$
3. $(\forall x)(Px)$

The test for validity requires a truth-functional expansion in a domain of 2^n individuals, where n represents the number of different monadic predicates appearing in the argument. Since the argument contains two such predicates, the domain must contain four individuals.

The argument's truth-functional expansion for a four-member universe is:

Premise 1: $(Sa_1 \& Pa_1)$v$(Sa_2 \& Pa_2)$v$(Sa_3 \& Pa_3)$v$(Sa_4 \& Pa_4)$
Premise 2: $(\sim Sa_1 \& Pa_1)$v$(\sim Sa_2 \& Pa_2)$v$(\sim Sa_3 \& Pa_3)$v$(\sim Sa_4 \& Pa_4)$
Conclusion: Pa_1 & Pa_2 & Pa_3 & Pa_4

A partial truth-table test on these expansions would prove that the argument is invalid. If Sa_1, Pa_1, and Pa_2 are assigned **T**, and Sa_2 and Pa_3 are assigned **F**, then the premises are true and the conclusion is false. This shows that the argument is invalid.

Here are two more examples. Consider the argument:

1.   (∀x) ~(Hx & ~Rx)/ (∀x) ~(Rx & ~Hx)

The premise expansion is:

~(Ha & ~Ra) & ~(Hb & ~Rb) & ~(Hc & ~Rc) & ~(Hd & ~Rd)

The conclusion expansion is:

~(Ra & ~Ha) & ~(Rb & ~Hb) & ~(Rc & ~Hc) & ~(Rd & ~Hd)

If Ha, Hb, Hc, and Hd are assigned **F**, and Ra, Rb, Rc, and Rd are assigned **T**, then the premise is true and the conclusion false, which shows that the argument is invalid.

Consider next:

2.   (∃x) (Hx & Gx)/ (∀x) (Hx & Gx)

The premise expansion is:

(Ha & Ga) v (Hb & Gb)

The conclusion expansion is:

(Ha & Ga) & (Hb & Gb)

If Ha is assigned **T**, Ga is assigned **T**, and Hb and Gb are assigned **F**, then the premise is true and conclusion is false.

EXERCISE 9.2

Construct truth-functional expansions showing that the following arguments are invalid.

1.   (∃x) (Ax & Px)/ (∀x)~(Ax & ~Px)
2.   (∃x) (Ax & Jx)/ (∃x) (Ax & ~Jx)
3.   (∀x) (Hx)/ (∃x) (Hx & Gx)

---

3. QUANTIFIED NATURAL DEDUCTION

Valid quantified arguments may be proven valid using another method, the method of natural deduction. In this section, all of the truth-functional inference and replacement rules from Chapter 3 will be presupposed. In addition, quantificational rules will be added to produce a set of rules for quantificational natural deduction.

QUANTIFIER PROOFS USING ONLY
TRUTH-FUNCTIONAL RULES

Many arguments involving quantifiers can be proven valid using just the truth-functional natural deduction rules from Chapter 3. In the following proofs singular sentences such as 'Pa' and 'Pab' express one complete sentence and so will be handled the way sentence constants such as P and Q were handled in Chapter 3. Quantified expressions such as $\forall x(Fx \supset Gx)$ and $(\exists x)(Fx \& Gx)$ also express complete sentences, so they will also be handled as sentence constants were handled in Chapter 3. If you understood the proofs in Chapter 3, the following should also make sense to you:

1. $(\forall x)(Fx \supset Gx) \vee \sim(\exists x)(Hx)$
2. $(\exists x)(Hx)$
3. $(\forall x)(Fx \supset Gx) \supset Fab$
4. $\underline{\sim Fab \vee Ae}$ /Ae
5. $(\forall x)(Fx \supset Gx)$ DS 1, 2
6. Fab   MP 3, 5
7. Ae   DS 4, 6

1. $Fab \supset Gst$
2. $Gst \supset Hab$
3. $\underline{(Fab \supset Hab) \supset (\forall x)(Fx)}$/ $(\forall x)(Fx)$
4. $Fab \supset Hab$ HS 1, 2
5. $(\forall x)(Fx)$ MP 3, 4

1. $(\forall x)(Hx) \supset (\exists x)(Fx)$
2. $Ab \vee Be$
3. $\sim Be$
4. $\underline{Ab \supset (\forall x)(Hx)}$/ $(\exists x)(Fx)$
5. Ab   DS 2, 3
6. $(\forall x)(Hx)$ MP 4, 5
7. $(\exists x)(Fx)$ MP 1, 6

These arguments were proven valid using only truth-functional rules of inference. However, many valid quantificational arguments cannot be proven valid using only truth-functional rules. For such arguments, we need special quantificational rules. In this section, five such rules are introduced.

UNIVERSAL INSTANTIATION

A wff is called a *universal* quantification if and only if it begins with a universal quantifier whose scope covers the entire wff. Examples include the following:

$$(\forall x)(Fx) \quad (\forall y)(Fy) \quad (\forall z)(Fz) \quad (\forall x)(Fx \supset Gx) \quad (\forall y)(Fy \vee Gy)$$

If we remove the quantifier from a universal quantification and uniformly replace *each* occurrence of the variable it binds with a constant, we produce an *instantiation* or *substitution instance* of the universal quantification. For each universal quantification, there are an infinite number of possible instantiations. Here are some examples. If we begin with the universal quantification

$$(\forall x)(Wx \supset Mx)$$

and remove the quantifier and uniformly replace the variable it binds, we may produce the following instantiations:

$$Wn \supset Mn$$
$$Wf \supset Mf$$

If we begin with the universal quantification

$$(\forall x)(Gx)$$

then the following are instantiations or substitution instances:

$$Gm$$
$$Gp$$
$$Gs$$

The distinction just drawn between a universal quantification and its substitution instances has its roots in ordinary English discourse. We sometimes draw a generalization and, when asked for an example, say, "For instance . . . ." Suppose, for instance, that in the middle of a political discussion, Pat says, "All communist countries place a high value on human rights." When challenged, Pat might say, "For instance, in Cuba, the constitution guarantees free speech, freedom of the press, freedom of religion, and freedom from arbitrary arrest."

We are now ready to formulate an inference rule dealing with universal quantifiers. A universal quantification makes a claim about all things in the universe of discourse. For instance, $(\forall x)(Gx)$ might be used to assert that all things are good. So, if a universal quantification is true, then its claim is true of each and every thing in the universe of discourse. Therefore, if a universal quantification is true, then each and every substitution instance of it is true as well. This gives us our first quantificational inference rule:

Universal Instantiation (UI)

From a universal quantification, one may infer any substitution instance.

Here are four applications of UI:

$(\forall x)(Fx)$	$(\forall x)(Fx \supset Gx)$	$(\forall x)(\forall y)(Kxy)$
Fa	Fc $\supset$ Gc	$(\forall y)(Kay)$

$(\forall x)(\sim Fx)$	$(\forall x)\sim(Fx \supset Gx)$
$\sim$Fs	$\sim(Fb \supset Gb)$

In each of these four examples, the sentence above the line is a universal quantification and the sentence below the line is a substitution instance of the universal quantification. Remember that when UI is applied, each occurrence of the variable must be uniformly replaced by the constant.

Here is natural deduction proof employing UI:

1. (∀x)(Fx⊃Gx)
2. (∀x)(Gx⊃Hx) / Fs⊃Hs
3. Fs⊃Gs   UI 1
4. Gs⊃Hs   UI 2
5. Fs⊃Hs   HS 3, 4

It is important to remember that UI may be applied only to a universal quantification. In a universal quantification, the universal quantifier appears at the start of the sentence and has the entire sentence as its scope. Thus, UI may not be applied to sentences such as the following:

a. ~(∀x)(Fx⊃Gx)
b. (∀x)(Fx)⊃(∀x)(Gx)

UI cannot be applied to sentences (a) and (b) because in each case the scope of the universal quantifier does not cover the entire sentence. Accordingly, the following inferences are not permissible:

$$\frac{\sim(\forall x)(Fx \supset Gx)}{\sim(Fa \supset Ga)} \qquad \frac{(\forall x)(Fx) \supset (\forall x)(Gx)}{Fa \supset Ga}$$

● EXISTENTIAL GENERALIZATION

Let us say that a sentence of QL is singular if and only if it contains no quantifiers and no variables. If we begin with a singular sentence containing one or more constants, and if we prefix to this an existential quantifier and uniformly replace *one or more* occurrences of a constant with the variable occurring in the existential quantifier, we produce the *existential generalization* of that singular sentence. In each example below, the sentence above the line is a singular sentence and the sentence below the line is an existential generalization of the singular sentence:

$$\frac{Fa}{(\exists x)(Fx)} \qquad \frac{Hb\&Rb}{(\exists x)(Hx\&Rx)} \qquad \frac{GsvFs}{(\exists y)(GyvFy)} \qquad \frac{Hbe}{(\exists x)(Hbx)}$$

A singular sentence containing one constant asserts that one particular thing—that thing designated by the constant—has a certain property. The existential generalization of such a sentence simply asserts that something or other has the property. Certainly if a specific individual designated by a singular term has a certain property, then it follows that something has that property. So, if a singular sentence is true, then the existential generalization of that sentence must be true as well.

For example, from 'George is sad', it certainly follows that 'Something is sad'. And from 'The Space Needle is yellow' it certainly follows that 'Something is yellow'. This simple and obviously valid pattern of reasoning is reflected in the next rule.

Existential Generalization (EG)

> From a singular sentence **P** containing one or more constants, you may infer any existential generalization of the sentence provided that the variable you insert does not already appear in **P**.

Here are a number of examples:

$\underline{Fa}$	$\underline{Fb}$	$\underline{\sim Fc}$	$\underline{FdvHd}$	$\underline{Srb}$
$(\exists x)(Fx)$	$(\exists x)(Fx)$	$(\exists x)(\sim Fx)$	$(\exists x)(FxvHx)$	$(\exists x)(Sxb)$

In each example above, the sentence above the line is a singular sentence and the sentence below the line is an existential generalization of the singular sentence.

The following proof employs EG:

1. $(\exists x)(Fx) \supset (\forall x)(Hx \supset Gx)$
2. Fa
3. $\underline{Hs\qquad}$ /Gs
4. $(\exists x)(Fx)$ EG 2
5. $(\forall x)(Hx \supset Gx)$ MP 1,4
6. $Hs \supset Gs$ UI 5
7. Gs   MP 3,6

EXISTENTIAL INSTANTIATION

In everyday reasoning, people sometimes have to reason about an unnamed individual. For example, the police might learn that someone robbed a bank Tuesday and they might have a lot of information about the robber even though they don't know the robber's name. In such cases, we often assign a temporary name to the unknown individual so that we can reason about the individual more easily. For example, if the police learn that the robber had red hair, they might decide to call the robber 'Red'. Then they could refer to him using just the name 'Red' instead of the more cumbersome 'The guy who robbed the bank Tuesday'. In the next inference rule, we are going to employ this practice of assigning a temporary name to an unnamed individual for the purpose of reasoning about that individual.

The following line of reason involves the assignment of a temporary name to an unnamed individual. Suppose we begin with the following premises:

1. There exists a singer.
2. All singers are happy.
3. If something is happy, then George is happy.

And suppose we want to prove from this that:

> George is happy.

Let us reason as follows. First, call the unnamed singer 'Sue'. So it follows from the premises that

> 4. Sue is a singer.

Since we know from (2) that all singers are happy, it follows that

> 5. If Sue is a singer then Sue is happy.

From (4) and (5) it follows that

> 6. Sue is happy.

Since Sue is happy it follows that:

> 7. Something is happy.

From 3 and 7 it follows that

> 8. George is happy.

Notice that at (4), we introduced a name new to the argument—Sue. That is, the name 'Sue' does not appear in either the premises or the conclusion. By using a name new to the argument, we assumed nothing improper. We just gave an unused name to an unnamed individual so that we could more conveniently reason about that individual. If we hadn't used a new name—if, in other words, we had used 'George'—we would have gotten our conclusion by simply assuming it true, which would not have been valid reasoning.

The reasoning by which we derived step (4) is contained within the next inference rule, *Existential Instantiation*. First, if a wff begins with an existential quantifier whose scope is the entire wff, the wff is called an *existential quantification*. If the existential quantifier is removed from an existential quantification and *each* of the variables bound by that quantifier is uniformly replaced by a constant, the result is an *instantiation* or *substitution instance* of the existential quantification. The inference rule Existential Instantiation may now be formulated as follows:

Existential Instantiation (EI)

From an existential quantification, you may infer any instantiation, provided that the constant you instantiate with does not appear in the premises or on any earlier line of the proof and does not appear in the conclusion.

Here are some applications of Existential Instantiation:

$(\exists x)(Fx)$	$(\exists x)(Fx)$	$(\exists x)(Hx \lor Gx)$	$(\exists x)(\exists y)(Kxy)$
Fa	Fb	Ha $\lor$ Ga	$(\exists y)(Kay)$

In each case above, the sentence above the line is an existential quantification and the sentence below the line is an instantiation of the existential quantification. Remember that each occurrence of the variable must be replaced uniformly when the instantiation is written.

It is crucial that a constant introduced through EI be new to the proof. In order to see why we must impose this restriction, consider what we could do if the restriction were dropped. Suppose the "new constant" restriction has been dropped and we can apply EI using any constant, including one already in the premises, one already in the conclusion, and one used in a previous instance of EI. Then the three following invalid arguments could be proven valid:

1.   There is a millionaire.

     So, Pat is a millionaire.

In QL, this is:

$$(\exists x)(Mx)/Mp$$

If we could apply EI with any constant, then we could use EI to prove this argument valid in one step:

    1. $(\exists x)(Mx)/Mp$
    2.   Mp EI 1   (Incorrect EI)

2.   There exists an octogenarian.

     There exists a teenager.

     So, there exists someone who is both a teenager and an octogenarian.

In QL, this is:

$$(\exists x)(Ox)$$
$$(\exists x)(Tx)/(\exists x)(Ox \& Tx)$$

If we could apply EI using any constant—with no restrictions—then we could instantiate twice with the same constant and prove this argument valid:

    1. $(\exists x)(Ox)$
    2. $(\exists x)(Tx)/(\exists x)(Ox \& Tx)$
    3.   Oa   EI 1   (Incorrect)
    4.   Ta   EI 1   (Incorrect)

5.   Oa&Ta   Conj 3, 4
6.   (∃x)(Ox&Tx)   EG 5

3.   Bob is an octogenarian.

    <u>Something is a teenager.</u>

    So, Bob is a teenager.

In QL:

Ob

<u>(∃x)(Tx)</u>/Tb

Again, if we could use any constant when applying EI, then we could construct the following proof:

1.   Ob
2.   <u>(∃x)(Tx)</u>/Tb
3.   Tb   EI 2 (Incorrect)

    Each of these three arguments is invalid, of course. And each proof involves a violation of EI's "new constant" restriction. Thus, in the first, we instantiated the existential quantification using a constant appearing in the conclusion. In the second, we instantiated using a constant that had been used in a previous existential instantiation. In the third, we instantiated using a constant that appears in the premises. In order to avoid invalid inferences such as the three above, we must make sure that when we apply EI, we use a constant that is new to the proof.

    The following proof employs a correct application of EI:

1.   (∀x)(Fx⊃Gx)
2.   <u>(∃x)(Fx)</u>/(∃x)(Gx)
3.   Fa   EI 2
4.   Fa⊃Ga   UI 1
5.   Ga   MP 3, 4
6.   (∃x)(Gx)   EG 5

    Notice that we applied EI before we applied UI. When you apply EI you must use a constant new to the proof while you may universally instantiate using any constant. For this reason, it is generally a good practice to apply EI before applying UI.

●— UNIVERSAL GENERALIZATION

If we take a singular sentence containing one or more constants and prefix to it a universal quantifier term, and uniformly replace *each* occurrence of a constant in the sentence with the variable contained in the quantifier, the result is the *universal generalization* of the singular sentence. Here are some examples:

Fa	Fa⊃Ha	~(Fa&Ga)
(∀x)(Fx)	(∀x)(Fx⊃Hx)	(∀x)~(Fx&Gx)

In each case above, the sentence above the line is a singular sentence and the sentence below the line is a universal generalization of the singular sentence.

We may now formulate our fourth rule:

Universal Generalization (UG)

> From a sentence **P** containing one or more constants, you may infer any universal generalization of **P** provided that: (a) the constant in **P** that is replaced by a variable does not appear in the premises and does not occur in any previous line that was obtained by E.I. and (b) the variable you insert does not already occur within **P**.

Later on, when we construct conditional and indirect proofs, an additional restriction for UG will be given.

In the following example, the English steps in the argument are given on the right and the QL symbolizations are given on the left. From the premises

	1.	(∀x)(Gx)	All things are good.
	2.	(∀x)(Cx)	All things are created by God.

we may apply UI:

	3.	Ga	UI 1	This apple is good.
	4.	Ca	UI 2	This apple was created by God.

By Simplification, we get:

	5.	Ga&Ca	This apple is good and was created by God.

Since the constant 'a' does not appear in the premises and was not introduced by EI, we may next apply Universal Generalization and infer:

	6.	(∀x)(Gx&Cx)	All things are good and created by God.

The rationale for this rule is the following: If we correctly apply UG to a singular statement containing one or more constants, then the constant generalized upon was not introduced by EI and it does not appear in the premises. How then did the constant enter the proof? It could have been introduced only through an application of UI. But then what is true in the case of this constant is true not only in the case represented by the constant but in all cases. That is, what is true of that which the constant designates is true of all things. Therefore, after we have finished reasoning about that which this constant represents, we may generalize our results and assert that what is true of that designated by the constant is true of all things.

The following proof employs UG:

1. $(\forall x)(Fx \supset Gx)$
2. $(\forall x)(Gx \supset Sx)$
3. <u>$(\forall x)(Sx \supset Rx)$</u> / $(\forall x)(Fx \supset Rx)$
4. $Fa \supset Ga$ UI 1
5. $Ga \supset Sa$ UI 2
6. $Sa \supset Ra$ UI 3
7. $Fa \supset Sa$ HS 4, 5
8. $Fa \supset Ra$ HS 6, 7
9. $(\forall x)(Fx \supset Rx)$ UG 8

Notice that we are allowed to universally generalize line (8) because the constant—'a'—does not appear in the premises and was not introduced by EI.

●— QUANTIFIER EXCHANGE

In Chapter 8, we saw how to exchange one quantifier for another. That discussion is the basis for the following replacement rule:

Quantifier Exchange (QE)

> If **P** contains either a universal or an existential quantification, **P** may be replaced by or may replace a sentence that is exactly like **P** except that one quantifier has been switched for the other quantifier in accord with (a) and (b) below.
>
> (a) switch one quantifier for the other
> (b) take the opposite of both sides of the quantifier

For example, if we apply QE to

$$(\forall x)(Fx)$$

we get

$$\sim(\exists x)(\sim Fx)$$

And if we apply QE to

$$(\exists x)(Fx)$$

we get

$$\sim(\forall x)(\sim Fx)$$

Further examples include:

1. ~(∀x)(Fx) . . . (∃x)(~Fx)
2. ~(∃x)(Fx) . . . (∀x)(~Fx)
3. (∀x)[~(Fx⊃Gx)] . . . ~(∃x)(Fx⊃Gx)
4. ~(∃x)(Fx&Gx) . . . (∀x)[(~(Fx&Gx)]
5. (∀x)(Fx⊃Gx) . . . ~(∃x)[~(Fx⊃Gx)]

The following proofs illustrate the use of Quantifier Exchange.

1. (∀x)(~Hx)⊃(∀x)(Fx⊃Gx)
2. ~(∃x)(Hx)
3. <u>Fa</u> / (∃x)(Gx)
4. (∀x)(~Hx)   QE 2
5. (∀x)(Fx⊃Gx)   MP 1, 4
6. Fa⊃Ga   UI 5
7. Ga   MP 3, 6
8. (∃x)(Gx)   EG 7

1. (∃x)(Fx)⊃(∃x)(Gx)
2. <u>(∀x)(~Gx)</u> /~(∃x)(Fx)
3. ~(∃x)(Gx)   QE 2
4. ~(∃x)(Fx)   MT 1, 3

Here are some further examples of proofs. In the next problem, we begin by removing the universal quantifiers. We then apply a truth-functional rule and end by attaching the existential quantifier:

1. (∀x)(Hx⊃Bx)
2. <u>Hg</u> /(∃x)(Bx)
3. Hg⊃Bg   UI 1
4. Bg   MP 2, 3
5. (∃x)(Bx)   EG 4

Notice that we instantiated premise (1) using the constant g. It is not a coincidence that this is the same constant that appears in line (2). We could have used any constant when we instantiated (1), but if we had instantiated (1) using a constant other than g, then we could not have applied Modus Ponens to (2) and (3).

Examine the first line of the following argument:

1. (∃x)(Jx)⊃(∀x)(Kx⊃Lx)
2. <u>Jb & Kb</u> /Lb

Since line (1) does not begin with a quantifier whose scope extends over the entire line, we cannot instantiate (1). Neither UI nor EI applies in this type of case.

Therefore, our only hope is to break (1) up into its two quantified components:

    3. Jb   Simp 2
    4. (∃x)(Jx)   EG 3
    5. (∀x)(Kx⊃Lx)   MP 1, 4

In the above steps, we derived the antecedent of (1) and then brought down (1)'s consequent by Modus Ponens. Now we can instantiate (5) and get rid of its quantifier:

    6. Kb⊃Lb UI 5
    7. Kb   Simp 2
    8. Lb   MP 6, 7

We instantiated (6) with the constant b so that we could apply Modus Ponens later in the proof.

In the last proof, we used something derived from line (2) to break up line (1). The following argument calls for the same strategy:

    1. (∀x)(Bx⊃Gx)⊃(∃x)(Ax&Hx)
    2. <u>(∀x)(Bx⊃Hx)&(∀x)(Hx⊃Gx)/(∃x)(Hx)</u>

Since (1) contains two quantified formulas joined by a horseshoe, we cannot apply to it either UI or EI. So, our first job is to somehow break (1) up into parts that can be instantiated. We therefore begin with the only move available—we simplify (2):

    3. (∀x)(Bx⊃Hx)   Simp 2
    4. (∀x)(Hx⊃Gx)   Simp 2

Next, we remove the quantifiers:

    5. Bc⊃Hc   UI 3
    6. Hc⊃Gc   UI 4

Notice that these two lines fit the requirements for the Hypothetical Syllogism Rule. Thus:

    7. Bc⊃Gc   HS 5, 6

From this, we derive the antecedent of (1):

    8. (∀x)(Bx⊃Gx)   UG 7

Now we can bring down the consequent of (1):

    9. (∃x)(Ax&Hx)   MP 1, 8

Finally:

10.   Aa&Ha   EI 9
11.   Ha   Simp 10
12.   (∃x)(Hx) EG 11

Consider the following argument:

1.   ~(∃x)(Bx&~Qx)
2.   ~(∀x)(~AxvQx) / (∃x)(~Bx)

Since neither of the two premises begins with a quantifier whose scope is the entire premise, neither premise can be instantiated. Our only option is to apply the Quantifier Exchange rule:

3.   (∀x)~(Bx&~Qx) QE 1
4.   (∃x)~(~AxvQx) QE 2

Now we can apply the instantiation rules:

5.   ~(~AavQa) EI 4
6.   ~(Ba&~Qa) UI 3

Next, using DeMorgan, we get:

7.   Aa&~Qa   DM 5
8.   ~BavQa   DM 6

Finally:

9.   ~Qa   Simp 7
10.   ~Ba   DS 8, 9
11.   (∃x)(~Bx) EG 10

●— CONDITIONAL and INDIRECT QUANTIFIED PROOFS

When you are constructing either a conditional or an indirect quantified proof, you will have to obey a special restriction on Universal Generalization if you are to avoid making an invalid inference. The restriction is this:

Conditional and Indirect Proof Restriction for UG

Within an indented CP or IP sequence, you cannot apply UG to a line containing a constant introduced in an assumed premise.

In order to see why we need this restriction, consider the following argument:

$$(\forall x)(Ax) \supset (\forall x)(Bx)/(\forall x)(Ax \supset Bx)$$

The premise tells us that if everything is an A, then everything is a B. The conclusion is that all A's are B's. This is clearly invalid. However, without the restriction on UG just given, we could construct a proof of this argument:

1. $(\forall x)(Ax) \supset (\forall x)(Bx)/(\forall x)(Ax \supset Bx)$

2. | Aa AP

3. | $(\forall x)(Ax)$ UG 2: Incorrect: Violates the restriction.

4. | $(\forall x)(Bx)$ MP 1, 3

5. | Ba UI 4

6.   Aa $\supset$ Ba CP 2–5

7. $(\forall x)(Ax \supset Bx)$ UG 6

Within the following conditional proof, the use of UG does not violate the restriction, for UG is applied to a line whose constant was not introduced within an assumed premise:

1. $(\forall x)(Ax \supset Bx) / (\forall x)(Ax) \supset (\forall x)(Bx)$

2. | $(\forall x)(Ax)$ AP

3. | Aa UI 2

4. | Aa $\supset$ Ba UI 1

5. | Ba MP 3, 4

6. | $(\forall x)(Bx)$ UG, 5 (allowable)

7. $(\forall x)(Ax) \supset (\forall x)(Bx)$ CP 2–6

Except for the restriction on the use of UG, the construction of quantified conditional and indirect proofs requires nothing beyond the conditional and indirect proof techniques covered in Chapter 3, for quantified conditional and indirect proofs follow the same general principles that truth-functional conditional and indirect proofs follow. Here is another example:

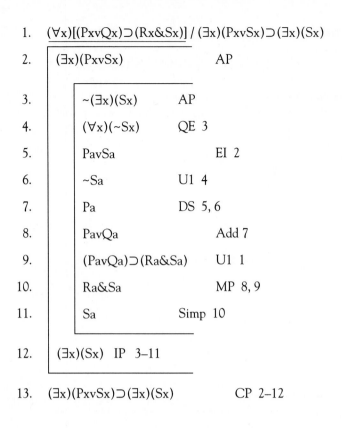

1. $(\forall x)[(Pxv Qx)\supset(Rx\&Sx)] / (\exists x)(PxvSx)\supset(\exists x)(Sx)$

2. $(\exists x)(PxvSx)$ AP

3. $\sim(\exists x)(Sx)$ AP

4. $(\forall x)(\sim Sx)$ QE 3

5. PavSa   EI 2

6. $\sim$Sa   U1 4

7. Pa   DS 5, 6

8. PavQa   Add 7

9. $(PavQa)\supset(Ra\&Sa)$ U1 1

10. Ra&Sa   MP 8, 9

11. Sa   Simp 10

12. $(\exists x)(Sx)$ IP 3–11

13. $(\exists x)(PxvSx)\supset(\exists x)(Sx)$ CP 2–12

It is also acceptable to employ UG in the following way:

1. $(\forall x)[Ax\supset(Bx\supset Cx)]$

2. $(\forall x)(Bx)/(\forall x)(Ax\supset Cx)$

3. $Aa\supset(Ba\supset Ca)$ UI 1

4. Ba   UI 2

5. Aa   AP

6. $Ba\supset Ca$ MP 3, 5

7. Ca   MP 4, 6

8. $Aa\supset Ca$ CP 5–7

9. $(\forall x)(Ax\supset Cx)$ UG 8

In this proof, it was permissible to use UG at step (9)—on line (8)—because although the constant in (8) was introduced within an assumed premise on line (5), that assumed premise was discharged at line (7) and so in effect was removed from the proof at that point. If we had used UG at line (6), however, we would have violated the restriction.

Here is one more example of a quantified conditional proof:

1. $(\forall x)(Ax \supset Bx) / (\exists x)(\sim Bx) \supset (\exists x)(\sim Ax)$

2. $(\exists x)(\sim Bx)$ AP

3. $\sim Ba$ EI 2

4. $Aa \supset Ba$ UI 1

5. $\sim Aa$ MT 3, 4

6. $(\exists x)(\sim Ax)$ EG 5

7. $(\exists x)(\sim Bx) \supset (\exists x)(\sim Ax)$ CP 2–6

PROOFS INVOLVING MULTIPLE QUANTIFIERS

When dealing with arguments containing adjacent or overlapping quantifiers, you may apply EI, EG, and UI in the usual way. In the following examples, quantifiers are removed one at a time:

1. $(\forall x)(\forall y)(Kxy) / Kab$
2. $(\forall y)(Kay)$ UI 1
3. Kab UI 2

1. $(\forall x)(\exists y)(Kxy)$
2. $(\exists y)(Kay) \supset Ga / Ga$
3. $(\exists y)(Kay)$ UI 1
4. Ga MP 2, 3

1. $(\exists x)(\exists y)(Fxy)/(\exists y)(\exists x)(Fxy)$
2. $(\exists y)(Fay)$ EI 1
3. Fab EI 2
4. $(\exists x)(Fxb)$ EG 3
5. $(\exists y)(\exists x)(Fxy)$ EG 4

1. $(\forall x)(\forall y)(Hxy)/(\forall y)(\forall x)(Hxy)$
2. $(\forall y)(Hay)$ UI 1
3. Hab UI 2
4. $(\forall x)(Hxb)$ UG 3
5. $(\forall y)(\forall x)(Hxy)$ UG 4

The last two proofs illustrate a point made in Chapter 8, namely, that when two quantifiers of the same type appear next to each other, their order doesn't affect the truth-value of the sentence. However, as we saw in Chapter 8, if universal and existential quantifiers appear next to each other, the order cannot arbitrarily be altered, for the order affects the truth-value of the sentence. That is:

$$(\forall x)(\exists y)(Fxy)$$

does not imply

$$(\exists y)(\forall x)(Fxy)$$

In order to rule out a proof that $(\forall x)(\exists y)(Fxy)$ implies $(\exists y)(\forall x)(Fxy)$, the following restriction on the use of UG must be emphasized:

> UG may not be applied to a line containing a constant if the constant appears in a line containing a constant introduced by EI.

If we didn't have this restriction, someone could construct a proof of the following invalid argument:

1. $(\forall x)(\exists y)(Fxy)/(\exists y)(\forall x)(Fxy)$
2. $(\exists y)(Fay)$ UI 1
3. Fab   EI 2
4. $(\forall x)(Fxb)$ UG 3   (Incorrect—violates restriction)
5. $(\exists y)(\forall x)(Fxy)$ EG 4

When you are working with sentences containing overlapping quantifiers—sentences in which one quantifier appears within the scope of another quantifier—the Quantifier Exchange rule is applied in the usual way. The following proof illustrates the proper procedure:

1. $\sim(\forall x)(\exists y)(Hxy)/(\exists x)(\forall y)(\sim Hxy)$
2. $(\exists x)\sim(\exists y)(Hxy)$ QE 1
3. $(\exists x)(\forall y)\sim(Hxy)$ QE 2

Notice that as the tilde "passes" over each quantifier, it converts the quantifier it passes over into the "opposite" quantifier.

In Chapter 8 the properties of reflexivity, transitivity, and symmetry were introduced. Let's put our understanding of transitivity to work. Suppose we wish to prove the following argument valid:

1. Carol's car is heavier than Pete's car.
2. Pete's car is heavier than Katie's car.
3. So, Carol's car is heavier than Katie's car.

If we let c designate Carol's car, p designate Pete's car, and k designate Katie's car, this argument goes into QL as:

1. Hcp
2. Hpk /Hck

The English argument is obviously valid. However, the QL argument can't be proven valid with our rules. The problem is that the argument is only valid if we assume that *being heavier than* is a transitive relation. We subconsciously add this assumption to the English version of the argument when we evaluate that, which is why the English version seems obviously valid even though the QL version cannot be proven valid. Let us fix up the QL version by adding to the argument a premise stating that *is heavier than* is a transitive relation:

1. Hcp
2. Hpk
3. $(\forall x)(\forall y)(\forall z)[(Hxy \& Hyz) \supset Hxz]$ /Hck

The third premise says that if one thing is heavier than a second thing, and if the second thing is heavier than a third thing, then the first is heavier than the third. Now, the argument can be proven valid:

4. $(\forall y)(\forall z)[(Hcy \& Hyz) \supset Hcz]$ UI 3
5. $(\forall z)[(Hcp \& Hpz) \supset Hcz]$ UI, 4
6. $(Hcp \& Hpk) \supset Hck$ UI, 5
7. $(Hcp \& Hpk)$ Conj 1, 2
8. Hck   MP 6, 7

The following argument provides another illustration involving multiple quantification:

Goober knows Betty. Betty does not know Goober. So, it's not the case that if one person knows another, then the second person knows the first person.

In QL, this is represented:

1. Kgb
2. ~Kbg / ~$(\forall x)(\forall y)(Kxy \supset Kyx)$

The proof follows.

1. Kgb

2. ~Kbg / ~(∀x)(∀y)(Kxy ⊃ Kyx)

3. | (∀x)(∀y)(Kxy ⊃ Kyx) AP
4. | (∀y)(Kgy ⊃ Kyg)   UI, 3
5. | Kgb ⊃ Kbg   UI, 4
6. | Kbg   MP 1, 5
7. | Kbg & ~Kbg   Conj 2, 6

8. ~(∀x)(∀y)(Kxy ⊃ Kyx) IP 3–7

●─ PROVING LOGICAL TRUTHS

Recall that within truth-functional logic, a formula is proven tautological or truth-functionally true if it can be proven using a premise-free proof. Formulas proven true with premise-free proofs are sometimes called "logical truths" or are said to be "logically true" because we can prove them true using purely logical procedures and without relying on scientific experiments or observations of the physical world.

The procedures for proving logical truths within quantified logic are essentially the same as the procedures found within truth-functional and modal logic. If a quantified formula can be proven using a premise-free proof, it is a logical truth of quantificational logic. It is also said to be *quantificationally true*. A premise-free proof must start, of course, with either a conditional proof assumption or an indirect proof assumption.

Here is a proof that (∀x)(Fx)⊃(∃x)(Fx) is quantificationally true:

_____/(∀x)(Fx)⊃(∃x)(Fx)

1. | (∀x)(Fx)   AP
2. | Fa   UI 1
3. | (∃x)(Fx)   EG 2

4. (∀x)(Fx)⊃(∃x)(Fx)   CP 1–3

The formulas (∃x)(Ax⊃Ax) and (∀x)(Ax⊃Ax) are pretty obviously logically true. Notice how similar their proofs are:

_____/ (∃x)(Ax⊃Ax)

1. | ~(∃x)(Ax⊃Ax)   AP

2. | (∀x)~(Ax⊃Ax)   QE 1

3. | ~(Aa⊃Aa)   UI 2

4. | ~(~AavAa)   Imp 3

5. | Aa&~Aa   DM 4

6. (∃x)(Ax⊃Ax) IP 1–5

_____/ (∀x)(Ax⊃Ax)

1. | ~(∀x)(Ax⊃Ax)   AP

2. | (∃x)~(Ax⊃Ax)   QE 1

3. | ~(Aa⊃Aa)   EI 2

4. | ~(~AavAa)   Imp 3

5. | Aa&~Aa   DM 4

6. (∀x)(Ax⊃Ax) IP 1–5

•— NATURAL DEDUCTION WITH IDENTITY

In order to work with arguments containing identity signs, we will need natural deduction rules applicable to '='. The rules for identity will be based on the following two principles:

The Principle of Self-Identity

Each and every thing is identical with itself. That is:

$$(\forall x)(x=x)$$

This principle needs no argument. It is pretty obviously a necessary truth.

The Principle of the Indiscernability of Identicals

If x is identical with y, then whatever is true of x is true of y and whatever is true of y is true of x.

For example, since Robert Zimmerman simply *is* Bob Dylan, if it is true that Bob Dylan sang with Joan Baez, then it is true that Robert Zimmerman sang with Joan Baez. If it is true that Robert Zimmerman was born in Minnesota, then it is true that Bob Dylan was born in Minnesota, and so on.

The first rule for identity is based on the self-identity principle:

Identity A ("Id A")

At any step in a proof, you may assert $(\forall x)(x=x)$.

Typically, after you have used this rule, you will want to use UI and instantiate $(\forall x)(x=x)$ using the appropriate constant. Here is an example. Suppose you want to prove that the following argument is valid:

1. Pat is a neighbor of Chris and Bob is a neighbor of Chris.

2. Therefore, someone is a neighbor of Chris and that person is Bob.

The proof, using obvious abbreviations, employs Identity A:

1. Npc & Nbc/$(\exists x)[(Nxc)\&(x=b)]$
2. Nbc   Simp 1
3. $(\forall x)(x=x)$ Id A
4. b=b   UI 3
5. Nbc & (b=b)   Conj 2, 4
6. $(\exists x)[(Nxc)\&(x=b)]$ EG 5

Notice that we instantiated line (3) in order to derive b=b.

The second rule for identity is based on the indiscernability principle.

Identity B ("Id B")

*If **c** and **d** are two constants and a line of a proof asserts that the individual designated by **c** is identical with the individual designated by **d**, then you may carry down and rewrite any available line of the proof replacing any or all occurrences of **c** with **d** or any or all occurrences of **d** with **c**. A line of a proof is available unless it is within the scope of a discharged assumption.*

Formally:

1. c=d
2. P
3. P[c//d]

where **P** is an available line of the proof containing one or more occurrences of one of the constants and **P[c//d]** is exactly like **P** except that one or more occurrences of **c** have been replaced with **d** or one or more occurrences of **d** have been replaced with **c**.

Here is an application of Identity B.

1. $(\forall x)(Ax \supset Bx)$
2. At
3. $(\forall x)[Bx \supset (x=g)]$ /Ag&Bg
4. At$\supset$Bt   UI 1
5. Bt   MP 2, 4
6. Bt$\supset$(t=g)   UI 3
7. t=g   MP 5, 6
8. Ag   Id B, 2 (g replaced t)
9. Bg   Id B, 5 (g replaced t)
10. Ag&Bg   Conj 8, 9

The Identity B Rule may be informally summed up as follows:

> *Given an identity statement asserting that the constants* **c** *and* **d** *designate the same individual, one constant may replace the other in any available line.*

Notice that when we use Identity B, we can, if we wish, replace only some occurrences of a constant. For example:

1. $\underline{a=b}$ / b=a
2. $(\forall x)(x=x)$ Id A
3. a=a   UI 2
4. b=a   Id B 1,3

In this proof, the identity statement on line (1) gives us the right—specified in the Identity B rule—to replace 'b' with 'a' or 'a' with 'b' on any line. Therefore, we derived (4) by replacing only the left side occurrence of 'a' in (3) with 'b'.

Here is another argument employing Id B:

Robert Zimmerman was a singer and Ed Sullivan was not a singer. Robert Zimmerman is identical with Bob Dylan. So, Ed Sullivan is not Bob Dylan.

1. Sr&~Se

2. $\underline{r=b}$/b≠e

3. Sr   Simp 1

4. ~Se   Simp 1

5.   b=e   AP

6.   ~Sb   Id B, 5, 4

7.   Sb   Id B, 2, 3

8. b≠e   IP 5–7

We can now use our natural deduction rules to prove that the identity relation is reflexive, transitive, and symmetric. First, identity is *reflexive*. That is, each and every thing is identical with itself. The claim that everything is identical with itself is expressed within QL as:

$$(\forall x)(x = x)$$

The proof of this follows:

1.	~$(\forall x)(x = x)$	AP
2.	$(\forall x)(x = x)$	Id A
3.	$(\forall x)(x = x)$	IP 1–2

Next, the relation of identity is symmetric. In QL, the claim that identity is a symmetrical relation is:

$$(\forall x)(\forall y)[(x=y) \supset (y\backslash = x)]$$

The proof of this follows.

1.	~$(\forall x)(\forall y)[(x = y) \supset (y = x)]$	AP
2.	$(\exists x)$ ~$(\forall y)\,[(x = y) \supset (y = x)]$	QE 1
3.	$(\exists x)(\exists y)$ ~$[(x = y) \supset (y = x)]$	QE 2
4.	$(\exists y)$ ~$[(a = y) \supset (y = a)]$	EI 3
5.	~$[(a = b) \supset (b = a)]$	EI 4
6.	~$[$~$(a = b) \lor (b = a)]$	Imp 5
7.	$(a = b) \,\&\,$ ~$(b = a)$	DM 6
8.	$a = b$	Simp 7
9.	~$(b = a)$	Simp 7
10.	~$(a = a)$	Id B
11.	$(\forall x)(x = x)$	Id A
12.	$a = a$	UI 11
13.	$(\forall x)(\forall y)\,[(x = y) \supset (y = x)]$	IP 1–12

Finally, the identity relation is transitive. That is,

$$(\forall x)(\forall y)(\forall z)\{[(x = y) \ \& \ (y = z)] \supset (x = z)\}$$

The proof is as follows.

1.	$\sim(\forall x)(\forall y)(\forall z)\{[(x = y) \ \& \ (y = z)] \supset (x = z)\}$	AP
2.	$(\exists x)\sim(\forall y)(\forall z)\{[(x = y) \ \& \ (y = z)] \supset (x = z)\}$	QE 1
3.	$(\exists x)(\exists y)\sim(\forall z)\{[(x = y) \ \& \ (y = z)] \supset (x = z)\}$	QE 2
4.	$(\exists x)(\exists y)(\exists z)\sim\{[(x = y) \ \& \ (y = z)] \supset (x = z)\}$	QE 3
5.	$(\exists y)(\exists z)\sim\{[(a = y) \ \& \ (y = z)] \supset (a = z)\}$	EI 4
6.	$(\exists z)\sim\{[(a = b) \ \& \ (b = z)] \supset (a = z)\}$	EI 5
7.	$\sim\{[(a = b) \ \& \ (b = c)] \supset (a = z)\}$	EI 6
8.	$\sim\{\sim[(a = b) \ \& \ (b = c)] \ v \ (a = c)\}$	Imp 7
9.	$[(a = b) \ \& \ (b = c)] \ \& \sim(a = c)$	DM 8
10.	$(a = b) \ \& \ (b = c)$	Simp 9
11.	$\sim(a = c)$	Simp 9
12.	$a = b$	Simp 10
13.	$b = c$	Simp 10
14.	$a = c$	Id B, 12, 13
15.	$(\forall x)(\forall y)(\forall z)\{[(x = y) \ \& \ (y = z)] \supset (x = z)\}$	IP 1–14

THE SYSTEM QD

The natural deduction system we have been using may be called 'QD' (for 'quantificational deduction'). QD consists of two parts:

i. The language QL

ii. The natural deduction rules of this chapter.

A *proof in QD* is a sequence of sentences of QL, each of which is either a premise or an assumption or follows from one or more previous sentences according to a QD deduction rule, and in which (i) every line (other than a premise) has a justification and (ii) any assumptions have been discharged. An argument is *valid in QD* if and only if it is possible to construct a proof in QD whose premises are the premises of the argument and whose conclusion is the conclusion of the argument. An argument is *invalid in QD* if and only if it is not valid in QD.

A proof whose last line is **P** and that contains no premises is called a *premise-free proof* of **P**. If a premise-free proof of **P** can be constructed in QD, then **P** is quantificationally true. In such a case, **P** is also said to be a *theorem of QD*. In this case, **P** is also called logically true, or quantificationally true since **P** can be proven true using only the methods of quantificational logic, without investigating the physical world.

4. QUANTIFIED TRUTH-TREES

•— INTRODUCTION

This section assumes a working knowledge of the truth-functional tree rules, the rationales behind those rules, and the general rationale underlying the truth-tree method as presented in Chapter 4. The first rule involves the universal quantifier. Since the reasoning behind the tree rules is basically the same as the reasoning underlying several of the natural deduction rules just presented, and since we cannot assume that every reader covers both the natural deduction section and the tree rules section of this chapter, much of this section must go over ground already covered in the natural deduction section.

•— UNIVERSAL INSTANTIATION

A wff is called a *universal* quantification if and only if it begins with a universal quantifier term whose scope covers the entire wff. Examples include the following:

$$(\forall x)(Fx) \qquad (\forall y)(Fy) \qquad (\forall z)(Fz) \qquad (\forall x)(Fx \supset Gx) \qquad (\forall y)(Fy \lor Gy)$$

If we remove the quantifier term from a universal quantification and uniformly replace *each* occurrence of the variable it binds with a constant, we produce an *instantiation* or *substitution instance* of the universal quantification. For each universal quantification, there are an infinite number of possible instantiations. Here are a number of examples:

a. Universal quantification:

English: All whales are mammals

$$\text{QL:} \quad (\forall x)(Wx \supset Mx)$$

Some instantiations:

If Namu is a whale, then Namu is a mammal.

$$Wn{\supset}Mn$$

If Fred is a whale, then Fred is a mammal.

$$Wf{\supset}Mf$$

b. Universal Quantification:

English: All things are good.

QL: $(\forall x)(Gx)$

Some instantiations:

The moon is good.

$$Gm$$

Pat is good

$$Gp$$

The Space Needle is good.

$$Gs$$

c. Universal Quantification

English: All persons have rights.

QL: $(\forall x)(Px{\supset}Rx)$

Some instantiations:

If Pat is a person, then Pat has rights.

$$Pp{\supset}Rp$$

If Chris is a person, then Chris has rights.

$$Pc{\supset}Rc$$

Universal Instantiation Left   We are now ready to formulate a tree rule dealing with universal quantifiers. A universal quantification makes a claim about *all* things in the universe of discourse. For instance, $(\forall x)(Gx)$ above was used to assert that all things are good. That is, according to $(\forall x)(Gx)$, Gx applies to every thing in the universe. So, if a universal quantification is true, then its claim is true of each and every thing in the universe of discourse. Therefore, if a universal quantification is true, each and every substitution instance of it is true as well. Thus, from a universal quantification situated on the true side of a tree, we may validly infer—on the true side—any substitution instance. This gives us our first quantificational tree rule:

Universal Instantiation Left

From a universal quantification situated on the true side of a tree, you may infer, on the true side, any substitution instance.

Here are some applications of this rule:

* $(\forall x)(Gx)$	* $(\forall x)(Wx \supset Mx)$	*$(\forall x)(\forall y)(Fxy)$	* $(\forall x)(\exists y)(Fxy)$
Gb	Wc⊃Mc	$(\forall y)(Fcy)$	$(\exists y)(Fay)$

We place a star instead of a check by a universal quantification situated on the left of a tree. The star signifies that we may come back and apply the rule to the formula again as many times as we wish.

For example, consider the following symbolized argument:

1. $(\forall x)(Fx \supset Gx)$
2. $\underline{(\forall x)(Gx \supset Hx)}$ / Fs⊃Hs

The premises are numbered and the conclusion follows the slash. In order to prove this argument valid on a tree, we first place the premises on the true side and the conclusion on the false side of a tree:

1. $(\forall x)(Fx \supset Gx)$	
2. $(\forall x)(Gx \supset Hx)$	Fs⊃Hs

Next, we apply Universal Instantiation Left to each of the universal quantifications on the left. We may use any constant. For obvious reasons, let us instantiate with s:

* $(\forall x)(Fx \supset Gx)$	
* $(\forall x)(Gx \supset Hx)$	Fs⊃Hs
Fs⊃Gs	
Gs⊃Hs	

Next we apply Horseshoe Right:

* $(\forall x)(Fx \supset Gx)$	
* $(\forall x)(Gx \supset Hx)$	Fs⊃Hs ✔
Fs⊃Gs	
Gs⊃Hs	
Fs	Hs

Now, after we apply Horseshoe Left to the remaining conditionals, the tree closes:

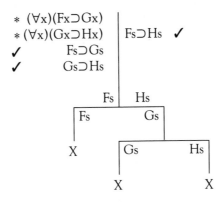

Universal Instantiation Right In everyday reasoning, people sometimes have to reason about an unnamed individual. For example, some people who work together might know that someone played a joke on the boss, but they don't know who the person was. In such cases, we often assign a temporary name to the unknown individual so that we can reason about the individual more easily. For example, if the boss is trying to figure out who played the joke on her, she might temporarily name the person "Dogbrain." Then, instead of saying, "I think that the person who played the joke on me worked the night shift," the boss could more simply say, "I think Dogbrain worked the night shift." (When the police assign a temporary name to an unknown individual, they often use a name such as John Doe or Jane Doe.) In the next tree rule, we are going to assign a temporary name to an unnamed individual for the purpose of reasoning about that individual.

Suppose a universal quantification is situated on the right side of a tree. For example:

$$\mid (\forall x)(Gx)$$

where Gx abbreviates 'x is good'. What follows? First, the assumption is that $(\forall x)(Gx)$ is false. That is, the assumption is that it's false that all things are good. If we assume that not all things are good, then it follows that there is at least one thing that is not good. Now, since we do not have a name for this unknown thing that is purportedly not good, and since we may need to reason about it later on, let us assign it a temporary name not used elsewhere in the tree. Call it 'Ozzie'. We will abbreviate this with 'o'. Then our inference is:

$$\mid \begin{array}{l} (\forall x)(Gx) ✔ \\ Go \end{array}$$

If we are working with a tree containing constants, we must be careful that the constant we assign an unknown individual is not a constant already on the tree. If we assign a constant that is already assigned to a known individual, then we are assuming that the unknown individual is the same individual as the named individual, which is an unjustified assumption. For example, suppose we begin with the assumption

$$Gd \mid (\forall x)(Gx)$$

We know that the individual designated by d is good and also that not all things are good. If we now instantiate the universal quantification using the constant 'd', we are assuming that the unnamed individual who is not good is the same individual as that designated by the 'd' on the left side. Such an assumption is completely unjustified. Hence, when we instantiate a right-side universal quantification, we must use a constant new to the tree:

$$Gd \mid (\forall x)(Gx) \; \checkmark$$
$$Ge$$

In general then, if it's not true that all things have a certain property, it follows that something doesn't have the property. This unknown "something" can be given a name so long as the name isn't already in use. This reasoning is contained in the following tree rule.

Universal Instantiation Right

> If a universal quantification is situated on the false side of a tree, you may infer on the false side any instantiation, provided that the constant used in the instantiation does not already appear elsewhere on the tree.

Here are some examples.

$$Gd \mid (\forall x)(Hx \supset Sx) \; \checkmark \qquad Hb \mid (\forall x)(\forall y)(Fxy) \; \checkmark \qquad Ra \mid (\forall x)(\exists y)(Hxy) \; \checkmark$$
$$Hc \supset Sc \qquad (\forall y)(Fsy) \qquad (\exists y)(Hdy)$$

Consider the following argument:

1. $(\forall x)(Fx \supset Gx)$
2. $(\forall x)(Gx \supset Sx)/ \quad (\forall x)(Fx \supset Sx)$

In order to prove this valid, we first place the premises on the truth side and the conclusion on the false side:

$$(\forall x)(Fx \supset Gx)$$
$$(\forall x)(Gx \supset Sx) \mid (\forall x)(Fx \supset Sx)$$

Next, we apply Universal Instantiation Right using a constant new to the tree:

$$(\forall x)(Fx \supset Gx)$$
$$(\forall x)(Gx \supset Sx) \mid (\forall x)(Fx \supset Sx) \; \checkmark$$
$$Fa \supset Sa$$

In the next step, Horseshoe Right is applied, and then Universal Instantiation Left is applied:

$$
\begin{array}{c|c}
* \quad (\forall x)(Fx \supset Gx) & \\
(\forall x)(Gx \supset Sx) & (\forall x)(Fx \supset Sx) \; ✔ \\
& Fa \supset Sa \; ✔ \\
Fa & Sa \\
Fa \supset Ga &
\end{array}
$$

Next, Universal Instantiation Left is applied to the remaining universal quantification:

$$
\begin{array}{c|c}
* \quad (\forall x)(Fx \supset Gx) & \\
* \quad (\forall x)(Gx \supset Sx) & (\forall x)(Fx \supset Sx) \; ✔ \\
& Fa \supset Sa \; ✔ \\
Fa & Sa \\
Fa \supset Ga & \\
Ga \supset Sa &
\end{array}
$$

Finally, the tree closes when Horseshoe Left is applied to the two conditionals:

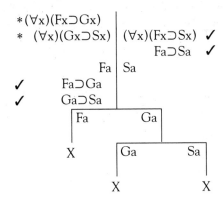

EXISTENTIAL INSTANTIATION

If a wff begins with an existential quantifier whose scope is the entire wff, the wff is called an *existential quantification*. If the existential quantifier is removed from an existential quantification and all the variables bound by that term are uniformly replaced by a constant, the result is an *instantiation* or *substitution instance* of the existential quantification. In each example below, the formula above the line is an existential quantification and the formula below the line is an instantiation of the existential quantification.

$$
\begin{array}{cccc}
\underline{(\exists x)(Fx)} & \underline{(\exists x)(Fx)} & \underline{(\exists x)(Hx \lor Gx)} & \underline{(\exists x)(\forall y)(Hxy)} \\
Fa & Fb & Ha \lor Ga & (\forall y)(Hay)
\end{array}
$$

Suppose we begin with the assumption that

$$\mid (\exists x)(Fx)$$

where Fx abbreviates 'x is funny'. The assumption is that it's not the case that at least one thing is funny. If we assume that it's not true that at least one thing is funny, then it follows that nothing is funny. So, for anything you pick—even something picked at random—that thing is not funny. Thus, we may use any constant and infer that the individual named by that constant is not funny. For instance, we may infer:

$$\begin{array}{l} \mid (\exists x)(Fx) \\ Fa \end{array}$$

Or, we may infer:

$$\begin{array}{l} \mid (\exists x)(Fx) \\ Fb \end{array}$$

Or we may infer:

$$\begin{array}{l} \mid (\exists x)(Fx) \\ Fc \end{array}$$

And so on.

In general, if it's not the case that at least one thing has a certain property, then nothing has that property, and it follows that any individual you pick does not have the property. The following rule reflects this reasoning pattern:

Existential Instantiation Right

> *From an existential quantification situated on the false side of a tree, you may infer on the false side any substitution instance.*

Here are a number of examples.

$$\begin{array}{lll} \mid (\exists x)(Hxv Gx)^* & \mid (\exists x)(Hxv Gx)^* & \mid (\exists x)(Hxv Gx)^* \\ Hav Ga & Hbv Gb & Hcv Gc \end{array}$$

The star beside each existential quantification indicates that we may come back to the formula and apply the rule again if we need to.

The next rule applies to existential quantifications situated on the other side of the tree. Suppose $(\exists x)(Fx)$ is true:

$$(\exists x)(Fx) \mid$$

This assumption indicates that some unnamed thing has the predicate F. In order to take account of this assumption, we will again employ the practice of assigning an unused name to an unknown thing for the purpose of reasoning about that thing. So, supposing that the constant b does not appear on the tree, we may legitimately infer:

$$(\exists x)(Fx) \quad \Big|$$
$$Fb$$

If the constant b had previously been used on the tree, then we could not have legitimately used b to instantiate $(\exists x)(Fx)$. This reasoning is reflected in the following rule:

Existential Instantiation Left

> *From an existential quantification situated on the true side of a tree, you may infer—on the true side—any substitution instance, provided that the constant used in the substitution instance is new to the tree.*

For example:

$$\checkmark(\exists x)(Fx)\Big|\ Hc \qquad \checkmark\ (\exists x)(\exists y)(Hxy)\Big|\ Ge \qquad \checkmark\ (\exists x)(\forall y)(Gxy)\Big|\ Jd$$
$$Fb \qquad\qquad\qquad (\exists y)(Hay) \qquad\qquad\qquad (\forall y)(Gcy)$$

Consider the following symbolized argument:

1. $(\forall x)(Fx \supset Gx)$
2. $\underline{(\exists x)(Fx)}/(\exists x)(Gx)$

In order to prove this valid, we place the premises and conclusion on a tree in the usual way:

$$(\forall x)(Fx \supset Gx) \quad \Big|$$
$$(\exists x)(Fx) \qquad\quad (\exists x)(Gx)$$

Next, let us apply Existential Instantiation Left using a constant new to the tree:

$$(\forall x)(Fx \supset Gx) \quad \Big|$$
$$\checkmark\ (\exists x)(Fx) \qquad\quad (\exists x)(Gx)$$
$$Fa$$

Now we may apply Universal Instantiation Left. On this move, we are allowed to use any constant:

$$*\ (\forall x)(Fx \supset Gx) \quad \Big|$$
$$\checkmark\ (\exists x)(Fx) \qquad\quad (\exists x)(Gx)$$
$$Fa$$
$$Fa \supset Ga$$

Next, using any constant, we may apply Existential Instantiation Right. For obvious reasons, we will use the constant a:

$$* \ (\forall x)(Fx \supset Gx)$$
$$✔ \quad (\exists x)(Fx) \quad | \quad (\exists x)(Gx) \ *$$
$$Fa$$
$$Fa \supset Ga$$
$$Ga$$

Finally, the tree closes when we apply Horseshoe Left to Fa⊃Ga:

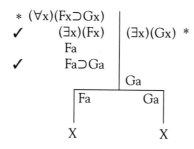

STRATEGY

Although tree rules may be applied to a problem in any order, the resulting tree will usually be simpler and easier to build if you follow two suggestions:

    i. Apply rules that do not split the tree before you apply rules that do.

    ii. Close off branches as soon as they self-contradict.

When building quantificational trees, two more suggestions will help you simplify things:

    iii. Apply rules that require new constants as soon as you possibly can in a tree.

    iv. Whenever you can, use constants that are already on the tree.

For example, consider the following argument:

    All baby aardvarks are cute.
    All things that are cute are precious.
    <u>Something is an aardvark.</u>
    So, something is precious.

In QL, this becomes:

    1. $(\forall x) \ (Ax \supset Cx)$
    2. $(\forall x) \ (Cx \supset Px)$
    3. <u>$(\exists x) \ (Ax)$</u> / $(\exists x)(Px)$

If we neglect suggestion (iii), the following tree results:

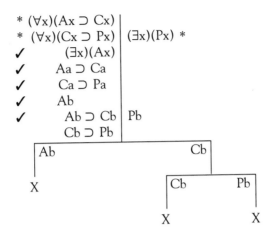

Notice that we began by applying Universal Instantiation Left to the two universal quantifications at the top of the tree. We then applied Existential Instantiation Left to $(\exists x)(Ax)$. However, when we applied Existential Instantiation Left to $(\exists x)(Ax)$, we had to use a constant new to the tree. This produced Ab. After doing this, we had to go back and apply Universal Instantiation Left all over again, using the constant b, in order to close the tree.

However, if we note suggestion (iii) and apply Existential Left before we apply Universal Left, the resulting tree is two steps shorter:

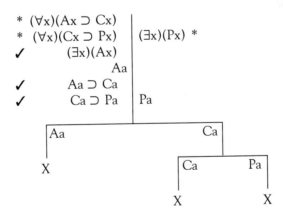

Both trees prove the same thing in the end. However, the second tree is simpler than the first tree.

Consider the following argument:

All cats are mammals.

Some cats are orange.

So, some orange things are mammals.

In QL, the argument is:

$$(\forall x)\,(Cx \supset Mx)$$
$$\underline{(\exists x)\,(Cx \,\&\, Ox)}/\,(\exists x)\,(Ox \,\&\, Mx)$$

If we follow suggestion (iv), then we will apply Existential Right and Universal Left using constants that are already on the tree. The resulting tree is:

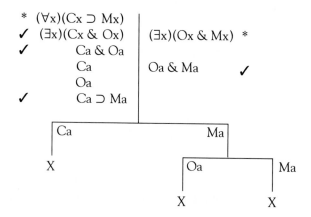

Generally, the fewer constants we use, the simpler the tree.

Here are two additional trees. Each illustrates an aspect of the quantificational tree technique of this chapter. The first tree proves the following argument valid:

1. $(\forall x)[(Ax \lor Bx) \supset Gx]$
2. $\underline{(\exists x)(Ax)}\quad /(\exists x)(Ax \,\&\, Gx)$

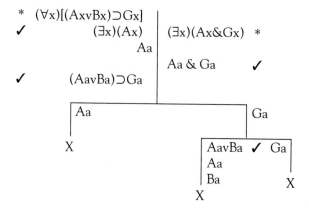

On this tree, notice that Existential Instantiation Left was applied first, using a constant new to the tree, before Universal Instantiation Left was applied. By applying Existential Instantiation Left first, we were able to use the same constant for every instantiation. Remember: The fewer the constants, the simpler the tree.

The second tree shows that the argument below is valid:

1. (∃x)(Fx&Gx)
2. (∀x)(Gx⊃Hx) / (∃x)(Fx&Hx)

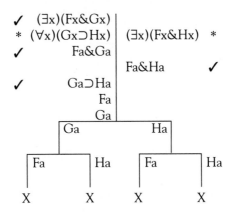

Again, by applying Existential Instantiation Left before applying Universal Instantiation Left, we were able to use the same constant on every instantiation.

TREES CONTAINING MULTIPLE QUANTIFIER FORMULAS

Is the following argument valid?

1. (∀x)(∀y)(Fxy⊃Gxy)
2. Fab/Gab

In order to test the argument, we place it on a tree as follows.

| (∀x)(∀y)(Fxy⊃Gxy) | |
| Fab | Gab |

Next, we apply Universal Instantiation Left and remove the first universal quantifier:

* (∀x)(∀y)(Fxy⊃Gxy)	
Fab	Gab
(∀y)(Fay⊃Gay)	

The next step involves another application of Universal Instantiation Left:

* (∀x)(∀y)(Fxy⊃Gxy)	
Fab	Gab
* (∀y)(Fay⊃Gay)	
Fab⊃Gab	

Next we apply Horseshoe Left:

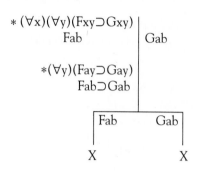

Since the tree closes, the argument is indeed valid.

●─ PROVING THEOREMS OF QUANTIFICATIONAL LOGIC

Consider the sentence $(\forall x)(Fx) \supset (\exists x)Fx$. Is there any interpretation on which this is false? Let us place it on the false side of a tree:

$$(\forall x)(Fx) \supset (\exists x)(Fx)$$

This represents the assumption that the formula is false. Next, we apply Horseshoe Right:

$$
\begin{array}{c|c}
& (\forall x)(Fx) \supset (\exists x)(Fx) \; ✔ \\
(\forall x)(Fx) & (\exists x)(Fx)
\end{array}
$$

The next move is an application of Existential Instantiation Right:

$$
\begin{array}{c|c}
& (\forall x)(Fx) \supset (\exists x)(Fx) \; ✔ \\
(\forall x)(Fx) & (\exists x)(Fx) \; * \\
& Fa
\end{array}
$$

Finally, when we apply Universal Instantiation Left, the tree closes:

$$
\begin{array}{c|c}
& (\forall x)(Fx) \supset (\exists x)(Fx) \; ✔ \\
* \;(\forall x)(Fx) & (\exists x)(Fx) \\
& Fa \\
Fa & \\
& \text{X}
\end{array}
$$

This sentence cannot possibly be false. If a sentence can be proven true using only the quantificational truth-tree method, the sentence is said to be a *theorem of quantificational logic*. It is also said to be logically true or quantificationally true, since it can be proven true using only the methods of quantificational logic—without investigating the physical world.

Next, consider the sentence $(\forall x)\{[(AxvBx)\&{\sim}Bx]{\supset}Ax\}$. Let us assume the sentence is false:

$$(\forall x)\{[(AxvBx)\&{\sim}Bx]{\supset}Ax\}$$

Universal Instantiation Right produces:

$$(\forall x)\{[(AxvBx)\&{\sim}Bx]{\supset}Ax\} \checkmark$$
$$[(AavBa)\&{\sim}Ba]{\supset}Aa$$

Next we apply Horseshoe Right:

	$(\forall x)\{[(AxvBx)\&{\sim}Bx]{\supset}Ax\}$ ✔
	$[(AavBa)\&{\sim}Ba]{\supset}Aa$ ✔
$(AavBa)\&{\sim}Ba$	Aa

Now, Ampersand Left gives us:

	$(\forall x)\{[(AxvBx)\&{\sim}Bx]{\supset}Ax\}$ ✔
	$[(AavBa)\&{\sim}Ba]{\supset}Aa$ ✔
✔ $(AavBa)\&{\sim}Ba$	Aa
$AavBa$	
${\sim}Ba$	

The tree closes once we apply Tilde Left and then Wedge Left:

	$(\forall x)\{[(AxvBx)\&{\sim}Bx]{\supset}Ax\}$ ✔
	$[(AavBa)\&{\sim}Ba]{\supset}Aa$ ✔
✔ $(AavBa)\&{\sim}Ba$	Aa
✔ $AavBa$	
✔ ${\sim}Ba$	
	Ba

$$\begin{array}{cc} Aa & Ba \\ X & X \end{array}$$

Thus, $(\forall x)\{[(AxvBx)\&{\sim}Bx]{\supset}Ax\}$ can't possibly be false.

● IDENTITY RULES

In order to work with arguments containing identity signs, we will need tree rules applicable to the identity sign. The rules for identity will be based on the following two principles:

The Principle of Self-Identity

Each and every thing is identical with itself. That is:

$$(\forall x)(x=x)$$

This principle is clearly a necessary truth.

The Principle of the Indiscernability of Identicals

If x is identical with y, then whatever is true of x is true of y and whatever is true of y is true of x.

For example, since Muhammad Ali simply is Cassius Clay, if it is true that Muhammud Ali was the greatest heavyweight boxer in the world, then it is true that Cassius Clay was the greatest heavyweight boxer in the world.

The first rule for identity is based on the self-identity principle. Let us say that a sentence of the form c=c, where c is any constant, is a *statement of self-identity*. Examples of self-identity statements would include: a=a, b=b, c=c, d=d, and so on. Clearly, any statement of self-identity is necessarily true. So, if a statement of self-identity shows up on the right side of a branch, a contradiction has been reached and the branch closes. Thus:

Identity Right

If a statement of self-identity appears on the right side of a branch, the branch closes.

The second rule for identity is based on the indiscernability principle.

Identity Left

If c and d are two constants, and a formula on the left side of a tree asserts that the individual designated by c is identical with the individual designated by d, then you may carry down and rewrite any formula on the tree, replacing any or all occurrences of c with d or any or all occurrences of d with c—as long as the formula remains on the same side of the tree after it is rewritten. More formally:

$$
\begin{array}{c|}
c=d \\
P \\
P[c//d]
\end{array}
$$

where **P** is a formula containing one or more occurrences of one of the constants and **P[c//d]** is exactly like **P** except that one or more occurrences of **c** have been replaced with **d** or one or more occurrences of **d** have been replaced with **c**.

Here are some examples. Suppose you want to prove that the following argument is valid:

1. (∀x)(Hx⊃Sx)
2. Hb
3. <u>b=e</u>/ Se

First, the premises must be placed on the true side and the conclusion must be placed on the false side.

(∀x)(Hx⊃Sx)	Se
Hb	
b=e	

Next, we apply Universal Instantiation Left:

*(∀x)(Hx⊃Sx)	Se
Hb	
b=e	
Hb ⊃Sb	

Now, Identity Left permits us to replace b with e in Hb⊃Sb:

*(∀x)(Hx⊃Sx)	Se
Hb	
b=e	
Hb ⊃Sb	
He⊃Se	

Applying Identity Left again, we replace b with e in Hb:

*(∀x)(Hx⊃Sx)	Se
Hb	
b=e	
Hb ⊃Sb	
He⊃Se	
He	

Finally, the tree closes when we apply Horseshoe Left:

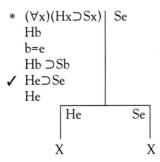

In Chapter 8 we examined the properties of reflexivity, symmetry, and transitivity. The following three trees show that identity is reflexive, symmetric, and transitive. First, identity is reflexive. That is, each thing is identical with itself. We begin by assuming this false:

$$(\forall x)(x = x)$$

The tree closes once we apply Universal Instantiation Right and Identity Right:

$$(\forall x)(x = x) \checkmark$$
$$e = e$$

X

Identity is also symmetrical. That is, if one thing is identical with a second thing, then the second thing is identical with the first. We first assume this false:

$$(\forall x)(\forall y)[(x = y) \supset (y = x)]$$

The tree closes in a few steps:

$$(\forall x)(\forall y)[(x = y) \supset (y = x)] \checkmark$$
$$(\forall y)[(a = y) \supset (y = a)] \checkmark$$
$$(a = b) \supset (b = a) \checkmark$$

$a = b$

$b = a$
$a = a$

X

The identity statement on the left allowed us to replace b with a in b = a. This allowed us to reach a = a on the right, which closed the tree.

Finally, identity is transitive. That is, if one thing is identical with a second and the second is identical with a third, then the first is identical with the third. The tree follows. In order to save space, all three quantifiers will be instantiated in one step:

$$(\forall x)(\forall y)(\forall z)\{[(x = y)\&(y = z)] \supset (x = z)\} \checkmark$$
$$[(a = b)\&(b = c)] \supset (a = c) \checkmark$$

$\checkmark$ (a = b)&(b = c)   a = c
$a = b$
$b = c$
$a = c$

X

●— A SUMMARY OF QL TREE TESTS

Let us say that a truth-tree constructed according to the rules of this chapter is a 'QL tree'. We may summarize a set of QL tree tests with a concise set of principles. In the following, '**P**' and '**Q**' are metavariables ranging over sentences of QL:

1. **P** is *quantificationally true* if and only if it is the case that if **P** is placed on the false side of a QL tree, the resulting completed tree closes.

2. **P** is *quantificationally false* if and only if it is the case that if **P** is placed on the true side of a QL tree, the resulting completed tree closes.

3. **P** is *quantificationally contingent* if and only if it is the case that if **P** is placed on the false side of a QL tree the resulting completed tree is open and it is the case that if **P** is placed on the true side of a QL tree, the resulting completed tree is open.

4. An argument is *quantificationally valid* if and only if it is the case that if the premises are placed on the true side of a QL tree and the conclusion is placed on the false side, the resulting completed tree closes. An argument is quantificationally invalid if and only if it is not quantificationally valid.

5. **P** and **Q** are *quantificationally inconsistent* if and only if it is the case that if the two are placed on the true side of a QL tree, the resulting completed tree closes.

6. **P** and **Q** are *quantificationally consistent* if and only if it is the case that if the two are placed on the true side of a QL tree, the resulting completed tree is open.

7. **P** *quantificationally implies* **Q** if and only if it is the case that if **P** is placed on the true side of a QL tree and **Q** is placed on the false side, the tree resulting completed closes.

8. **P** and **Q** are *quantificationally equivalent* if and only if it is the case that whenever the two are placed on opposite sides of a QL tree, the resulting completed tree closes.

● EXERCISES ●

Exercise 9.3

If you covered the section on natural deduction, supply justifications for the following natural deduction proofs:

1.
   1. $(\forall x)\,[Fx \supset (Gx\&Sx)]$

   2. $(\forall x)[(Gx\&Sx) \supset (HxvRx)]/(\forall x)[Fx \supset (HxvRx)]$

   3. $Fa \supset (Ga\&Sa)$ _____

   4. $(Ga\&Sa) \supset (HavRa)$ _____

   5. $Fa \supset (HavRa)$ _____

   6. $(\forall x)[Fx \supset (HxvRx)]$ _____

2. 1. (∀x) [Px⊃(SxvHx)]
    2. (∃x)(Px&~Sx)/ (∃x)(Px&Hx)
    3. Pc&~Sc _____
    4. Pc _____
    5. Pc⊃(ScvHc) _____
    6. ScvHc _____
    7. ~Sc _____
    8. Hc _____
    9. Pc & Hc _____
    10. (∃x)(Px&Hx) _____

3. 1. (∃x)(Fx)⊃(∀x)(Ax⊃Bx)
    2. Fc
    3. Ag/(∃y)(By)
    4. (∃x)(Fx) _____
    5. (∀x)(Ax⊃Bx) _____
    6. Ag⊃Bg _____
    7. Bg _____
    8. (∃y)(By) _____

4. 1. (∀x)(Hx)⊃(∀x)(Sx⊃Px)
    2. ~(∀x)(Sx⊃Px)
    3. ~(∀x)(Hx)⊃(∀x)(Mx)/(∀x)(Mx)
    4. ~(∀x)(Hx) _____
    5. (∀x)(Mx) _____

Exercise 9.4

Use either trees or natural deduction to show that the following are valid.

1. 1. (∀x)(Wx⊃Sx)
    2. (∀x)(Sx⊃Px)/(∀x)(Wx⊃Px)

2. 1. (∀x)(Hx⊃Jx)
    2. (∃x)(Hx)/(∃x)(Jx)

3. 1. (∀x)(Sx⊃Gx)
    2. Sa/Ga

4.   1.   (∀x)(Ax⊃Bx)

      2.   ~Bc/~Ac

5.   1.   (∀x)(Mx)

      2.   Hg/Hg&Mg

6.   1.   (∃x)(Fx&~Mx)

      2.   (∀x)(Fx⊃Hx)/ (∃x)(Hx&~Mx)

7.   1.   Sa⊃(∀x)(Fx)

      2.   HavSa

      3.   ~Ha

      4.   (∀x)(Fx⊃Gx)/(∀x)(Gx)

8.   1.   (∀x){Sx⊃[Px⊃(Bx&Qx)]}

      2.   (∃x)(Sx)/(∃x)(Px⊃Qx)

9.   1.   (∀x)(Axv~Bx)

      2.   Hs/(∃x)[(Hx&Ax)v(Hx&~Bx)]

10.   1.   (∀x)(Hx&Sx)

      2.   (∃x)(Hx)⊃(∃x)(Gx)/ (∃x)(Sx)&(∃x)(Gx)

11.   1.   (∃x)(Jx)⊃(∀x)(HxvSx)

      2.   (∀x)(Fxv~Sx)

      3.   (∀x)(Jx)/(∃x)(HxvFx)

12.   1.   (∀x)(Hx)⊃(∃x)(Sx)

      2.   (∀x)(Fx)v(∀x)(Hx)

      3.   ~(∀x)(Fx)__/(∃x)(Sx)

13.   1.   (∀x)(Ax⊃Bx)

      2.   (∀x)(Jx⊃Fx)

      3.   (∀x)(AxvJx)/ (∀x)(BxvFx)

14.   1.   (∀x)(Ax⊃Bx)

      2.   (∀x)(Ax)/(∀x)(Bx)

15.   1.   (∀x)(Hx⊃Gx)

      2.   (∀x)(~Gx)/(∀x)(~Hx)

16.   1.   (∀x)(Sx)⊃(∀x)(Gx)

      2.   ~(∀x)(Gx)/~(∀x)(Sx)

17. 1. (∃x)(Bx)⊃(∀x)(Hx⊃Gx)

    2. Bb&Hb/Gb

18. 1. (∀x)[Hx⊃(SxvGx)]

    2. ~Sb&~Gb/~Hb

19. 1. (∃y)(Sy&Hy)

    2. (∀x)[(GxvRx)⊃~Sx]/ (∃x)(~Gx)

20. 1. (∀x)[Hx⊃(Bx&Wx)]

    2. (∃y)(~By)/(∃z)(~Hz)

21. 1. (∀x)[Hx⊃(SxvGx)]

    2. (∃y)(~Sy&~Gy)/(∃x)(~Hx)

22. 1. (∀x)(Fx⊃Sx)

    2. Fa & Fb/Sa & Sb

23. 1. (∀x)(Hx⊃Qx)

    2. Ha v Hb/Qa v Qb

24. 1. (∃x)(Sx)⊃(∀x)(Hx⊃Gx)

    2. Sp&Hp/Gp

25. 1. (∃x)(Fx)⊃(∀x)(Sx)

    2. (∃x)(Hx)⊃(∀x)(Gx)

    3. Fs & Hg/(∀x)(Sx&Gx)

26. 1. (∃x)(Gx)⊃(∀x)(Hx⊃Sx)

    2. (∃x)(Gx v Sx)

    3. (∀x)(Sx⊃Gx)/(∀x)(~Gx⊃~Hx)

Exercise 9.5

Use either trees or natural deduction to show that each of the following is valid. If you use natural deduction, the Quantifier Exchange Rule will be necessary.

1. 1. ~(∀x)(Sx)/(∃x)(Sx⊃Px)

2. 1. (∀x)[(BxvCx)⊃Wx]

    2. ~(∀x)(Hxv~Bx)/(∃x)(Wx)

3. 1. (∀x)(Hx)⊃(∃x)(Sx)

    2. (∀x)(~Sx)/(∃x)(~Hx)

4. 1. (∃x)(~Hx)v(∃x)(~Sx)

    2. (∀x)(Sx)/~(∀x)(Hx)

5. 1. ~(∃x)(Ax)/(∀x)(Ax⊃Bx)

6. 1. (∃x)(Gx)v(∃x)(Hx&Sx)

   2. ~(∃x)(Hx)/(∃x)(Gx)

7. 1. (∀x)(Sx&Gx)v(∀x)(Qx&Hx)

   2. ~(∀x)(Qx)/(∀x)(Gx)

8. 1. (∀x)(Jx)⊃(∃x)(~Sx)

   2. ~(∀x)(Sx)⊃(∃x)(~Hx)/(∀x)(Hx)⊃(∃x)(~Jx)

9. 1. (∃x)(PxvGx)⊃(∀x)(Hx)

   2. (∃x)(~Hx)/(∀x)(~Px)

10. 1. ~(∃x)(Mx&~Gx)

   2. ~(∃x)(Mx&~Jx)/ (∀x)[Mx⊃(Gx&Jx)]

11. 1. (∀x)[(Px&Qx)⊃Rx]

   2. ~(∀x)(Px⊃Rx)/(∃x)(~Qx)

12. 1. (∃x)(~Hx)⊃(∀x)(Ax⊃Bx)

   2. ~(∀x)(HxvBx)/(∃x)(~Ax)

Exercise 9.6

Use either trees or natural deduction to show that the following are valid. If you use natural deduction, use either conditional or indirect proof.

1. 1. (∀x)(Ax⊃~Bx)

   2. (∀x)[Bx⊃(Hx&Ax)]/(∃x)(~Bx)

2. 1. (Ax)(Hx⊃Sx)

   2. (∀x)(Sx⊃Gx)/ (∀x)[Hx⊃(Sx&Gx)]

3. 1. (∀x)[Ax⊃(Bx&Cx)]/ (∀x)(Ax⊃Cx)

4. 1. (∀x)[Ax⊃(BxvCx)]/ (∃x) (Ax) ⊃(∃x)(BxvCx)

5. 1. (∀x)(Hx⊃Qx)

   2. (∀x)(Hx⊃Rx)/(∀x)[Hx⊃(Qx&Rx)]

6. 1. (∀x)[Hx⊃(Qx&Sx)]

   2. (∀x)[Px⊃(Rx&Mx)]/(∀x)(Sx⊃Px)⊃(∀x)(Hx⊃Mx)

7. 1. (∀x)[Px⊃(Hx&Qx)]/(∀x)(Sx⊃Px)⊃(∀x)(Sx⊃Qx)

8. 1. (∀x)~(Hx&~Bx)

   2. (∀x)~(Bx&~Gx)/(∀x)~(Hx&~Gx)

9. 1. (∀x)[(FxvGx)⊃Px]

    2. (∃x)(~FxvSx)⊃(∀x)(Rx)/(∀x)(Px)v(∀x)(Rx)

10. 1. (∃x)(Px)v(∃x)(Qx&Rx)

    2. (∀x)(Px⊃Rx)/(∃x)(Rx)

11. 1. (∃x)(Fx)⊃(∀x)(Gx⊃Sx)

    2. (∃x)(Hx)⊃(∀x)(~Sx)/(∀x)[(Fx&Hx)⊃~Gx]

12. 1. (∀x)(Jx⊃Px)

    2. (∀x)(Hx⊃Mx)/(∃x)(JxvHx)⊃(∃x)(PxvMx)

13. 1. (∀x)[(HxvPx)⊃Qx]

    2. (∀x)[(QxvMx)⊃~Hx]/(∀x)(~Hx)

14. 1. (∃x)(Qx)⊃(∀x)(Sx)

    2. Qa⊃~Sa/~Qa

15. 1. (∃x)(PxvJx)⊃~(∃x)(Px)/(∀x)(~Px)

16. 1. (∃x)(Hx)⊃(∃x)(Sx&Fx)

    2. ~(∃x)(Fx)/(∀x)(~Hx)

17. 1. (∃x)(Sx)⊃(∃x)(Hx&Jx)

    2. (∀x)(Fx⊃Sx)/(∃x)(Fx)⊃(∃x)(Hx)

18. 1. (∃x)(Px)⊃(∃x)(Qx&Sx)

    2. (∃x)(SxvHx)⊃(∀x)(Gx)/(∀x)(Px⊃Gx)

Exercise 9.7

Using either trees or natural deduction, prove the following theorems.

1. _____ /(∀x)[(HxvPx)v~Px]

2. _____ /(∀x)(PxvQx)v(∃x)(~Pxv~Qx)

3. _____ /~(∀x)(Ax&Bx)⊃(∃x)(~Axv~Bx)

4. _____ /(∃x)(Px)v(∀x)(~Px)

5. _____ /(∀x)(Px)⊃~(∃x)(~Px)

6. _____ /~(∃x)(Px&~Px)

7. _____ /(∃x)(Fx&Gx)⊃[(∃x)(Fx)&(∃x)(Gx)]

8. _____ /[(∀x)(Fx)&(∀x)(Gx)]⊃(∀x)(Fx&Gx)

9. _____ /(∀x)(Sx)v(∃x)(~Sx)

10. _____ /(∀x)[(Sx&Hx)⊃(SxvHx)]

11. _____ /(∀x)(Hx)⊃(∃x)(Hx)

12. _____ /(∀x)(Sx⊃Px)⊃[(∃x)(Sx)⊃(∃x)(Px)]

Exercise 9.8

Prove the following arguments valid using either trees or natural deduction.

1. 1. (∀x)(∀y)(Hxy⊃~Hyx)

    2. (∃x)(∃y)(Hxy)/(∃x)(∃y)(~Hyx)

2. 1. (∃x)(∀y)(Sxy)/(∀y)(∃x)(Sxy)

3. 1. (∀x)(Fx⊃Bx)

    2. (∀x)(∃y)(Sxyv~Bx)/(∃x)(∃y)(Sxyv~Fx)

4. 1. (∃x)(∀y)(Hyx⊃~Hxy)/~(∀x)(Hxx)

5. 1. (∀x)(∃y)(Sx&Py)/(∃y)(∃x)(Sx&Py)

6. 1. (∀x)(∃y)(JxvRy)/(∃y)(∃x)(JxvRy)

7. 1. (∀x)(∃y)(Hxy)⊃(∀x)(∃y)(Sxy)

    2. (∃x)(∀y)(~Sxy)/(∃x)(∀y)(~Hxy)

8. 1. (∃x)(∀y)(Pxy⊃Sxy)

    2. (∀x)(∃y)(~Sxy)/ ~(∀x)(∀y)(Pxy)

9. 1. (∀x)(∃y)(Px⊃Wy)/ (∀x)(Px)⊃(∃y)(Wy)

10. 1. (∃x)(∃y)(Kxy)/(∃y)(∃x)(Kxy)

Exercise 9.9

These arguments require the identity rules.

1. 1. Aa⊃Ha

    2. ~Ha

    3. a=b/~Ab

2. 1. Aa⊃Ha

    2. ~(Ab⊃Hb)/~(a=b)

3. 1. (∀x)(Ax⊃Px)

    2. (∀x)(Px⊃Hx)

    3. Aa&~Hb/~(a=b)

4. 1. Ab⊃Bb

    2. Rd⊃Sd

    3. Ab&Rd

    4. b=d/Bd&Sb

Exercise 9.10

Symbolize the following arguments and then prove each valid using either trees or natural deduction.

1. Every cat is a mammal. No fish is a mammal. So, no fish is a cat.

2. No cat is a reptile. Some pets are reptiles. So, some pets are not cats.

3. Some cats are orange. Every cat is a mammal. So, some mammals are orange.

4. Every dog is a mammal. No airplane is a mammal. If no airplane is a dog, then no airplane barks at cats. So, no airplane barks at cats.

5. Every musician is an artist. Every artist is a dreamer. Some high school dropouts are musicians. So, some high school dropouts are dreamers.

6. Charlie's car has a personality. Anything that has a personality is a person. So, Charlie's car is a person.

7. Cats and dogs reason, learn, and love. Any creature that reasons, learns, and loves possesses an immortal soul and goes to be with God when it dies. Therefore, cats and dogs possess immortal souls and will go to be with God when they die.

8. Every person loves that which he or she makes. Each person makes his or her own enemies. So, each person loves his or her enemies.

9. Anyone who loves hamburgers isn't a vegetarian. Wimpy loves hamburgers. So, Wimpy isn't a vegetarian.

10. Each event in one's life possesses eternal significance. A person's birth is one event in his or her life. So, a person's birth is an event of eternal significance.

11. Every hamburger sold by Dag's has Dag's special sauce on it. No other burger joint puts Dag's special sauce on its burgers. The burger Joe is eating does not have Dag's special sauce on it. Therefore, the burger Joe is eating is not a Dag's burger.

12. No member of the Revolutionary Communist Party is a Republican. Some Marxists are members of the Revolutionary Communist Party. So, some Marxists are not Republicans.

13. All who hang out at the Hasty Tasty admire Mao Tse-tung. All who admire Mao Tse-tung belong to the Progressive Labor Party. Stephanie does not belong to the Progressive Labor Party. So, Stephanie does not hang out at the Hasty Tasty.

    (During the 1960s, the Hasty Tasty, located on "the Ave" in Seattle's University District, was a hangout for student radicals and political activists. It was also a great place to get a good greasy burger any time of the day or night.)

14. Anyone who frequents the Blue Moon Tavern reads the Helix. If Deane sells the Helix, then Deane reads the Helix. Deane sells the Helix but doesn't frequent the Blue Moon. (He only goes in there once

in a blue moon.) So, some who read the Helix don't frequent the Blue Moon.

15. Dragons live forever. Nothing that lives forever is to be feared. Puff is a dragon. So, Puff is not to be feared.

16. All songs written by either Lennon or McCartney are rock 'n' roll songs. Therefore, there is no song written by Lennon or McCartney that is not a rock 'n' roll song.

17. If that piece of varnished sewer sludge is a work of art, then anything is a work of art. That piece of varnished sewer sludge is a work of art. So this glazed pile of dead bugs is a work of art.

18. If rocking horse people eat marshmallow pies, then newspaper taxis are waiting to take you away. If the girl with kaleidoscope eyes meets you, then rocking horse people eat marshmallow pies. The girl with kaleidoscope eyes will meet you. So, newspaper taxis are waiting to take you away. (Hint: Do you need to use quantifiers to symbolize this?)

19. Either all things are created by God or all things are material. If all things are material, then life has no transcendent meaning. If God does not exist, then it's not the case that all things are created by God. So, if God does not exist, then life has no transcendent meaning. (Do you need to use quantifiers when you symbolize this argument?)

APPENDIX ONE: MEASURING PROBABILITIES

I nductive strength is a matter of degree, but deductive validity is not. When students are introduced to the concept of inductive strength, they often want to know how the degree of inductive strength is measured or calculated. Unfortunately, logicians have not yet discovered a standard of measurement that will measure degrees of probability in all inductive arguments. For example, consider the following argument:

> Argument A
> 1. Sue graduated from MIT with honors in electrical engineering.
> 2. Sue did excellent work at Boeing.
> 3. So, Sue will perform well in her new engineering job.

With what degree of probability does the conclusion follow? 60 percent? 70 percent? 80 percent? With this argument, nobody knows any way to measure the probability in any precise, objective sense. However, consider the next argument:

> Argument B
> 1. A card was drawn at random from a perfectly shuffled poker deck.
> 2. Therefore, the card drawn was an ace.

In this case, we can measure the probability that the conclusion is true, given the truth of the premise. (That probability is 4/52 or 1/13.)

So, in the case of some inductive arguments, we can measure the probabilities involved; in the case of other inductive arguments, we have no way to measure probabilities. As far as anyone knows, there exists no overall standard of measurement that would measure the degree of probability involved in every inductive argument.

The Classical Theory of Probability In the seventeenth century, Blaise Pascal and Pierre Fermat developed the beginnings of a theory of probability now known as the *classical* theory. Pascal and Fermat wanted to develop an exact method of measuring the odds in games of chance. Suppose two assumptions can reasonably be made:

1. All possible outcomes can be counted.
2. Each outcome is equally likely.

Then, according to the classical theory, the probability of an event E occurring is:

$$P(E) = \frac{f}{t}$$

where 'P(E)' abbreviates 'the probability of event E occurring', f is the number of favorable outcomes, and t is the total number of possible outcomes.

For example, suppose you reach into a perfectly shuffled poker deck and draw a card at random. What is the probability that the card drawn is a queen? Since the deck contains 52 cards, the total number of possible outcomes is 52. The deck contains four queens, so there are four favorable possible outcomes. Therefore, assuming it's reasonable to suppose that all possible outcomes have been counted and each is equally likely, the probability of drawing a queen from a shuffled poker deck is 4/52 or 1/13.

PROBLEMS

Use the classical theory to compute the probabilities of the following events:

a. A six-sided die is rolled and comes up four.
b. A six-sided die is rolled and lands showing an odd number.
c. A card drawn from a poker deck is red.

During the eighteenth century, mathematicians in England developed a theory of probability that could be used to calculate rates and values on insurance policies. Whereas Pascal and Fermat were concerned with the odds at the roulette table and other games of chance, the English were concerned with questions such as: What is the probability that a 40-year-old person in good health will live to be 50? The problem was that when trying to figure a person's probable life span, the two assumptions required by the classical theory simply could not be made. (Given any 40-year-old, there is no way to count all the possible outcomes of the person's life; there is no way to determine how many of these are favorable to the person living 50 years; and there is no reason to suppose each outcome is equally likely.)

The English developed what is called the *relative frequency theory* of probability. According to this theory, the probability of an event E, when event E belongs to a category of events that occur again and again, is:

$$P(E) = \frac{f_o}{T_o}$$

where f_o is the number of observed favorable outcomes and T_o is the total number of observed outcomes.

To use this theory to determine the probability that a 40-year-old in good health will live to be 50, a statistician might observe a group of, say, 1,000 healthy 40-year-olds. If, 10 years later, 950 are still alive, then the probability that a healthy 40-year-old will live to be 50 is 950/1000 or 95 percent.

Measuring the Probabilities of Compound Events   The two theories introduced above were developed to provide a way to assign a probability value to a *single* event. Another problem within probability theory is this: Once probabilities have been assigned to single events, how are we to assign probabilities to *compound* arrangements of events? For example, we earlier used the classical theory to assign a probability to the event of drawing a queen on one draw from a shuffled poker deck. What is the probability of drawing two cards in a row and both are queens? To determine such a compound probability, we need a set of principles known as the *probability calculus*.

If the occurrence of an event E has no effect on the occurrence of an event E', then E and E' are called *independent* events. For example, if someone draws a card and replaces it before drawing a second card, the first draw has no effect on the second draw and so the two draws are independent events. If two events cannot both occur at the same time, the two events are called *mutually exclusive* events. For example, if a normal coin is tossed, it's not possible that it will come up both heads and tails at the same time, and so the heads outcome and the tails outcome are mutually exclusive events.

In the following, let p and q be variables ranging over events, and let 'P(q)' abbreviate 'the probability of event q occurring'. We will briefly examine seven fundamental rules of the probability calculus.

— **Rule 1** —•

The probability of an event that necessarily must happen is, by convention, 1. Thus:

> If p necessarily must happen, then P(p)=1

For example, the probability that it will either snow tomorrow or it will not snow tomorrow is exactly 1.

— **Rule 2** —•

The probability of an event that necessarily cannot happen is 0. Thus:

> If p necessarily cannot happen, P(p)=0.

For example, the probability that it will snow tomorrow in Atlanta at noon and yet not snow tomorrow in Atlanta at noon is 0.

— **Rule 3   The Restricted Conjunction Rule** —•

This principle is used to compute the probability of two events occurring together when the two events are independent of each other. If p and q are independent events, then the probability that p and q both happen is equal to the probability of p multiplied by the probability of q. That is,

$$P(p \text{ and } q)=P(p) \times P(q)$$

where p and q are independent.

Here is an example. We have already seen that the probability of drawing a queen from a shuffled poker deck is 1/13. What is the probability of drawing a queen on the first draw, and then, after placing that card back in the deck, drawing a queen again on the second draw? According to the Restricted Conjunction Rule, the probability is figured this way:

$$P(\text{queen 1 and queen 2})=P(\text{queen 1}) \times P(\text{queen 2})$$
$$= 1/13 \times 1/13 = 1/169$$

where 'queen 1' abbreviates 'a queen is drawn on the first draw' and 'queen 2' abbreviates 'a queen is drawn on the second draw'.

For another example, what is the probability that you toss two quarters in the air and both land tails up? Since the first tails has no effect on the second tails, the two events are independent. The probability of a fair coin landing tails is obviously 1/2. So, the probability that both land tails is calculated as follows:

$$P(\text{tails 1 and tails 2})=P(\text{tails 1}) \times P(\text{tails 2})$$
$$= 1/2 \times 1/2 = 1/4$$

What is the probability of rolling two threes with a standard pair of dice? Since each die has six sides, the probability that a single die is rolled and comes up three is 1/6. Since the outcome of one die's roll doesn't effect the outcome of the other's roll, the probability that a pair of dice comes up two threes is calculated according to the Restricted Conjunction Rule as follows:

$$P(\text{Three 1 and Three 2})=P(\text{Three 1}) \times P(\text{Three 2})$$
$$= 1/6 \times 1/6 = 1/36$$

— **Rule 4   The General Conjunction Rule** —•

This rule is used to compute the probability of two events occurring together in cases where the two events are not independent:

If the occurrence of p affects the occurrence of q, then the probability of both p and q happening is equal to the probability of p times the probability that q occurs where q's probability is calculated on the assumption that p occurred. That is:

$$P(p \text{ and } q) = P(p) \times P(q \text{ given that p occurred})$$

For example, what is the probability of drawing two cards in a row without re-placing any cards drawn, and both are aces? The probability that the first card is an ace is, of course, 4/52. The probability that the second card is an ace—given that the first card was an ace and was not replaced—is 3/51, for at the second draw, only three aces remain among the 51 remaining cards. The probability of drawing two aces in a row without replacement is therefore:

$$P(\text{ace 1 and ace 2}) = P(\text{ace 1}) \times P(\text{ace 2 given ace 1})$$
$$= 4/52 \times 3/51 = 12/2652 = 1/221$$

The General Conjunction Rule may also be used if the two events are indepen-dent. In such a case, the value of P(q given p) will be the same as P(q). So, when the two events are independent, the General Conjunction Rule gives the same result as the Restricted Conjunction Rule.

PROBLEM

A bowl contains eight apples, three oranges, and six peaches.

1. What is the probability of drawing at random two apples in a row with replacement? What is the probability of drawing two apples in a row without replacement?

2. What is the probability of randomly drawing an orange and then a peach, with replacement? What is the probability of drawing an or-ange and then a peach without replacement?

— Rule 5   The Restricted Disjunction Rule —•

This rule is used to compute the probability that one or the other of two events oc-curs when the two events are mutually exclusive. The rule is:

> *If p and q are mutually exclusive events, then the probability that either p or q occurs is equal to the probability of p plus the probability of q. That is:*
>
> $$P(p \text{ or } q) = P(p) + P(q)$$
>
> *where p and q are mutually exclusive.*

For example, what is the probability of drawing just one card from a poker deck and the card is either a jack or an ace? The probability that a drawn card turns out an ace is 4/52. The probability that a drawn card turns out to be a jack is also 4/52. Therefore, the probability that a single card drawn is either a jack or an ace is:

$$P(\text{jack or ace}) = P(\text{jack}) + P(\text{ace})$$
$$= 4/52 + 4/52 = 8/52 = 2/13$$

PROBLEM

A bowl contains eight grapes, five cherries, and three hamburgers. What is the probability that someone randomly selects one item from the bowl and the item is either a grape or a cherry? What is the probability one item is randomly drawn and it is not a grape?

Combining the Rules Let us now use the Restricted Disjunction Rule in combination with the Restricted Conjunction Rule. Assume we are reaching into the bowl mentioned in the exercise immediately above. What is the probability of randomly drawing out either a grape or a cherry and then, after replacing what was drawn, drawing again either a grape or cherry? That is, what is the value of:

P[(grape or cherry)1 and (grape or cherry)2]

The probability of drawing either a grape or a cherry on a single draw is, according to the Restricted Disjunction Rule:

P(grape or cherry) = P(grape) + P(cherry)
= 8/16 + 5/16
= 13/16

According to the Restricted Conjunction Rule, the probability that this happens twice is calculated by multiplying the probability of the first by the probability of the second. We may now insert this into the format required by the Restricted Conjunction Rule as follows:

P[(grape or cherry)1 and (grape or cherry)2]
=P(grape or cherry)1 × P(grape or cherry)2
=(8/16 + 5/16) × (8/16 + 5/16)
= 13/16 × 13/16

── **Rule 6 The General Disjunction Rule** ──•

This rule is used to compute the probability that one or the other of two events occurs, whether or not the two events are mutually exclusive events. In this discussion, we will limit ourselves to the simpler case in which the two events are independent. The rule in this case is:

The probability that either p or q occurs— whether or not p and q are mutually exclusive events— is equal to the probability of p plus the probability of q minus the probability that p and q both occur. That is:

P(p or q) = P(p) + P(q) – P(p and q)

where p and q are mutually exclusive events.

Since p and q are independent, this is equal to:

$$P(p \text{ or } q) = P(p) + P(q) - [P(p) \times P(q)]$$

What is the probability of drawing at least one ace on two draws from a deck of cards, with replacement? That is, what is the probability of drawing either an ace on the first draw, or on the second draw, or on both draws—with replacement? Notice that this is an inclusive disjunction of events. Since the events are not mutually exclusive, the Restricted Disjunction Rule does not apply and we must use the General Disjunction Rule. According to that rule, in this case:

$$
\begin{aligned}
P(\text{ace 1 or ace 2}) &= P(\text{ace 1}) + P(\text{ace 2}) - P(\text{ace 1 and ace 2}) \\
&= (4/52 + 4/52) - (4/52 \times 4/52)
\end{aligned}
$$

What is the probability of getting tails on either or both of two tosses of a quarter? According to the General Disjunction Rule:

$$
\begin{aligned}
P(\text{tails 1 or tails 2}) &= P(\text{tails 1}) + P(\text{tails 2}) - P(\text{tails 1 and tails 2}) \\
&= (1/2 + 1/2) - (1/2 \times 1/2) \\
&= 1 - 1/4 \\
&= 3/4
\end{aligned}
$$

For one more example, what is the probability of getting at least one 3 when rolling a pair of 10-sided dice, where the sides are numbered 1 through 10? The General Disjunction Rule tells us:

$$
\begin{aligned}
P(\text{three 1 or three 2}) &= P(\text{three 1}) + P(\text{three 2}) - P(\text{three 1 and three 2}) \\
&= (1/10 + 1/10) - (1/10 \times 1/10) \\
&= (1/10 + 1/10) - 1/100 \\
&= 19/100
\end{aligned}
$$

PROBLEM

1. You have been blindfolded and placed in front of two buckets of hot greasy burgers. The first bucket contains three hamburgers, four cheeseburgers, and five fishburgers. The second bucket contains two hamburgers, three cheeseburgers, and six fishburgers. You are to reach into the first bucket and pull out two burgers in a row, without replacement. You are next to reach into the second bucket and do the same thing. If one or more of the buckets yields first a cheeseburger and then a fishburger, you win the grand prize. (You get to eat the burgers you pull out!) What is the probability that you will win? That is, what is the probability you will draw first a cheeseburger then a fishburger from one or the other or from both of the buckets? (Hint: You will have to combine both the General Conjunction and General Disjunction rules.)

2.   Assume you have in front of you the two buckets of burgers from the previous problem. What is the probability of drawing one burger from each bucket, getting either a cheeseburger or a fishburger? That is, what is the probability that either the first bucket yields either a cheeseburger or a fishburger or the second bucket yields either a cheeseburger or a fishburger? (Hint: You will have to use both the General Disjunction and the Restricted Disjunction rules.)

— Rule 7   The Negation Rule ⟶•

The Negation Rule is used to compute the probability of an event when the probability of the event *not* happening is known. The rule is:

$$P(p) = 1 - P(\text{not } p).$$

The proof of this rule is fairly obvious. According to the Restricted Disjunction Rule:

$$P(p \text{ or not } p) = P(p) + P(\text{not } p)$$

Since either p or else not p must happen, Rule 1 tells us:

$$P(p \text{ or not } p) = 1$$

Therefore:

$$1 = P(p) + P(\text{not } p)$$

If we subtract P (not p) from each side, we get:

$$1 - P(\text{not } p) = P(p).$$

For example, what is the probability of tossing two dimes and getting tails on at least one? The probability of this *not* happening is the probability of getting heads on both tosses, which is $1/2 \times 1/2 = 1/4$. So, the probability of at least one tail on two tosses is:

$$1 - 1/4 = 3/4.$$

———————————• PROBLEMS •———————————

Part I   Figure the following probabilities:

1.   P(six or two on a single roll of a die)

2.   P(heads on 4 successive tosses of a coin)

3.   P(king or ace on one draw)

4. P(at least one king or queen on 2 draws, with replacement)

5. P(two kings in two draws) if:

 a. The first card drawn is replaced.

 b. The first card is NOT replaced.

Part II If a pair of dice are rolled, what is the probability that the points add up to:

1. 3?

2. 4?

3. 5?

4. 6?

Part III Imagine two bowls of marbles. One has three green, six blue, and seven yellow marbles, and the other has two blue, four green, and five yellow marbles. If a single marble is drawn from each, what is the probability that:

1. Both are green?

2. At least one is blue?

3. One is blue, the other is green?

4. At least one is either blue or yellow?

5. Both are the same color?

Part IV Imagine a jar containing five purple, six orange, and seven gold marbles. If two marbles are drawn without replacement, what is the probability that:

1. Both are the same color?

2. Both are gold?

APPENDIX TWO: QUANTIFIED MODAL LOGIC

Arguments that contain both modal operators and quantifiers are called *quantified modal arguments*. *Quantified modal logic* is the branch of modal logic concerned with the study and evaluation of such arguments.

Since it contains both quantifiers and modal operators, a quantified modal argument obviously cannot adequately be symbolized within QL alone or within ML alone. In order to symbolize a quantified modal argument, we need a formal language that combines the symbols of QL and ML into one formal language. The language combining QL and ML will be called QML (for Quantified Modal Logic).

We won't take the space to set out formally the syntax for QML. However, you will get an idea of how the language is used if we symbolize some ordinary quantified modal sentences. As we saw earlier, *materialism* is the view that all things are composed of nothing but matter. A *materialist* is one who believes that materialism is true. Now, in QL, materialism may be symbolized as:

$$(\forall x)(Mx)$$

where Mx abbreviates 'x is material'.

Next, two types of materialism may be distinguished. Some materialists claim that materialism is a necessary truth, but others claim it is merely a contingent truth. To claim that materialism is necessarily true is really to claim two things:

1. all things in the actual world are material;
2. a nonmaterial thing is not even a logical possibility, that is, in all worlds all things are material.

To claim that materialism is contingently true is also to claim two things:

3. all things in the actual world are material;

4. a nonmaterial thing is a logical possibility, that is, in some possible worlds some things are nonmaterial.

We must use QML if we are going to symbolize either (2) or (4).

How shall we symbolize the claim that materialism is necessarily true? There are two possibilities:

5. □(∀x)(Mx)
6. (∀x) □(Mx)

Sentence (5) abbreviates 'It is necessarily true that all things are material', which is to say that in all worlds it's true that all things are material. Sentence (6) symbolizes 'each and every thing in the actual world is necessarily material', which is to say that each actual thing is such that in every world it has the property of being material.

Let us now compare (5) and (6). Notice that in (5), the modal operator takes the entire sentence as its scope. During the Middle Ages, the philosopher Thomas Aquinas called such sentences *de dicto* modal sentences because the modal operator applies to the whole sentence (*dictum*). A *de dicto* modal sentence or modality attributes a modal property to a *proposition*.

However, in (6), it is the *quantifier* that takes the whole sentence as its scope and the modal operator applies only to the part to its right, Mx. Thus, in (6), the box indicates that x possesses the property represented by M—the property of being material—in every world. That is, x possesses the property necessarily. Aquinas called such sentences *de re* modalities because the modal operator applies in such cases to a thing (*res*), or, more specifically, to a thing's possession of a property.

Before we conclude something from this, it will be instructive to consider an example that mixes a modal operator and an existential quantifier. Atheism is the view that there exists no such being as God. Traditional theism, on the other hand, is the view that there exists a being—often called God—who is omnipotent, omniscient, omnibenevolent, and the creator of the world. In QL, atheism may be abbreviated as:

7. ~(∃x)(Ox)

where 'Ox' abbreviates 'x is omnipotent, omniscient, omnibenevolent, and the creator of the world.'

However, two types of atheism may be distinguished. Some atheists claim that atheism is necessarily true:

8. □~(∃x)(Ox)

According to (8), it is necessarily true that there is no such thing as a being who is omnipotent, omniscient, omnibenevolent, and the creator of the universe. Sentence (8) is, of course, a *de dicto* modality.

To claim that atheism is necessarily true is really to claim two things:

9. there is no such being as God;

10. the existence of a being such as God is not even a logical possibility.

On this view, the very concept of God is contradictory or unintelligible; no possible world contains a being such as God. So, not only is there no God, but there could not even possibly have been such a being.

On the other hand, some atheists claim that it is only contingently true that there is no such being as God. To claim this is to claim two things:

11. there is no such being as God;

12. there could have been such a being—the existence of God is at least a logical possibility.

The first part of this view—(11)—is symbolized simply as $\sim(\exists x)(Ox)$. How shall we symbolize the second part, namely (12)? If the modal operator governs the entire sentence, then we have the *de dicto* modality:

$$13. \quad \Diamond(\exists x)(Ox)$$

According to this, it's possibly true that there is a being such as God, that is, in some possible world there is an omnipotent, omniscient, omnibenevolent, creator of the universe. However, if the quantifier governs the entire sentence, we get the *de re* modality

$$14. \quad (\exists x)\Diamond(Ox)$$

This sentence expresses a very strange claim. According to (14), there actually exists an individual who could have been God. That is, there is an actually existing being who, in some other possible world, is omnipotent, omniscient, omnibenevolent, and creator of the universe. Which of (13) and (14) do you think is the more plausible construal of the contingent form of atheism?

There is considerable agreement among philosophers that a satisfactory semantical theory for *de dicto* quantified modalities has been worked out. However, when we turn to the case of *de re* quantified modal sentences, it's a different story altogether. No widespread agreement exists on the semantics of such sentences, and a number of logicians, most notably W. V. O. Quine, argue that an acceptable semantical theory for *de re* sentences is an impossibility. We only have space to peek a little ways into this issue.

The goal of a semantical theory for *de re* modalities is to specify the truth-conditions for *de re* sentences. Consider a claim associated with the ancient Greek philosophers:

All persons are necessarily rational.

The *de re* symbolization of this is:

$$15. \quad (\forall x)(Px \supset \Box Rx)$$

Literally, this tells us that all persons in the actual world are—in every world—rational. That is, all persons have the property of rationality necessarily. Now, if you think about it, the truth of this claim seems to require that persons in this world also exist in other possible worlds as well. In other words, the truth of this *de re* sentence seems to presuppose that individual things exist in more than one world. This implies that in some cases, individuals in one world are identical with—are the same individuals as—individuals in other possible worlds. Thus, if we are going to formulate a semantics for *de re* sentences, it looks as if we will have to suppose individuals exist in more than one world—that individuals in one world are the same individuals as individuals in other worlds.

The view that in some cases things in one world are identical with things in other worlds is known as the *doctrine of transworld identity* (or 'TWI' for short). It appears that the truth of a *de re* modality requires the truth of TWI.

What is wrong with that? The problem is that a number of philosophers have argued that the concept of transworld identity—the concept of an individual in one world being one and the same individual as an individual in another possible world—is either contradictory or, at least, unintelligible. But if the concept of transworld identity is contradictory, or if it just makes no sense at all, and if any semantical theory for *de re* sentences must presuppose the possibility of transworld identity, then no acceptable semantical theory for *de re* sentences is going to be found.

A view opposed to TWI is the view that each individual thing exists in exactly one world. This has been named the *doctrine of world-bound individuals* (or 'WBI' for short). If this view is correct, then no thing exists in more than one world. Leibniz held a view of this sort. The most prominent contemporary defender of WBI is the philosopher David Lewis. And so there is a debate within modal logic between those who favor transworld identity and those who deny the very possibility of it.

The defenders of TWI have developed ingenious arguments in support of transworld identity and powerful criticisms of the theory of world-bound individuals. Unfortunately for those who like clear-cut answers to things, the defenders of WBI have developed ingenious arguments in favor of WBI and powerful criticisms of TWI. Some philosophers reject the very concept of a possible world as problematic or unintelligible, and along with that they reject both TWI and WBI.

Which view is correct? The arguments for and against the various views are extremely complex, and an examination of the issue would take us beyond the scope of an introductory symbolic logic text. The adventurous student who wants to explore the issue may consult the Further Reading section at the end of the book.

Some philosophers have argued that we can make good sense of *de re* modal sentences if we interpret them—specify a semantics for them—in terms of the theory known as *Aristotelian essentialism*. According to this theory, things have both *necessary* (or essential) properties and *contingent* (or accidental) properties. Roughly, x has the property P essentially just in case x has P and it's not possible that x exist but lack P. In terms of possible worlds, if P is an essential property of x, then x has P in every world in which x exists. For example, philosophers in ancient Greece argued that a capacity for rational thought was an essential property of any human being. An individual x has the property P accidentally or contingently just in case x has P in some possible circumstances but lacks P in other possible circumstances. In terms of possible worlds, x has P accidentally or contingently, if and only if x has P in some worlds and x lacks P in other worlds. The length of a person's hair or the color of a person's eyes are examples of accidental properties of persons.

Next, let '□Px' abbreviate 'x has P essentially' and let '◇Px' abbreviate 'in some world or other x has P'. Then we can abbreviate the claim that all human beings are essentially rational as:

$$16. \quad (\forall x)(Hx \supset \Box Rx)$$

And we can abbreviate the claim that Socrates is only accidentally long-haired as:

$$17. \quad \Diamond Ls \& \Diamond \sim Ls$$

According to essentialism, the collection consisting of those properties essential to an individual constitutes that individual's essence. As we follow an individual in our minds from possible world to possible world, the individual's accidental properties may vary, but the essence remains the same.

There is considerable debate within philosophy over the truth of Aristotelian essentialism. The reader wishing to pursue the matter may consult the references in the Further Reading section.

Let us consider some issues any semantical theory of quantified modal logic must resolve. Should the semantics for QML count the following as a theorem?

$$18. \quad (\forall x)\Box(Fx) \supset \Box(\forall x)(Fx)$$

According to this, if everything that actually exists is necessarily F, then in every world everything is F. This formula is known as the Barcan Formula because Ruth Barcan, one of the pioneers of contemporary modal logic, first discussed its modal status. Equivalent to the Barcan Formula is:

$$19. \quad \Diamond(\exists x)(Fx) \supset (\exists x)(\Diamond Fx)$$

This says that if there's a possible world in which something is F then there's actually a thing in this world that is possibly F.

Should the semantics of QML count the formula known as the *Converse Barcan Formula* as necessarily true? That formula is:

$$20. \quad \Box(\forall x)(Fx) \supset (\forall x)\Box(Fx)$$

According to (20), if in all worlds all things are F, then all things are necessarily F. Equivalent to this is:

$$(\exists x)(\Diamond Fx) \supset \Diamond(\exists x)(Fx)$$

This says that if there actually exists something that is possibly F, then there's a world in which something is F.

The Homogeneity Assumption The set of objects existing in a possible world constitutes the *domain* of that world. Suppose we assume that each possible world has exactly the same domain, that is, the domain of objects does not change from world to world. Then each world contains exactly the same set of objects. What varies from world to world is only the *properties* things have and the *relations* things bear to each other. Call this assumption the *homogeneity assumption*. On this assumption, each thing in existence exists in every world, which is to say that each thing in existence has necessary existence.

The Barcan and Converse Barcan formulas turn out to be logical truths only if we make the homogeneity assumption. You can verify this for yourself if you reflect upon the two formulas and on the homogeneity assumption. So the Barcan and Converse Barcan formulas require that each thing in existence has necessary existence.

However, the homogeneity assumption contradicts some of our deepest modal intuitions. For example, it seems plausible to suppose that many things in existence are contingent beings, which is to say that they might not have existed if things had been different. Contingent beings do not exist in every possible world. It also seems plausible to suppose that there might have been more things than there actually are. For instance, it seems plausible to suppose that John F. Kennedy could have had one more brother than he actually had, that General Motors might have made one more car in 1992 than it actually did, and so on. But if these statements are true, then there are possible worlds containing objects that don't exist in the actual world, for to say (for example) that John F. Kennedy could have had one more brother is merely to say that there is a possible world in which JFK had one more brother than he actually had. These intuitions contradict the homogeneity assumption.

Let us call the view that the domain varies from world to world the heterogeneity assumption. If we suppose that the domain changes from world to world, that is, that some worlds contain objects that don't exist in other worlds, then the Barcan Formula counts as logically false. This can be seen if you reflect upon the heterogeneity assumption and the Barcan Formula.

Why would someone reject the heterogeneity assumption and favor the homogeneity assumption? Some philosophers have argued that if we make the heterogeneity assumption, then we must suppose that there exist possible but nonactual objects and they go on to argue that the concept of a possible but nonactual object is contradictory or incoherent. The student who wishes to investigate this question should consult the Further Readings section.

PROBLEMS

1. Symbolize in a *de dicto* form and then in a *de re* form:

   a. All singers necessarily have voices.

   b. All bicyclists are necessarily two-legged.

   c. Some people might possibly go swimming.

2. Should a semantics for QML count the following formula as necessarily true?

$$\Diamond(\forall x)(Ax) \supset (\forall x)(\Diamond Ax)$$

Give a reason for your answer.

3. In Meditation VI, Descartes gives several arguments for mind–body dualism. Mind–body dualism is the view that the mind is an entity separate from the body. One of his arguments may be summarized thus:

   a. The body is essentially divisible.

   b. The mind is essentially indivisible.

   c. Therefore the body and the mind are two distinct entities. (Something is divisible just in case it is composed of parts; something is indivisible if it is not composed of parts.)

   Symbolize this argument in QML. Do you find the argument valid or invalid? Give a reason for your answer.

4. Should the semantics for QML count the following as necessarily true?
   $(\forall x)(\Diamond Fx) \supset \Diamond(\forall x)(Fx)$

   Give a reason for your answer.

FURTHER READING

The following list of books is not meant to be complete. It is offered merely to give you some places to start if you want to pursue a particular topic.

CHAPTER ONE

The most important reference work in philosophy is the eight-volume *Encyclopedia of Philosophy* edited by Paul Edwards. This contains articles on just about every topic in logic and in philosophy. The individual articles also provide references for further research. If a particular topic interests you, this encyclopedia is a good place to begin your search.

On ancient philosophy:
Copleston [1962] is a multivolume history of philosophy. Volume I is a readable and short introduction to ancient philosophy. For other introductions to the topic see Matson [1987] volume I, Robinson [1968], Armstrong [1977], and Barnes [1982]. Also, Wilbur and Allen [1979] is a good anthology of articles on the topic. Aristotle's view of human nature is developed in his *Nichomachean Ethics*.

On inductive reasoning:
Giere [1991] is an excellent introduction to inductive reasoning as that type of reasoning is carried on within science. Skyrms [1966] is an excellent introduction to inductive logic. The first appendix to this text concerns the measurement of probability within inductive logic.

On the teleological argument for God's existence:
Swineburne [1979] contains a presentation and defense of the argument. Mackie [1982] contains a critique of the argument. Dawkins [1987] is a book-length critique, by a biologist, of the argument. Burrill [1967] contains several good introductory articles.

Other:
Burrill [1967] contains several good articles on the cosmological argument. Swineburne [1979] contains a defense of the argument, and Mackie [1982] contains a critique. Gale [1991] contains a detailed discussion of several versions of the argument.

CHAPTER TWO

On the development of modern truth-functional logic:
Bertrand Russell was one of the pioneers of modern truth-functional logic. Sainsbury [1979] and Ayer [1972] are good studies of Russell's philosophy. Whitehead and Russell [1950], originally published in 1910, is the book that founded modern truth-functional logic; if you are interested, see volume I.

Other:
For further reading on validity, see Haack [1978] Chapter 2 and Sainsbury [1991]. For further reading on the sentence connectives, see Haack [1978] Chapter 3. For further reading on the horseshoe see Haack [1978] Chapter 3 and Sainsbury [1991] Chapter 2.

CHAPTERS THREE and FOUR

If you would like to look at an advanced treatment of truth-functional logic, Hunter [1973] contains an account of the metatheory of truth-functional logic.

CHAPTER FIVE

On the general topic of possible worlds:
Loux [1979] is a marvelous introduction to the subject. This book contains an excellent and comprehensive introduction, a very extensive bibliography, articles critical of the possible worlds framework, articles in defense of that framework, and articles that make use of that framework to explore philosophical issues. Coburn [1990] contains two excellent chapters—4 and 5—dealing with some interesting theoretical difficulties associated with the possible worlds framework of ideas. Plantinga [1974] Chapter IV presents one philosopher's account of the nature of a possible world. For a very different account, see Lewis [1986].

On philosophy of logic:
Haack [1978] and Sainsbury [1991] are excellent introductions to the philosophy of logic. Sainsbury [1991] includes an extensive discussion of possible worlds semantics. Putnam [1971] and Quine [1986] are two shorter introductions. Putnam argues for the existence of abstract entities. Copi and Gould [1967] is an anthology containing interesting articles on the topic.

On truth:
Chisholm [1977], Haack [1978], and Martinich [1985] contain introductions to the topic. Martinich [1985] also contains a number of important articles on theories of truth. Rescher [1970] is a good introduction to the coherence theory. On the

pragmatic theory, see "Pragmatism's Conception of Truth" in James [1948]. Coburn [1990] Chapter 8 is an interesting discussion of truth. Also recommended is Haack [1978] Chapter 7.

On the nature of a proposition:
For an excellent introduction to the topic see Bradley and Swartz [1979] Chapter 2. Bradley and Swartz defend the *sui generis* account. Salmon and Soames [1988] contains articles dealing with the issue. See also Haack [1978] Chapter 6. Compare the views developed in Putnam [1971] and Quine [1986].

On abstract entities:
Putnam [1971] provides a strong argument for the existence of abstract entities. Compare Putnam's discussion with the discussion in Quine [1986]. Staniland [1972] is a good introduction to the subject.

On galaxies and Earth's place within the universe:
Hawking [1988], Ferris [1988], Weinberg [1977], and Gribbin [1986] are excellent introductions to the subject.

On philosophy of language:
Martin [1987] and Devit and Sterelny [1987] are good introductions. Zabeeh, Klemke, and Jacobsen [1974], Martinich [1985], and Harmon and Davidson [1973] are important anthologies.

Leibniz's philosophical writing:
For Leibniz's philosophical writings see Weiner [1951] and Leibniz [1985]. For a study of Leibniz's philosophy see Mates [1986].

CHAPTER SIX

On the nature of necessary truth:
Sleigh [1972] is an excellent introduction to the topic. Plantinga [1974] is a defense of the idea of necessary truth. Tomberlin and van Inwagen [1985] contains a number of articles on Plantinga's philosophy.

On ratiocinative and experiential knowledge:
Moser [1987] is an excellent introduction. Bradley and Swartz [1979] contains a nice introduction to the topic in Chapter 4. See also the discussion in Chisholm [1977]. Bonjour [1985] is an interesting defense of a particular theory of experiential knowledge and also treats the issue of ratiocinative knowledge.

On epistemology:
Audi [1988] and Chisholm [1977] are excellent introductions.

On the paradoxes of implication:
See Haack [1978] pp. 197–203 and Sainsbury [1991]. Anderson and Belnap [1975] is a presentation of logical theory that aims to do away with the paradoxes.

APPENDIX TO CHAPTER 6

On the history of logic:
Kneale and Kneale [1962] is a very comprehensive and important history of logic. Mates [1972] contains a nice appendix on the history of logic. Mates [1953] is an interesting account of Stoic logic. Rist [1978] contains interesting articles on Stoic logic and philosophy. Also, Hadas [1961] contains the main works of Stoic philosophy.

PHILOSOPHICAL INTERLUDE

On the concept of God:
Beaty [1990], Morris [1987b], and Urban and Walton [1978] contain excellent articles on the topic. The discussions in Davis [1983], Gale [1991], Morris [1987a], and Swineburne [1977] are also recommended.

On Anselm's ontological argument:
Adams [1985] contains a rigorous analysis of the argument and an interesting article on necessary existence. Deane [1962] contains Anselm's writings. Hick and McGill [1967] and Plantinga [1965] are collections of articles on the argument. Mackie [1982] contains a critique of the argument. Adams [1985] contains several excellent articles. Schufreider [1978] is a book-length study.

On Plantinga's ontological argument:
Plantinga [1974] Chapter X contains Plantinga's argument. A simplified version of the argument is presented in Plantinga [1975]. Mackie [1982] contains a critique of the argument. Gale [1991] also contains a critique of the argument. Also of interest is Plantinga's article "Necessary Being" in Burrill [1967].

On the problem of evil:
Kirwan [1989] is an excellent study of Augustine's thought. Plantinga [1974] provides an interesting version of the free will theodicy. Plantinga's version employs the possible worlds framework. A simplified version is presented in Plantinga [1975]. Mackie [1982] contains critiques of the main theodicies. Hick [1968] summarizes the Irenaean theodicy. Pike [1964] and Flew [1984] are also recommended.

On fatalism:
Cahn [1967] an excellent introduction. Taylor [1975] contains an interesting chapter on the topic. Van Inwagen [1983] contains a chapter devoted to a critique of fatalism. Gale [1968] contains an interesting article by Raymond Bradley on fatalism.

On the nature of time:
Coburn [1990] Chapter 6 contains an excellent introduction to the issue and an interesting discussion of the nature of time. See also Gale [1968], Taylor [1975], and Schlesinger [1980].

On determinism and free will:
O'Conner [1971] and Trusted [1984] are good introductions. Coburn [1990] Chapter 2 contains an excellent discussion. Van Inwagen [1983] and Dennett [1984] are

excellent book-length treatments. Entemen [1967], Fischer [1986], Lehrer [1966], and Hook [1958] are good anthologies on the topic. Also, Taylor [1975] contains a good chapter on the topic and Richman [1983] argues for a version of compatibilism. Nozick [1981] Chapter 4 contains an interesting exploration of the issue. Agent causation is defended by Roderick Chisholm in "Freedom and Necessity," reprinted in Lehrer [1966], and by Richard Taylor in Taylor [1975]. Coburn [1990] Chapter 2 contains a criticism of agent causation.

On dualism and materialism:
Armstrong [1968] defends a materialist view, and Foster [1991] is a recent defense of dualism. Campbell [1970] and Churchland [1984] are excellent introductions to the topic. Coburn [1990] Chapter 3 contains a good introduction of the issue and an interesting discussion. Daniel Dennett argues against dualism in Dennett [1991] Chapter 2. Flanagen [1984] is a good introduction to the general topic. Nagel [1979] contains a well-known argument for dualism in the chapter "What Is It Like to Be a Bat?" Popper and Eccles [1977] is also a defense of dualism. Tomberlin and van Inwagen [1985] contains a criticism of Plantinga's modal argument for dualism. Levin [1979] is a book-length argument for materialism.

On the nature of matter, quantum theory, and the modern scientific view:
There are an enormous number of "popular" books on these topics. Barrow [1988], Close [1987], Ferris [1988], Hawking [1988], Mulvey [1981], Trefle [1980], Gribbin [1986], and Weinberg [1977] are only a few of the many excellent accounts of the modern scientific view. Feynman [1985] and Gribbin [1984] introduce quantum theory to the general reader.

On the philosophy of religion:
Wainwright [1988] and Hick [1972] are excellent introductions. Rowe and Wainwright [1973] is an excellent anthology. Smith [1979] makes a case for atheism. Swineburne [1979] builds a cumulative case for theism. Mackie [1982] is a criticism of Swineburne's cumulative case. See also Gale [1991].

On many-valued logics:
Rescher [1969] is a good introduction. See also Haack [1978] Chapter 11.

CHAPTER SEVEN

Hughes and Cresswell [1968] is an extended treatment of the different systems of modal logic. Konyndyk [1986] and Chellas [1980] are excellent texts covering the major systems of modal logic. Loux [1979] contains an excellent introduction to the subject and important articles. Lewis and Langford [1932] helped lay the foundations of the modern treatment of the subject. See also Haack [1978] Chapter 9.

CHAPTER EIGHT

On the development of modern quantificational logic:
Bertrand Russell and Gotlob Frege were the pioneers of modern quantificational

logic. On Russell's contributions to the development of quantificational logic see Sainsbury [1979]. On Frege, see Kneale and Kneale [1962], Wright [1984], and Haack [1978] Chapter 4. For Russell's solution to the problem of nonreferring descriptions see the articles "Descriptions and Existence" in Russell [1973] and "On Denoting," which is reprinted in Copi and Gould [1967].

Other:
Sainsbury [1991], Haack [1978], Nagel and Newman [1958], and Martin [1987] contain philosophical treatments of issues in quantificational logic. Coburn [1990] Chapter 1 contains an interesting argument on the nature of personal identity. In his argument, Coburn refers to the properties of transitivity, symmetry, and reflexivity.

CHAPTER NINE

For an advanced treatment of quantificational logic:
Hunter [1973] contains a unit on the metatheory of quantificational logic.

APPENDIX ONE

Skyrms [1966] is a good introduction to inductive logic. For a possible worlds treatment see Bradley and Swartz [1979] Chapter 6.

APPENDIX TWO

On essentialism:
Theories of essentialism are developed and defended in Plantinga [1974] and Kripke [1980]. Linsky [1971] contains important articles on the topic. See also McCulloch [1989], Salmon [1981], and Brody [1980]. Coburn [1990] Chapter 5 discusses some of the philosophical problems associated with essentialism.

On quantified modal logic:
See the introduction in Loux [1979]. This book contains several articles on the topic, including David Lewis's important "Counterpart Theory and Quantified Modal Logic." Konyndyk [1986], and Forbes [1985] present introductory accounts. Plantinga [1974] and Lewis [1973] and [1986] are also recommended. For Quine's criticism of quantified modal logic, see Quine [1961], the appendix to Plantinga [1974], and Haack [1978] Chapter 10. Linsky [1977] is also of interest.

On TWI and WBI:
An excellent introduction to the issue may be found in the introduction to Loux [1979]. For Plantinga's treatment see Chapter VI of Plantinga [1974]. For an alternative view see Lewis [1974] and the chapter on the topic in Lewis [1986]. An interesting discussion of the issue is Peter van Inwagen's "Plantinga on Trans-world Identity" in Tomberlin and van Inwagen [1985]. Also recommended is Chisholm's "Identity Across Possible Worlds" in Loux [1979].

BIBLIOGRAPHY

Adams, Robert M. [1985]. *The Virtue of Faith*. London: Oxford University Press.

Anderson, A. R., and Belnap, N. D., Jr. [1975]. *Entailment*. Princeton: Princeton University Press.

Armstrong, A. H. [1977]. *An Introduction to Ancient Philosophy*. 3rd ed. Totowa, NJ: Rowman & Allanheld.

Armstrong, D. M. [1968]. *A Materialist Theory of the Mind*. London: Routledge & Kegan Paul.

Audi, Robert [1988]. *Belief, Justification, and Knowledge: An Introduction to Epistemology*. Belmont, CA: Wadsworth.

Ayer, A. J. [1972]. *Russell*. London: Wm. Collins.

Barnes, Jonathon [1982]. *The Pre-Socratic Philosophers*. London: Routledge & Kegan Paul.

Barrow, John D. [1988]. *The World Within the World*. Oxford: Clarendon Press.

Beaty, Michael D., ed. [1990]. *Christian Theism and the Problems of Philosophy*. South Bend, IN: University of Notre Dame Press.

Bonjour, Lawrence [1985]. *The Structure of Empirical Knowledge*. Boston: Harvard University Press.

Bradley, Raymond, and Swartz, Norman [1979]. *Possible Worlds*. Indianapolis: Hackett Publishing.

Brody, Baruch [1980]. *Identity and Essence*. Princeton: Princeton University Press.

Burrill, Donald R. [1967]. *The Cosmological Arguments*. Garden City, NY: Doubleday.

Cahn, Steven M. [1967]. *Fate, Logic, and Time*. New Haven: Yale University Press.

Campbell, Keith [1970]. *Body and Mind*. Garden City, NY: Doubleday.

Chellas, Brian [1980]. *Modal Logic*. Cambridge: Cambridge University Press.

Chisholm, Robert [1977]. *Theory of Knowledge*. Englewood Cliffs, NJ: Prentice-Hall.

Churchland, Paul [1984]. *Matter and Consciousness*. Cambridge: MIT Press.

Close, Frank; Marten, Michael; and Sutton, Christine [1987]. *The Particle Explosion*. New York: Oxford University Press.

Coburn, Robert [1990]. *The Strangeness of the Ordinary*. Savage, MD: Rowmand and Littlefield Publishers.

Copi, Irving, and Gould, J., eds. [1967]. *Contemporary Readings in Logical Theory*. New York: Macmillan.

Copleston, Frederick, S. J. [1962]. *A History of Philosophy*. Garden City, NY: Image Books, 8 volumes.

Davies, P. C. W., and Brown, J. R. [1986]. *The Ghost in the Atom. A Discussion of the Mysteries of Quantum Physics*. Cambridge: Cambridge University Press.

Davis, Stephen T. [1983]. *Logic & the Nature of God*. Grand Rapids, MI: William B. Eerdmans.

Dawkins, Richard [1987]. *The Blind Watchmaker*. New York: W.W. Norton & Co.

Deane, S. N., ed. [1962]. *St Anselm: Basic Writings*. La Salle, IL: Open Court Publishing.

Dennett, Daniel [1984]. *Elbow Room*. Cambridge: Harvard University Press.

——— [1978]. *Brainstorms. Philosophical Essays on Mind and Psychology*. Montgomery, VT: Bradford.

——— [1991]. *Conciousness Explained*. Boston: Little Brown.

Descartes, Rene [1960]. *Discourse on Method* and *Meditations*. New York: Bobbs-Merrill.

Devit, Michael, and Sterelny, Kim [1987]. *Language and Reality. An Introduction to the Philosophy of language*. Cambridge: MIT Press.

Entemen, Willard, ed. [1967]. *The Problem of Free Will*. New York: Charles Scribner's Sons.

Feldman, Fred [1986]. *A Cartesian Introduction to Philosophy*. New York: McGraw Hill.

Ferris, Timothy [1987]. *Galaxies*. New York: Harrison House.

——— [1988]. *Coming of Age in the Milky Way*. New York: William Morrow.

Feynman, Richard [1985]. *QED. The Strange Theory of Light and Matter*. Princeton: Princeton University Press.

Finnis, John [1982]. *Natural Law and Natural Rights*. London: Oxford University Press.

Fischer, John, ed. [1986]. *Moral Responsibility*. Ithaca: Cornell University Press.

Flanagen, Owen, Jr. [1984]. *The Science of the Mind*. Cambridge: Bradford Books.

Flew, Antony [1984]. *God, Freedom and Immortality. A Critical Analysis*. Buffalo, NY: Prometheus Books.

Forbes, Graeme [1985]. *The Metaphysics of Modality*. Oxford: Clarendon Press.

Foster, John [1991]. *The Immaterial Self. A Defense of the Cartesian Dualist Conception of the Mind*. New York: Routledge.

Gale, Richard [1991]. *On the Nature and Existence of God*. Cambridge: Cambridge University Press.

———, ed. [1968]. *The Philosophy of Time*. Garden City, NY: Anchor Books.

Giere, Ronald N.[1991]. *Understanding Scientific Reasoning*. 3rd ed. Fort Worth: Holt, Rinehart and Winston.

Glashow, Shelden [1988]. *Interactions*. New York: Warner Books.

Glover, Jonathon [1977]. *Causing Death and Saving Lives*. New York: Penquin Books.

Gribbin, John [1986]. *In Search of the Big Bang*. New York: Bantam Books.

―――― [1984]. *In Search of Schrodinger's Cat*. New York: Bantam Books.

Haack, Susan [1978]. *Philosophy of Logics*. Cambridge: Cambridge University Press.

Hacking, Ian [1975]. *Why Does Language Matter to Philosophy?* New York: Cambridge University Press.

Hadas, Moses [1961]. *Essential Works of Stoicism*. New York: Bantam Books.

Harmon, Gilbert, and Davidson, Donald [1973]. *The Semantics of Natural Language*. Dordrecht: Reidel.

Hawking, Stephen W. [1988]. *A Brief History of Time*. New York: Bantam Books.

Hick, John [1968]. *Evil and the God of Love*. London: Macmillan.

―――― [1972]. *Philosophy of Religion*. Englewood Cliffs, NJ: Prentice-Hall.

――――, and McGill, Arthur [1967]. *The Many Faced Argument*. New York: Macmillan.

Hook, Sidney, ed. [1958]. *Determinism and Freedom*. New York: Colleen Books.

Hughes, G. E., and Cresswell, M. J. [1968]. *An Introduction to Modal Logic*. London: Meuthen.

Hunter, Geoffrey [1973]. *Metalogic*. Los Angeles: University of California Press.

James, William [1948]. *Essays in Pragmatism*. New York: Macmillan.

Kirwan, Christopher [1989]. *Augustine*. New York: Routledge.

Kneale, William, and Kneale, Martha [1962]. *The Development of Logic*. Oxford: Clarendon Press.

Konyndyk, Kenneth [1986]. *Introductory Modal Logic*. South Bend, IN: University of Notre Dame Press.

Kripke, Saul [1980]. *Naming and Necessity*. Cambridge: Harvard University Press.

Lehrer, Keith, ed. [1966]. *Freedom and Determinism*. New York: Random House.

Leibniz, G. W. [1985]. *Theodicy*. La Salle, IL: Open Court Publishing.

Levin, Michael [1979]. *Metaphysics and the Mind–Body Problem*. Oxford: Oxford University Press.

Lewis, C. I., and Langford, C. H. [1932]. *Symbolic Logic*. New York: Dover.

Lewis, David [1986]. *On the Plurality of Worlds*. Oxford: Basil Blackwell.

―――― [1973]. *Counterfactuals*. Cambridge: Harvard University Press.

Linsky, Leonard, ed. [1971]. *Reference and Modality*. Oxford: Oxford University Press.

―――― [1977]. *Names and Descriptions*. Chicago: University of Chicago Press.

Loux, Michael, ed. [1979]. *The Possible and the Actual*. Ithaca: Cornell University Press.

Lovejoy, Arthur O. [1960]. *The Revolt Against Dualism*. La Salle, IL: Open Court Publishing.

Mackie, John [1982]. *The Miracle of Theism*. Oxford: Clarendon Press.

Martin, Robert [1987]. *The Meaning of Language*. Cambridge: Bradford Books.

Martinich, A. P., ed. [1985]. *The Philosophy of Language*. New York: Oxford University Press.

Mates, Benson [1953]. *Stoic Logic*. Los Angeles: University of California Press.

——— [1972]. *Elementary Logic*. New York: Oxford University Press.

——— [1986]. *The Philosophy of Leibniz*. Los Angeles: University of California Press.

Matson, Wallace I. [1987]. *A New History of Philosophy*. New York: Harcourt Brace Jovanovich.

McCulloch, Gregory [1989]. *The Game of the Name, Introducing Logic, Language and Mind*. Oxford: Clarendon Press.

Morris, Thomas V. [1987a]. *Anselmian Explorations*. South Bend, IN: University of Notre Dame Press.

———, ed. [1987b]. *The Concept of God*. Oxford: Oxford University Press.

Moser, Paul K., ed. [1987]. *A Priori Knowledge*. Oxford: Oxford University Press.

Mulvey, J. H., ed. [1981]. *The Nature of Matter*. Oxford: Oxford University Press.

Nagel, Thomas [1979]. *Mortal Questions*. New York: Cambridge University Press.

Nagel, Ernest, and Newman, James [1958]. *Gödel's Proof*. New York: New York University Press.

Nozick, Robert [1981]. *Philosophical Explanations*. Cambridge: Harvard University Press.

——— [1974]. *Anarchy, State, and Utopia*. New York: Basic Books.

O'Connor, D. J. [1971]. *Free Will*. New York: Doubleday.

Oursler, Fulton, and Oursler, Will [1956]. *Father Flanagan of Boys Town*. Garden City, NY: Doubleday.

Pike, Nelson, ed. [1964]. *God and Evil*. Englewood Cliffs, NJ: Prentice-Hall.

Plantinga, Alvin, ed. [1965]. *The Ontological Argument*. Garden City, NY: Doubleday.

——— [1974]. *The Nature of Necessity*. Oxford: Clarendon Press.

——— [1975]. *God, Freedom and Evil*. London: George Allen and Unwin.

Popper, Karl, and Eccles, John [1977]. *The Self and Its Brain*. London: Routledge & Kegan Paul.

Putnam, Hilary [1971]. *Philosophy of Logic*. New York: Harper.

Quine, W. V. O. [1986]. *Philosophy of Logic*. 2nd ed. Cambridge: Harvard University Press.

——— [1961]. *From a Logical Point of View*. 2d ed. Cambridge: Harvard University Press.

Rawls, John [1971]. *A Theory of Justice*. Boston: Harvard University Press.

Rescher, Nicholas [1970]. *The Coherence Theory of Truth*. London: Oxford University Press.

——— [1969]. *Many Valued Logic*. New York: McGraw Hill.

Richman, Robert J. [1983]. *God, Free Will, and Morality*. Boston: D. Reidel Publishing.

Rist, John, ed. [1978]. *The Stoics*. Los Angeles: University of California Press.

Robinson, Mansley [1968]. *Introduction to Early Greek Philosophy*. New York: Houghton Mifflin.

Rowe, William, and Wainwright, William [1973]. *Philosophy of Religion: Selected Readings*. New York: Harcourt Brace Jovanovich.

Russell, Bertrand [1945]. *A History of Western Philosophy*. New York: Simon and Schuster.

——— [1973]. *Essays in Analysis*. New York: George Braziller.

Sainsbury, Mark [1991]. *Logical Forms. An Introduction to Philosophical Logic.* Cambridge: Basil Blackwell.

—— [1979]. *Russell.* London: Routledge & Kegan Paul.

Salmon, Nathan [1981]. *Reference and Essence.* Princeton: Princeton University Press.

——, and Soames, Scott [1988]. *Propositions and Attitudes.* Oxford: Oxford University Press.

Schlesinger, George [1980]. *Aspects of Time.* Indianapolis: Hackett Publishing.

Schufreider, Gregory [1978]. *An Introduction to Anselm's Argument.* Philadelphia: Temple University Press.

Skyrms, Brian [1966]. *Choice and Chance.* Belmont, CA: Dickenson.

Sleigh, Robert, ed. [1972]. *Necessary Truth.* Englewood Cliffs, NJ: Prentice-Hall.

Smith, Peter, and Jones, O. R. [1986]. *The Philosophy of Mind.* New York: Cambridge University Press.

Smith, George [1979]. *Atheism. The Case against God.* Buffalo, NY: Prometheus Books.

Staniland, Hilary [1972]. *Universals.* Garden City, NY: Doubleday.

Swineburne, Richard [1977]. *The Coherence of Theism.* Oxford: Clarendon Press.

—— [1979]. *The Existence of God.* Oxford: Clarendon Press.

Taylor, Richard [1975]. *Metaphysics.* Englewood Cliffs, NJ: Prentice-Hall.

Tomberlin, James, and van Inwagen, Peter, eds. [1985]. *Alvin Plantinga: A Profile.* Dordrect: Reidel.

Trefil, James S. [1980]. *From Atoms to Quarks. An Introduction to the Strange world of Particle Physics.* New York: Charles Scribner's Sons.

Trusted, Jennifer [1984]. *Free Will and Responsibility.* New York: Oxford University Press.

Urban, Linwood, and Walton, Douglass, eds. [1978]. *The Power of God.* New York: Oxford Univerity Press.

Van Inwagen, Peter [1983]. *An Essay on Free Will.* Oxford: Clarendon Press.

Wainwright, William [1988]. *Philosophy of Religion.* Belmont, CA: Wadsworth.

Weinberg, Steven [1977]. *The First Three Minutes.* New York: Basic Books.

Weiner, Philip, ed. [1951]. *Leibniz Selections.* New York: Charles Scribner's Sons.

White, Morton [1978]. *The Philosophy of the American Revolution.* New York: Oxford University Press.

—— [1987]. *Philosophy, the Federalist, and the Constitution.* New York: Oxford University Press.

Whitehead, Alfred North, and Russell, Bertrand [1950]. *Principia Mathematica.* Cambridge: Cambridge University Press.

Wilbur, J. B., and Allen, H. J. [1979]. *The Worlds of the Early Greek Philosophers.* New York: Prometheus Books.

Wright, Crispin, ed. [1984]. *Frege. Tradition and Influence.* New York: Basil Blackwell.

Zabeeh, Farhang; Klemke, E. D.; and Jacobson, Arthur, eds. [1974]. *Readings in Semantics.* Chicago: University of Illinois Press.

SELECTED ANSWERS

Exercise 1.1

V: valid   I: invalid   U: unsound   S: sound

1. V,U	5. V,U	10. I,U	15. I,U

Exercise 1.2

C: consistent   I: inconsistent

1. C	10. C	20. I
5. I	15. C	

Exercise 1.3

pi: p implies q   qi: q implies p

1. Neither implies

10. pi, qi

5. Controversial. You decide for yourself.

15. Neither implies

Exercise 1.4

Not: not equivalent   E: equivalent

1. 1. Not

  5. Not

  10. Not

  15. Not

  20. Not

2. 1. Not

  5. You decide for yourself.

  10. E

  15. Not

Exercise 1.5

1. Strong   5. Strong

Exercise 1.7

1. F   5. T   10. F

Exercise 1.11

1. Deductive, valid. Indicator word: must

5. Deductive, invalid. Indicator word: must

Exercise 1.13

5. Conclusion: ". . .the gifts of fortune by their instability cannot ever lead to happiness."
    Premises: The previous sentences.

Exercise 1.14

1. V   10. I   20. V

5. I   15. I

SELECTED ANSWERS   CHAPTER 2

Exercise 2.2

1. (A & B) v D   10. ~M & I

5. D & (F v O)   15. (~J v ~T) v (~E & ~D)

Exercise 2.3

1. wff   10. Not a wff

5. wff   15. Not a wff

Exercise 2.4

1. Second ampersand from left   5. Ampersand

Exercise 2.5

1. F   10. T   20. T

5. T   15. T

Exercise 2.6

1. T	5. F	10. T	15. F

Exercise 2.7

Part I   Part II

1. T	10. T	1. F	10. T
5. T	15. T	5. T	15. T

Exercise 2.8

1. T	5. T	10. F	15. T

Exercise 2.9

1. F ⊃ (J v B)	5. ~B ⊃ (C v D)

Exercise 2.10

T: tautological   X: contingent   C: contradictory

1. T	15. C	30. X
5. T	20. T	35. C
10. C	25. X	40. X

Exercise 2.11

1. If **P** is a tautology, then **P** is true on every row of its table. The negation of **P** is false just in case **P** is true. So, if **P** is true on every row of the table, then the negation of **P** is false on every row. If a sentence is false on every row of its table, then it is a contradiction. So if **P** is a tautology, then its negation is a contradiction.

5. A conditional is true if its antecedent is false. If the antecedent is a contradiction, then the antecedent is false on every row of the table. So on every row of the conditional's table, the conditional is true—if its antecedent is contradictory. A sentence that is true on every row of its table is tautological. So, if the antecedent is contradictory, the conditional is a tautology.

Exercise 2.12

1. If the conclusion is a tautology, then the conclusion is true on every row of the argument's truth-table. It follows that there is no row of the table on which the premises are true and the conclusion is false. If there is no row of the table on which the premises are true and the conclusion is false, then the argument is valid.

Exercise 2.13

Part I

5.

AB	A ⊃ B	B ⊃ A
TT	T	T
TF	F	T
FT	T	F
FF	T	T

Invalid. Row 3 is an invalidating row.

10.

ABE	A ⊃ B	B ⊃ E	B v E	A v B
TTT	T	T	T	T
TTF	T	F	T	T
TFT	F	T	T	T
TFF	F	T	F	T
FTT	T	T	T	T
FTF	T	F	T	T
FFT	T	T	T	F
FFF	T	T	F	F

Invalid. Row 7 is an invalidating row.

Part II

5.

AB	(A ⊃ B)	⊃	(B ⊃ A)
TT	T	T	T
TF	F	T	T
FT	T	F	F
FF	T	T	T
		*	

The corresponding conditional is not a tautology. The argument is therefore invalid.

8.

ABC	[(A ⊃ B)	&	(B ⊃ C)]	⊃	(A ⊃ C)
TTT	T	T	T	T	T
TTF	T	F	F	T	F
TFT	F	F	T	T	T
TFF	F	F	T	T	F
FTT	T	T	T	T	T
FTF	T	F	F	T	T
FFT	T	T	T	T	T
FFF	T	T	T	T	T
				*	

The corresponding conditional is a tautology. The argument is therefore valid.

Exercise 2.14

Part I

I: implies D: Does not imply

1. D
5. I
10. I
15. I
20. I
25. I
30. D

Part II

C: consistent

1. C
5. C
10. C
15. C
20. C
25. C
30. C

Exercise 2.15

Part I

1. E   20. E
5. N   25. E
10. E   30. E
15. N

Part II

1. C   20. C
5. C   25. C
10. C   30. C
15. C

Exercise 2.16

Part I

1. S ⊃ (~J ⊃ R) /R ⊃ ~J
    TT FTT T/ T FFT

5. (A ⊃ R) R ⊃ (C v P) P ⊃ B/A ⊃ B
    T T T TT TTF FTFTFF

Part II

1. P ⊃ (Q v R) Q ⊃ S/P ⊃ S
    TT FTT   F TFTF F

5. ~(P & Q) ~P/Q
    TFF F TF F

10. ~(P & Q)/~P
     TT F F FT

15. P ⊃ Q R ⊃ S Q v S/P v R
     FT T FT TTTTFFF

20. P ⊃ Q Q ⊃ R S ⊃ R/P ⊃ S
     TT T T T TFT TTF F

Exercise 2.17

1.

BJ	B⊃J	~J	~B
TT		FT	FT
TF	TFF		FT
FT			TF
FF			TF

Valid. No invalidating row.

5.

GJD	(GvJ)⊃D	~D	~G&~J
TTT	T	F	F
TTF	F	T	F
TFT	T	F	F
TFF	F	T	F
FTT	T	F	F
FTF	F	T	F
FFT	T	F	T
FFF	T	T	T

Valid. No invalidating row.

Exercise 2.18

a.	All are instances	j.	2, 4
e.	1, 9	o.	5, 18

Exercise 2.19

1. Tautology (The sentence is an instance of **Pv~P.**)

Exercise 2.20

1.	I	10.	T	25.	I	15.	I
5.	T	20.	C	30.	T		

Exercise 2.21

1.	I	10.	V	20.	V	30.	V
5.	V	15.	V	25.	V	35.	I

SELECTED ANSWERS   CHAPTER 3

Exercise 3.1

Part I

1.	HS 1,2				DS 4,6		DS 2,5
	MP 3,5	5.	MT 3,4				MP 1,6

10.	MP 1,2	15.	HS 1,2
	MP 1,3		DS 3,5
	MT 1,4		MT 4,6

Part II

1. 1. (A & B) ⊃ S   2. A

    2. H ⊃ R   3. <u>~E</u>/ ~B

    3. (A & B) / S   4. B ⊃ E   MP 1,2

    4. S   MP 1,3   5. ~B   MT 3,4

5. 1. F ⊃ ~(H & B)   18. 1. A ⊃ B

    2. A ⊃ (H & B)   2. B ⊃ W

    3. <u>A</u> / ~F   3. ~W

    4. H & B   MP 2,3   4. P ⊃ Q

    5. ~F   MT 1,4   5. S ⊃ ~Q

10. 1. R ⊃ H   6. ~A ⊃ (P v Z)

    2. H ⊃ S   7. <u>S</u> / Z

    3. S ⊃ G   8. ~Q   MP 5,7

    4. <u>(R ⊃ G) ⊃ F</u> / F   9. ~P   MT 4,8

    5. R ⊃ S   HS 1,2   10. A ⊃ W   HS 1,2

    6. R ⊃ G   HS 3,5   11. ~A   MT 3,10

    7. F   MP 4,6   12. P v Z   MP 6,11

15. 1. A ⊃ (B ⊃ E)   13. Z   DS 9,12

Part III

1. 1. A⊃B   5. 1. R⊃A

    2. AvC   2. N⊃R

    3. <u>~B</u> / C   3. <u>~A</u> / ~N

    4. ~A   MT 1,3   4. ~R   MT 1,3

    5. C   DS 2,4   5. ~N   MT 2,4

Exercise 3.2

Part I

1. Simp 5	MP 2,8	Add 6
Add 6	Add 9	MP 4,12
MP 1,7	MP 3,10	Conj 11,13

5.   Simp 4   DS 2,9   CD 5,6,12

     Add 7   MP 3,10

     MP 1,8   Add 11

Part II

1.   1.   A ⊃ B   6.   (P & S) v F   CD 1,2,5

     2.   B ⊃ R   7.   G   MP 4,6

     3.   ~R / ~A & ~B   15.   1.   (A ⊃ B) & (S ⊃ Q)

     4.   ~B   MT 2,3   2.   P ⊃ R

     5.   ~A   MT 1,4   3.   (A v P) & (S v E) / B v R

     6.   ~A & ~B   Conj 4,5   4.   A ⊃ B   Simp 1

5.   1.   B ⊃ (S v R)   5.   A v P   Simp 3

     2.   S ⊃ P   6.   B v R   CD 2,4,5

     3.   R ⊃ G   20.   1.   (A & S) ⊃ Z

     4.   H & B / P v G   2.   ~P ⊃ A

     5.   B   Simp 4   3.   ~S ⊃ P

     6.   S v R   MP 1,6   4.   ~P & B / Z v Q

     7.   P v G   CD 2,3,6   5.   ~P   Simp 4

10. 1.   A ⊃ (P & S)   6.   A   MP 2,5

     2.   B ⊃ F   7.   S   MT 3,5

     3.   A   8.   A & S   Conj 6,7

     4.   [(P & S) v F] ⊃ G / G   9.   Z   MP 1,8

     5.   A v B   Add 3   10. Z v Q   Add 9

Exercise 3.3

Part I

1.   Dist 1   MP 3,8   5.   Simp 1

     Simp 4   Add 9   MP 4,10

     MP 2,5   DS 3,6   DN 11

     MP 2,6   Comm 7   DS 5,12

     DN 7

Part II

5. 1. ~(A v B)
    2. ~B ⊃ E
    3. <u>E ⊃ S /S</u>
    4. ~A & ~B DM 1
    5. ~B Simp 4
    6. E MP 2,5
    7. S MP 3,6

10. 1. (R ⊃ S) v ~G
    2. ~~G
    3. (R ⊃ S) ⊃ ~~P
    4. <u>H ⊃ ~P /~H</u>
    5. R ⊃ S DS 1,2
    6. ~~P MP 3,5
    7. ~H MT 4,6

15. 1. ~R ⊃ ~S
    2. ~~S
    3. ~~R ⊃ H
    4. <u>(H ⊃ A) & (Z ⊃ O) /A v O</u>
    5. ~~R MT 1,2
    6. H MP 3,5
    7. H v Z Add 6
    8. H ⊃ A Simp 4
    9. Z ⊃ O Simp 4
    10. A v O CD 7,8,9

20. 1. K
    2. ~(A & B) & ~(E & F)
    3. <u>(H & K) ⊃ [(A & B) v (E & F)] /~H</u>
    4. ~[(A & B) v (E & F)] DM 2
    5. ~(H & K) MT 3,4
    6. ~H v ~K DM 5
    7. ~H DS 1,6

25. 1. H ⊃ (E & P)
    2. A v (B & ~C)
    3. A ⊃ ~E
    4. <u>~C ⊃ ~P /~H</u>
    5. (A v B) & (A v ~C) Dist 2
    6. A v ~C Simp 5
    7. ~E v ~P CD 3,4,6
    8. ~(E & P) DM 7
    9. ~H MT 1,8

30. 1. A ⊃ (B & H)   6. (~A v H) v S   Add 5
     2. <u>(H v S) ⊃ Z</u>/A ⊃ Z   7. ~A v (H v S)   Assoc 6
     3. ~A v (B & H)   Imp 1   8. A ⊃ (H v S)   Imp 7
     4. (~A v B) & (~A v H)   Dist 3   9. A ⊃ Z   HS 2,8
     5. ~A v H   Simp 4

Exercise 3.4

Part I

1. Trans 1   5. Imp 1
   MP 2,5   MP 2,6
   Add 6   MT 3,7
   MP 3,7   MP 4,8
   Exp 8   Add 9
   MP 4,9   MP 5,10
   Trans 10   Taut 11

Part II

1. 1. H ⊃ ~S   3. A ⊃ (E ⊃ S)   Exp 1
    2. (~H v ~S) ⊃ F   4. A ⊃ F   HS 2,3
    3. <u>F ⊃ B</u>/B   15. 1. (A ⊃ B) & (E ⊃ B)
    4. ~H v ~S   Imp 1   2. <u>~(~A & ~E)</u>/B
    5. F   MP 2,4   3. A v E   DM 2
    6. B   MP 3,5   4. A ⊃ B   Simp 1

5. 1. ~(A & ~B) ⊃ C   5. E ⊃ B   Simp 1
    2. A ⊃ B   6. B v B   CD 3,4,5
    3. <u>H v ~C</u>/H   7. B   Taut 6
    4. ~A v B   Imp 2   20. 1. A ⊃ B
    5. ~(A & ~B)   DM 4   2. <u>A ⊃ ~B</u>/~A
    6. C   MP 1,5   3. B ⊃ ~A   Trans 2
    7. H   DS 3,6   4. A ⊃ ~A   HS 1,3

10. 1. (A & E) ⊃ S   5. ~A v ~A   Imp 4
    2. <u>(E ⊃ S) ⊃ F</u>/A ⊃ F   6. ~A   Taut 5

25. 1. P ⊃ (S & H)
    2. R ⊃ (~S & ~H)
    3. P v R/H ⊃ S
    4. (S & H) v (~S & ~H)   CD 1,2,3
    5. S ≡ H   Equiv 4
    6. (S ⊃ H) & (H ⊃ S)   Equiv 5
    7. (H ⊃ S)   Simp 6

30. 1. (P ⊃ Q) ⊃ (Q ⊃ P)
    2. (P ≡ Q) ⊃ ~(A & ~B)
    3. Q & A/A & B
    4. Q   Simp 3
    5. Q v ~P   Add 4
    6. ~P v Q   Comm 5
    7. P ⊃ Q   Imp 6
    8. Q ⊃ P   MP 1,7
    9. (P ⊃ Q) & (Q ⊃ P)   Conj 7,8
    10. P ≡ Q   Equiv 9
    11. ~(A & ~B)   MP 2,10
    12. ~A v B   DM 11
    13. A   Simp 3
    14. B   DS 12,13
    15 A & B   Conj 13,14

Exercise 3.5

1. 1. A ⊃ (B & C)/A ⊃ C
   2. | A AP
   3. | B&C MP 1, 2
   4. | C Simp 3
   5. A ⊃ C   CP 2-4

5. 1. ~Q v Z
   2. Z ⊃ A/Q ⊃ (Z & A)
   3. | Q AP
   4. | Z DS 1, 3
   5. | A MP 2, 4
   6. | Z&A Conj 4,
   7. Q ⊃ (Z & A)   CP 3-6

10. 1. A ⊃ (B v C)
     2. E ⊃ S
     3. B ⊃ C/A ⊃ C
     4. | A AP
     5. | B v C MP 1, 4
     6. | C v B Com 5

     7. | ~C ⊃ B Imp 6
     8. | ~C ⊃ C HS 3, 7
     9. | C v C Imp 8
     10. | C Taut 9
     11. A ⊃ C CP 4-10

Exercise 3.6

1. 1. P v (Q & E)
    2. P ⊃ E/E
    3. | ~E AP
    4. | ~P MT 2, 3
    5. | Q & E DS 1, 4
    6. | E Simp 5
    7. E IP 3-6

5. 1. (H v S) ⊃ (A & B)
    2. (B v F) ⊃ K
    3. H v F/K
    4. | ~K AP
    5. | ~(B v F) MT 2, 4
    6. | ~B & ~F DM 5
    7. | ~F Simp 6
    8. | H DS 3, 7
    9. | H v S Add 8
    10.| A & B MP 1, 9
    11.| B Simp 10
    12.| ~B Simp 6
    13. K IP 4-12

10. 1. (P & Q) v E
     2. ~E v Q/P ⊃ Q
     3. | ~(P ⊃ Q) AP
     4. | ~(~P v Q) Imp 3
     5. | P & ~Q DM 4
     6. | ~Q Simp 5
     7. | ~E DS 2, 6
     8. | P & Q DS 1, 7
     9. | Q Simp 8
     10. P ⊃ Q IP 3-9

15. 1. P ⊃ (~Q ⊃ S)
     2. (P ⊃ Q) ⊃ S/S
     3. | ~S AP
     4. | ~(P ⊃ Q) MT 2, 3
     5. | ~(~P v Q) Imp 4
     6. | P & ~Q DM 5
     7. | P Simp 6
     8. | ~Q ⊃ S MP 1, 7
     9. | ~Q Simp 6
     10.| S MP 8, 9
     11. S IP 3-10

Exercise 3.7

1. 1. P ⊃ (Q ⊃ R)
    2. <u>R ⊃ (Q ⊃ S)</u>/P ⊃ (Q ⊃ S)
    3. | P AP
    4. | | Q AP
    5. | | Q ⊃ R MP 1, 3
    6. | | R MP 4, 5
    7. | | Q ⊃ S MP 2, 6
    8. | | S MP 4, 7
    9. | Q ⊃ S CP 4–8
    10. P ⊃ (Q ⊃ S) CP 3-9

5. 1. H ⊃ (S & T)
    2. <u>B ⊃ (A & G)</u>/(T ⊃ B) ⊃ (H ⊃ G)
    3. | T ⊃ B AP
    4. | | H AP
    5. | | S & T MP 1, 4
    6. | | T Simp 5
    7. | | B MP 3, 6
    8. | | A & G MP 2, 7
    9. | | G Simp 8
    10. | H ⊃ G CP 4–9
    11. (T ⊃ B) ⊃ (H ⊃ G) CP 3-10

Exercise 3.8

5. 1. | ~[~A v (B v A)] AP
    2. | A & ~(B v A) DM 1
    3. | A Simp 2
    4. | ~(B v A) Simp 2
    5. | ~B & ~A DM 4
    6. | ~A Simp 5
    7. [~A v (B v A)] IP 1-6

10. 1. | ~[~(P & Q) v ~(~P & ~Q)] AP
     2. | (P & Q) & (~P & ~Q) DM 1
     3. | P & Q Simp 2
     4. | ~P & ~Q Simp 2
     5. | P Simp 3
     6. | ~P Simp 4

     7. [~(P & Q) v ~(~P & ~Q)]   IP 1-6

13. 1. | ~P AP
     2. | ~P v ~Q Add 1
     3. | ~Q v ~P Comm 2
     4. | ~(Q & P) DM 3
     5. ~P ⊃ ~(Q & P)   CP 1-4

17. 1. | A AP
     2. | | A ⊃ B AP
     3. | | B MP 1, 2
     4. | (A ⊃ B) ⊃ B CP 2-3
     5. A ⊃ [(A ⊃ B) ⊃ B]   CP 1-4

SELECTED ANSWERS   CHAPTER 4

Exercise 4.1

1.
```
        | ~P ⊃ (P ⊃ Q) ✔
  ✔ ~P  | P ⊃ Q ✔
        | P
    P   | Q
        x
     Tautology
```

5.
```
         | ~(P & ~P) ✔
  ✔ P & ~P |
         P |
  ✔    ~P |
           | P
         x
      Tautology
```

10. ✔ ~[P ⊃ (Q ⊃ P)]
```
          | P ⊃ (Q ⊃ P) ✔
       P  | Q ⊃ P ✔
       Q  | P
          x
     Contradiction
```

15. ✔ P & (Q & ~P)
```
             P
    ✔    Q & ~P
    ✔       Q
    ✔       ~P
                | P
                x
          Contradiction
```

20.
```
      | Q ⊃ (P v ~P) ✔
   Q  | P v ~P ✔
      | P
      | ~P ✔
   P  |
      x
   Tautology
```

25.

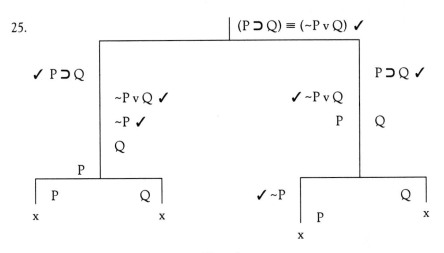

Tautology

30.

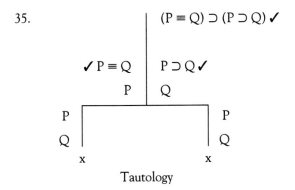

Tautology

35.

$(P \equiv Q) \supset (P \supset Q)$ ✓

✓ $P \equiv Q$ $P \supset Q$ ✓
    P   Q

P   P
Q   Q
   x   x

Tautology

40.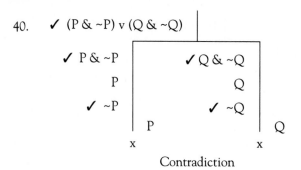

Exercise 4.2

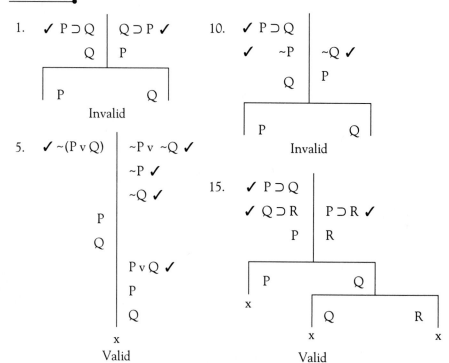

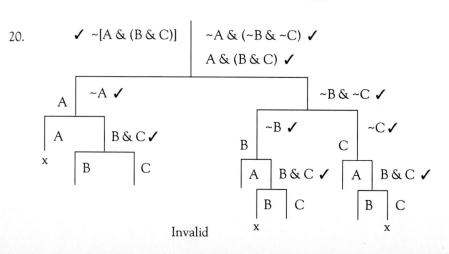

25.

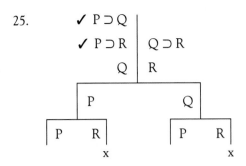

Invalid

30.

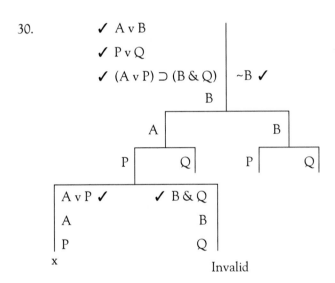

Invalid

35.

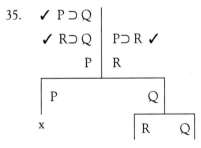

Invalid

40.
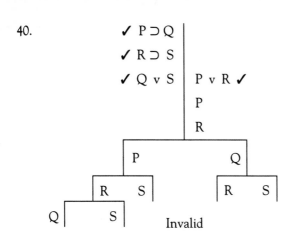

45.
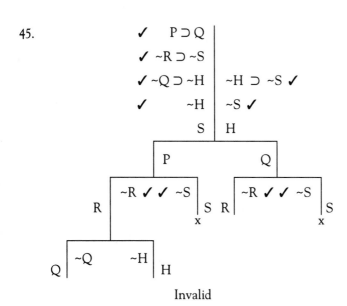

Exercise 4.3 A

1.
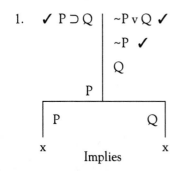

10.
```
        ✔ P & Q  │  ~(~P v ~Q) ✔
             P
             Q
        ✔ ~P v ~Q
    ┌────────────┴────────────┐
  ✔ ~P    │           ✔ ~Q   │
        P │                 Q │
    x                    x
            Implies
```

15.
```
        ✔ ~(P & Q)  │  ~P & ~Q ✔
                        P & Q ✔
        ┌───────────────┴──────┐
                P            │  Q
    ┌───────────┴──────┐
    ~P ✔          ~Q ✔
  P │          Q │
  x
            Does not imply
```

20.
```
     P ⊃ Q  │  ~Q ⊃ ~P ✔
    ✔ ~Q    │   ~P ✔
            │    Q
        ┌───┴───┐
      P │        Q │
    ┌───┴───┐
  P │      Q │
  x          x
      Implies
```

25.
```
    ✔ (P v Q) & ~P  │  Q
         ✔ P v Q
          ✔ ~P
                 ┌────┴────┐
                          P │
            ┌───────┴───────┐
          P │              Q │
          x      Implies    x
```

Exercise 4.3 B

1.
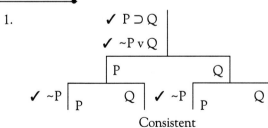
```
            ✔ P ⊃ Q  │
            ✔ ~P v Q
        ┌───────┴───────┐
          P            Q
    ┌─────┴─────┐    ┌─────┴─────┐
  ✔ ~P       Q │  ✔ ~P       Q │
      P │            P │
        Consistent
```

5.
```
      ✔ ~P   │
    ✔ P ⊃ Q
           P │
    ┌───────┴───────┐
    P             Q │
      Consistent
```

10. ✔ P & Q
   ✔ ~(~P v ~Q)

        ~P v ~Q ✔
     P
     Q
        ~P ✔
        ~Q ✔
     P
     Q
      Consistent

15. ✔ ~(P & Q)
    ✔ ~P & ~Q
      ✔ ~P
      ✔ ~Q

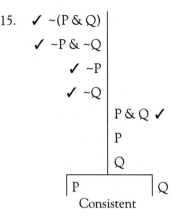

         P & Q ✔
         P
         Q
     P   Q
      Consistent

20.
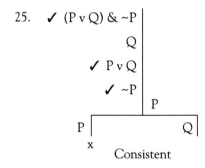

    ✔ P ⊃ Q
   ✔ ~Q ⊃ ~P

    P   Q
  ~Q ✔ ✔ ~P   ~Q ✔ ✔ ~P
 Q   P Q   P
      Consistent

25. ✔ (P v Q) & ~P
       Q
    ✔ P v Q
    ✔ ~P

          P
  P   Q
  x
      Consistent

Exercise 4.4 A

1.   ✔ P v Q | ~P & ~Q ✔
   P   Q
   ~P ✔ ~Q ✔ ~P ✔ ~Q ✔
 P   Q   P   Q
    Not equivalent

5.   P ⊃ P | P ≡ P ✔
  P   P
    P   P
   x   x
   P ≡ P | P ⊃ P ✔
    P | P
     x
   Equivalent

10.
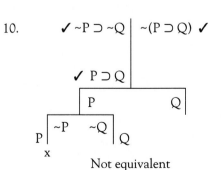

  ✔ ~P ⊃ ~Q | ~(P ⊃ Q) ✔

     ✔ P ⊃ Q
       P   Q
   ~P ~Q
 P   Q
 x
    Not equivalent

15.

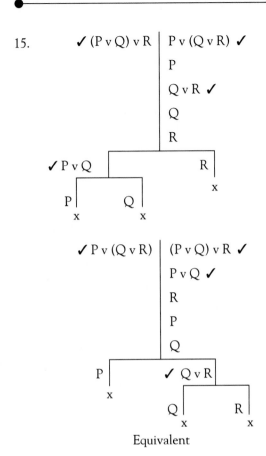

Equivalent

Exercise 4.4 B

1.

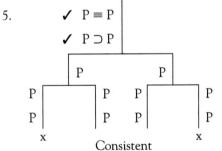

Inconsistent

5. ✔ P ≡ P
    ✔ P ⊃ P

Consistent

10.

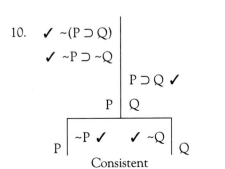

Consistent

Ex 4.4 A: 20.

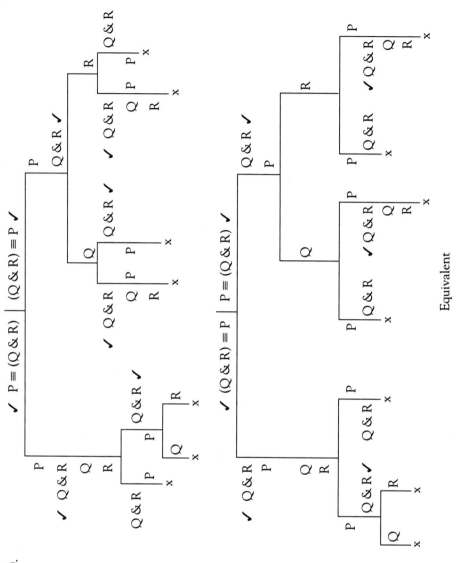

Equivalent

15.

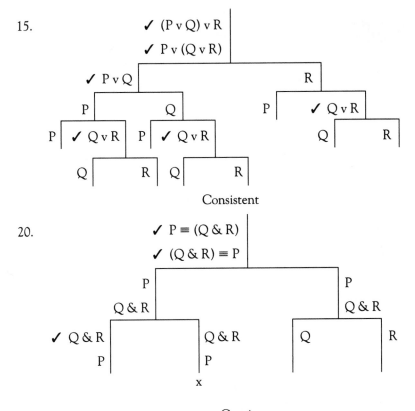

20.

Consistent

Exercise 5.1

LP: logically possible   P: physically possible

1. LP

5. Neither LP nor P

10. LP, P

15. Neither LP nor P

20. LP, P

Exercise 5.2

Part I
1. numeral

5. number, number

Part II
1. sentence token

5. sentence token, sentence token

SELECTED ANSWERS   CHAPTER 6

Exercise 6.1

5A

1. F   5. F

5B

1. Every necessary truth is possibly true and also not possibly false. So some possible truths are not possibly false. Therefore, 1 is false.

5. A necessary truth is a proposition that could not possibly be false. So 5 is false.

Exercise 6.2

2.	e.	T	3.	e.	F	4.	e.	T	5.	a.	T
	j.	T		j.	T		j.	F		e.	T
	n.	F		n.	T		n.	T		j.	T

Exercise 6.3

1. '∇P' abbreviates 'P is contingent.' When we assert that P is "contingently true", we assert more than this; we assert that P is actually true in addition to being contingent.

Exercise 6.4

See the answers for the corresponding exercises in Chapter One Selected Answers.

Exercise 6.5

Co	none	D↔	6	C→	all
A→	2	B→	2		

SELECTED ANSWERS   CHAPTER 7

Exercise 7.1

S5 Natural Deduction:

1.   1.   ◇ (H & S) ⊃ G   4.   H & S   DS 2,3

     2.   ~R   5.   ◇ (H & S)   M2 4

     3.   R v (H & S)/G   6.   G   MP 1,5

5. 1. H → G
    2. G → (M & N)
    3. [H → (M & N)] ⊃ S/S
    4. H → (M & N)   M5 1,2
    5. S   MP 3,4

10. 1. F → (□ A & □ B)
    2. G ⊃ F
    3. H & G/ □(□ A & □ B)
    4. G   Simp 3
    5. F   MP 2,4
    6. ◇F   M2, 5
    7. □(□ A & □ B)   M8 1,6

15. 1. □ (A v B)
    2. B ⊃ H
    3. ~R & ~H/A
    4. A v B   M1, 1
    5. ~ H   Simp 3
    6. ~ B   MT 2,5
    7. A   DS 4,6

20. 1. □ (A ⊃ B)
    2. □ (~ B v ~ A)
    3. G ⊃ ~□ ~ A/ ~ G
    4. │ □ (A ⊃ B)   S5 Reit
    5. │ □ (~ B v ~ A) S5 Reit
    6. │ A ⊃ B   M 1, 4
    7. │ ~ B v ~ A   M 1, 5
    8. │ B ⊃ ~ A   Imp 7
    9. │ A ⊃ ~ A   HS 6, 8
    10. │ ~ A v ~ A   Imp 9
    11. │ ~ A   Taut 10
    12. □ ~ A   S5 Nec 4-11
    13. ~ G   MT 3, 12

25.   1. $\square\,(Z \equiv M)$

      2. $\square\,Z$

      3.   <u>$\square\,[M \supset (S\,\&\,A)]$</u>$/\,Q \rightarrow (A \vee R)$

      4. | Q AP

      5. | $\square\,(Z \equiv M)$ S5 Reit

      6. | $\square\,Z$ S5 Reit

      7. | $\square\,[M \supset (S\,\&\,A)]$ S5 Reit

      8. | $Z \equiv M$ M1, 5

      9. | $(Z \supset M)\,\&\,(M \supset Z)$ Equiv 8

     10. | $Z \supset M$ Simp 9

     11. | Z M1, 6

     12. | M MP 10, 11

     13. | $M \supset (S\,\&\,A)$ M1, 7

     14. | $S\,\&\,A$ MP 12, 13

     15. | A Simp 14

     16. | $A \vee R$ Add 15

     17. $Q \rightarrow (A \vee R)$ MCP 4-16

30.   1.   <u>$\sim\!\lozenge A$</u>$/\,\lozenge\!\sim\!A$

      2. $\square\!\sim\!A$ DE 1

      3. $\sim\!A$ M1, 2

      4. $\lozenge\!\sim\!A$ M2, 3

35.   1. $\square\,(A \supset B)$

      2.   <u>$\lozenge\!\sim\!B$</u>$/\,\lozenge\!\sim\!A$

      3. | $\sim\!\lozenge\!\sim\!A$ AP

      4. | $\square\,A$ DE 3

      5. | | $\square\,(A \supset B)$ S5 Reit

      6. | | $\square\,A$ S5 Reit

      7. | | $A \supset B$ M1, 5

      8. | | A M1, 6

      9. | | B MP 7, 8

     10. | $\square\,B$ S5 Nec 5-9

     11. | $\sim\!\lozenge\!\sim\!B$ DE 10

     12. | $\lozenge\!\sim\!B\,\&\,\sim\!\lozenge\!\sim\!B$ Conj 2, 11

     13. $\lozenge\!\sim\!A$ IP 3-12

40. 1. $\Diamond A \lor \Diamond B / \Diamond (A \lor B)$

    2. $\sim \Diamond (A \lor B)$ AP

    3. $\Box \sim (A \lor B)$ DE 2

    4. $\Box \sim (A \lor B)$ S5 Reit 3

    5. $\sim (A \lor B)$ M1, 4

    6. $\sim A \,\&\, \sim B$ DM 5

    7. $\sim A$ Simp 6

    8. $\sim B$ Simp 6

    9. $\Box \sim A$ S5 Nec 4-8

   10. $\Box \sim B$ S5 Nec 4-8

   11. $\sim \Diamond A$ DE 9

   12. $\sim \Diamond B$ DE 10

   13. $\sim \Diamond A \,\&\, \sim \Diamond B$ Conj 11, 12

   14. $\sim (\Diamond A \lor \Diamond B)$ DM 13

   15. $\sim (\Diamond A \lor \Diamond B) \,\&\, (\Diamond A \lor \Diamond B)$ Conj 1, 14

   16. $\Diamond (A \lor B)$ IP 2-15

45. 1. $A \to B / A \to \Diamond B$

    2. A AP

    3. $A \to B$ S5 Reit

    4. B M3, 2, 3

    5. $\Diamond B$ M2, 4

    6. $A \to \Diamond B$ MCP 2-5

50. 1. $\Box (A \,\&\, B) / \Box A \,\&\, \Box B$

    2. $\Box (A \,\&\, B)$ S5 Reit

    3. $A \,\&\, B$ M1, 2

    4. A Simp 3

    5. B Simp 3

    6. $\Box A$ S5 Nec 2-5

    7. $\Box B$ S5 Nec 2-5

    8. $\Box A \,\&\, \Box B$ Conj 6, 7

55. 1. $P \to Q$

    2. $\Box P$

    3. $Q \to R / \Box R$

    4. $P \to R$ M5, 1, 3

    5. $P \to R$ S5 Reit

    6. $\Box P$ S5 Reit

    7. P M1, 6

    8. R M3, 5, 7

    9. $\Box R$ S5 Nec 4-8

60. 1. $\Diamond A / \Box \Box \Diamond A$

    2. $\Diamond A$ S5 Reit

    3. $\Box \Diamond A$ S5 Nec 2

    4. $\Box \Diamond A$ S5 Reit

    5. $\Box \Box \Diamond A$ S5 Nec 4

65. 1. $\square \lozenge A / \lozenge \lozenge A$

  2. $\lozenge A$ M1, 1

  3. $\lozenge \lozenge A$ M2, 2

S5 Trees:

1. ✓ $\lozenge$ (H & S) ⊃ G

  ✓ ~R

  ✓ Rv (H & S) | G

  | R

  R | H & S

  x | $\lozenge$ (H & S) ✓   G

    | (H & S)

    x   x

5.

        H → G

      G → (M & N)

    ✓ [H → (M & N)] ⊃ S | S

    | H → (M & N) ✓ | S

  H | M & N   x

  ✓ H → G

  ✓ G → (M & N)

  | H   G

  x | G   M & N

    x   x

10. ✓ F → ($\square$A & $\square$B)

  ✓ G ⊃ F

  ✓ H & G | $\square$A & $\square$B

  H

  G

  | G   F

  x | F $\square$A & $\square$B

    x   x

15.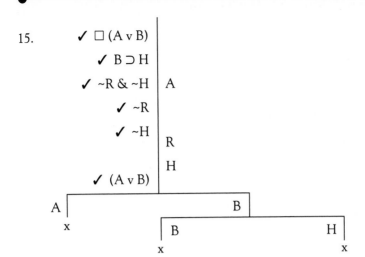

20.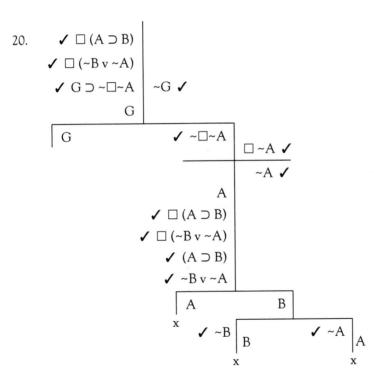

25.

✓ □(Z ≡ M)	
✓ □Z	
✓ □[M ⊃ (S & A)]	Q → (A v R) ✓

| Q | A v R ✓ |

✓ □(Z ≡ M)

✓ □Z

✓ □[M ⊃ (S & A)]

✓ Z ≡ M

Z

✓ M ⊃ (S & A)

| | A |
| | R |

	Z		Z
	M		M
M	✓ S & A	x
x	S	
	A	
	x	

30.

✔ ~◇ A	◇ ~ A ✔
	◇ A ✔
	A ✔
	~A ✔
A	
x	

40.

| ✓ ◇ A v ◇ B | ◇ (A v B) ✓ |

| ✓ ◇ A | ✓ ◇ B |

A	B
◇ (A v B) ✓	◇ (A v B) ✓
A v B ✓	A v B ✓
A	A
B	B
x	x

45. ✓ A → B | A → ◇ B ✓
     ─────────────────
         A | ◇ B ✓
     ✓ A → B
     ┌──────────────┐
     A   B
     x   │ B
                    x

50.   ✓ □ (A & B) | □ A & □ B ✓
     ┌──────────────────┐
         □ A ✓   □ B ✓
     ─────────   ─────────
             A   B
     ✓ □ (A & B)   ✓ □ (A & B)
             A   A
             B   B
             x   x

60. ✔ ◇ A | □ □ ◇ A ✔   65. ✔ □ ◇ A | ◇ ◇ A ✔
     ─────────────────   ◇ A | ◇ A
              □ ◇ A ✔   x
     ─────────────────
      ◇ A | ◇ A
        x

Exercise 7.3

S5 Natural Deduction:

1. 1. | ~P   AP
   2. | ~P v Q   Add 1
   3. | P ⊃ Q   Imp 2
   4. ~P ⊃ (P ⊃ Q)   CP 1-3
   5. □ [~P ⊃ (P ⊃ Q)]   Taut Nec 4

5. 1. | P ⊃ Q   AP
   2. | ~Q ⊃ ~P   Trans 1
   3. (P ⊃ Q) ⊃ (~Q ⊃ ~P)   CP 1-2
   4. □ [(P ⊃ Q) ⊃ (~Q ⊃ ~P)]   Taut Nec 3

10. 1. | (P v Q) & (~Q v S)   AP
     2. | P v Q   Simp 1
     3. | ~Q v S   Simp 1
     4. | ~P ⊃ Q   Imp 2
     5. | Q ⊃ S   Imp 3
     6. | ~P ⊃ S   HS 4, 5
     7. | P v S   Imp 6
     8. [(P v Q) & (~Q v S)] ⊃ (P v S)   CP 1-7
     9. □ {[(P v Q) & (~Q v S)] ⊃ (P v S)}   Taut Nec 8

15. 1. | P   AP
     2. P → P   MCP 1

20. 1. | ~□P   AP
     2. | ◇ ~P   DE 1
     3. ~□P → ◇ ~P   MCP 1-2

25. 1. | ~◇P   AP
     2. | □ ~P   DE 1
     3. ~◇P → □ ~P   MCP 1-2

30. 1. | P → Q   AP
     2. | □ (P ⊃ Q)   E1, 1
     3. | □ (~Q ⊃ ~P)   Trans 2
     4. | ~Q → ~P   E1, 3
     5. (P → Q) → (~Q → ~P)   MCP 1-4

35. 1. | □ (P ⊃ Q)   AP
     2. | P → Q   E1, 1
     3. □ (P ⊃ Q) → (P → Q)   MCP 1-2

S5 Trees:

1.
```
        | □ [~P ⊃ (P ⊃ Q)]  ✔
        | ~P ⊃ (P ⊃ Q)  ✔
   ~P   | P ⊃ Q  ✔
        | P
   P    | Q
      x
```

5.

$$\Box\,[(P \supset Q) \supset (\sim Q \supset \sim P)] \; ✓$$

$$(P \supset Q) \supset (\sim Q \supset \sim P) \; ✓$$

✓ $P \supset Q$ $\sim Q \supset \sim P \; ✓$

   ✓ $\sim Q$ $\sim P \; ✓$

       P | Q

    P   Q

    x   x

10.

$$\Box\,\{[(P \vee Q) \,\&\, (\sim Q \vee S)] \supset (P \vee S)\} \; ✓$$

$$[(P \vee Q) \,\&\, (\sim Q \vee S)] \supset (P \vee S) \; ✓$$

✓ $(P \vee Q) \,\&\, (\sim Q \vee S)$ $P \vee S \; ✓$

        ✓ $P \vee Q$

      ✓ $\sim Q \vee S$ P

                S

        P   Q

        x   ✓ $\sim Q$ S

                Q   x

                x

15. $P \rightarrow P$ ✔

   P | P

    x

20. $\sim \Box\,P \rightarrow \Diamond \sim P$ ✔

✔ $\sim \Box\,P$ | $\Diamond \sim P$

            $\Box\,P$ ✔

            P

            $\Diamond \sim P$ ✔

      P | $\sim P$ ✔

       x

25. $\sim \Diamond\,P \rightarrow \Box \sim P$ ✔

✔ $\sim \Diamond\,P$ | $\Box \sim P$ ✔

            $\Diamond\,P$

            $\sim P$ ✔

            $\Diamond\,P$ ✔

      P | P

       x

30.

$$(P \rightarrow Q) \rightarrow (\sim Q \rightarrow \sim P) \checkmark$$

✔ $P \rightarrow Q$	$\sim Q \rightarrow \sim P$ ✔
✔ $\sim Q$	$\sim P$ ✔
✔ $P \rightarrow Q$	

$$Q$$

$$P$$

P	Q
x	x

35.

$$\Box (P \supset Q) \rightarrow (P \rightarrow Q) \checkmark$$

✔ $\Box (P \supset Q)$	$P \rightarrow Q$
P	Q
✔ $\Box (P \supset Q)$	

$$P \supset Q$$

P	Q
x	x

Exercise 7.5

1.   1. $\sim \Diamond (S \& K)$

     2. $\sim S \rightarrow \sim M$

     3. $\underline{\sim K \rightarrow \sim M} / \sim \Diamond M$

     4. $\Box \sim (S \& K)$ DE 1

     5. $\Box \sim (S \& K)$ S5 Reit

     6. $\sim (S \& K)$ M1, 5

     7. $\sim S \lor \sim K$ DM 6

     8. $\sim S \rightarrow \sim M$ S5 Reit

     9. $\sim K \rightarrow \sim M$ S5 Reit

    10. $\Box (\sim S \supset \sim M)$ E1, 8

    11. $\Box (\sim K \supset \sim M)$ E1, 9

    12. $\sim S \supset \sim M$ M1, 10

    13. $\sim K \supset \sim M$ M1, 11

    14. $\sim M \lor \sim M$ CD 7, 12, 13

    15. $\sim M$ Taut 14

16. □ ~M   S5 Nec 5-15

17. ~ ◇ M   DE 16

S5 Tree:

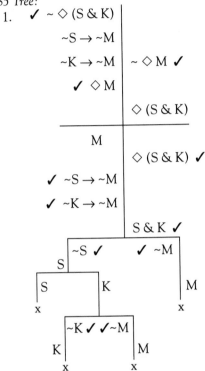

1. ✓ ~ ◇ (S & K)

~S → ~M

~K → ~M   ~ ◇ M ✓

✓ ◇ M

◇ (S & K)

―――――――

M

◇ (S & K) ✓

✓ ~S → ~M

✓ ~K → ~M

S & K ✓

~S ✓   ✓ ~M

S

S   K   M

x   x

~K ✓ ✓ ~M

K   M

x   x

SELECTED ANSWERS   CHAPTER 8

Exercise 8.1

1. Hb   b: Beth   Hx: x is happy

5. Af   f: The first person to step on the moon

Ax: x is an American astronaut.

Exercise 8.2

1. (∀x)(Ax ⊃ Cx)

5. (∀x)(W∀x ⊃ Bx)

10. (∀x)(Hx ⊃ ~Lx)

15. (∀x)(Rx ⊃ ~Ex)

20. (∀x)(Cx ⊃ Gx)

25. (∀x)(Sx ⊃ Ex)

Exercise 8.3

1. $(\forall x)[(Gx \lor Hx) \supset Cx]$

5. $(\forall x)[(Sx \& Ax) \supset Bx]$

Exercise 8.4

1. $(\forall x)(Mx) \supset Wd$

5. $(\forall x)(\exists y)(Lxy)$ (Domain: persons)

10. $(\forall x)[(Ex \lor Px) \supset Lx]$

15. $(\forall x)(Px \supset Lx)$ Lx: x likes a Dick's Deluxe burger

Px: x is a person

20. $(\exists x)(\exists y)(\exists z)[(Dx \& Py \& Pz) \& (Oxy \& Bxz)]$

Oxy: x is owned by y   Py: y is a person

Bxz: x barked at z   Dx: x is a dog

25. $(\forall x)(Ex \supset Lxm)$ m: Nathan's pet mouse

Ex: x is an elephant

30. $(\exists x)(Sx \& Lxs)$ s: the sun

Lxs: x is larger than the Sun

35. $(\forall x)(Hx \supset Px)$ Hx: x was hired

Px: x puts pepper in x's coffee

Domain: persons

Exercise 8.5

1. $(\forall x)[(Mx \& (x \neq s)) \supset Tsx] \& Ms$

Mx: x is a member of the team

Tsx: s is taller than x   s: Sue

5. Domain: numbers

$\sim(\exists x)(\forall y)[(x \neq y) \supset Gxy]$

10. $(\exists x)(\forall y)[Dx \& (Dy \supset (y=x))]$

15. $(\forall x)[(Bx \& Px) \supset (x=j)]$

Bx: x was a Beatle

Px: x belonged to the Plastic Ono Band

j: John Lennon

20. $(\exists x)(\exists y)(\forall z)\{[Sx \& Sy \& (x \neq y)] \& [Sz \supset (z=x \lor z=y)]\}$

Exercise 8.6

1. b = (ɿx)(Rx)   Rx: x is the richest person in Washington State.

b: Bill Gates

5. (∀x)(Px ⊃ Wxr)   r: The richest person in town.

Px: x is a person in town

Or:

{r = (ɿx){(∀y)[y≠x]⊃Rxy]}}&(∀y)[(y≠r)⊃Wyr]

Rxy: x is richer than y

Wyr: y wants to borrow money from r

Domain: People in town

10. j = (ɿx)(Axt)   t: the book A *Theory of Justice*

Axt: x authored t

j: John Rawls

Exercise 8.7

2.b.   nonsymmetric, transitive, irreflexive

1.   reflexive, symmetric, transitive

SELECTED ANSWERS   CHAPTER 9

Exercise 9.1

1. Domain: Numbers
Ax: x is even
Bx: x is rational

5. Domain: Persons
Ax: x is conscious at least once in x's life
Bx: x has never experienced consciousness

Exercise 9.3

1. UI   1
UI   2
HS 3,4
UG 5

Exercise 9.4

Natural Deduction:

1.   1.   (∀x)(Wx ⊃ Sx)   4.   Sa⊃Pa   UI 2

2.   (∀x)(Sx ⊃ Px)/ (∀x)(Wx ⊃ Px)   5.   Wa⊃Pa   HS 3,4

3.   Wa⊃Sa   UI 1   6.   (∀x)(Wx ⊃ Px)   UG 5

5.   1. $(\forall x)(Mx)$
     2.   Hg/Hg & Mg
     3.   Mg   UI 1
     4.   Hg & Mg   Conj 2,3

10. 1. $(\forall x)(Hx)$ & Sa
     2. $(\exists x)(Hx) \supset (\exists x)(Gx)/ (\exists x)(Sx)$ & $(\exists x)(Gx)$
     3. $(\forall x)(Hx)$ Simp 1
     4.   Ha   UI 3
     5. $(\exists x)(Hx)$ EG 4
     6. $(\exists x)(Gx)$ MP 2,5
     7.   Sa   Simp 1
     8. $(\exists x)(Sx)$ EG 7
     9. $(\exists x)(Sx)$ & $(\exists x)(Gx)$ Conj 6,8

15. 1. $(\forall x)(Hx \supset Gx)$
     2. $(\forall x)(\sim Gx)/(\forall x)(\sim Hx)$
     3.   Hs $\supset$ Gs   UI 1
     4. $\sim$Gs   UI 2
     5. $\sim$Hs   MT 3,4
     6. $(\forall x)(\sim Hx)$ UG 5

20. 1. $(\forall x)[Hx \supset (Bx\&Wx)]$
     2. $(\exists y)(\sim By)/(\exists z)(\sim Hz)$
     3. $\sim$Bc   EI 2
     4.   Hc$\supset$(Bc & Wc)   UI 1
     5. $\sim$Bc v $\sim$Wc   Add 3
     6. $\sim$(Bc & Wc)   DM 5
     7. $\sim$Hc   MT 4,6
     8. $(\exists z)(\sim Hz)$ EG 7

25. 1. $(\exists x)(Fx) \supset (\forall x)(Sx)$
     2. $(\exists x)(Hx) \supset (\forall x)(Gx)$
     3.   Fs & Hg/$(\forall x)(Sx$ & $Gx)$
     4.   Fs   Simp 3
     5. $(\exists x)(Fx)$ EG 4
     6. $(\forall x)(Sx)$ MP 1,5
     7.   Hg   Simp 3
     8. $(\exists x)(Hx)$ EG 7
     9. $(\forall x)(Gx)$ MP 8,2
     10. Sa   UI 6
     11. Ga   UI 9
     12. Sa & Ga   Conj 10,11
     13. $(\forall x)(Sx$ & $Gx)$ UG 12

Trees:

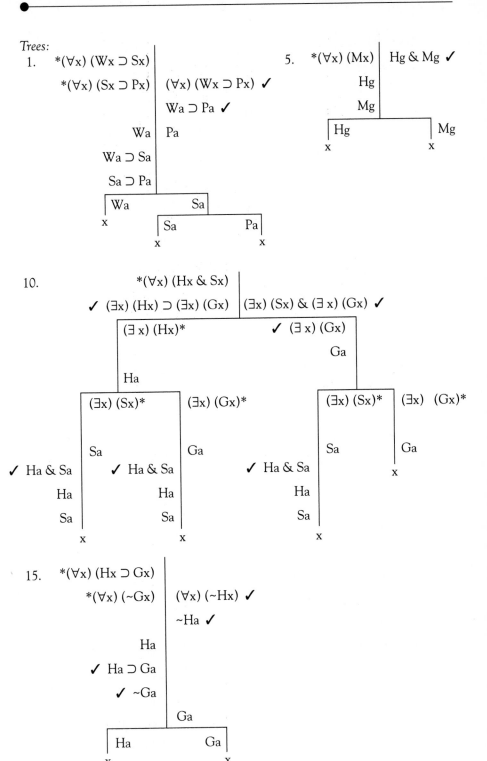

1. $*(\forall x)(Wx \supset Sx)$
       $*(\forall x)(Sx \supset Px)$ | $(\forall x)(Wx \supset Px)$ ✓
                                        $Wa \supset Pa$ ✓

                                  Wa | Pa
                              $Wa \supset Sa$
                              $Sa \supset Pa$
                        Wa | Sa
                        x Sa Pa
                               x   x

5. $*(\forall x)(Mx)$ | $Hg \& Mg$ ✓
                           Hg
                           Mg
                    Hg | | Mg
                    x   x

10. $*(\forall x)(Hx \& Sx)$
     ✓ $(\exists x)(Hx) \supset (\exists x)(Gx)$ | $(\exists x)(Sx) \& (\exists x)(Gx)$ ✓
               $(\exists x)(Hx)^*$ ✓ $(\exists x)(Gx)$
                                                           Ga
               Ha
          $(\exists x)(Sx)^*$ $(\exists x)(Gx)^*$ $(\exists x)(Sx)^*$ $(\exists x)(Gx)^*$

          Sa Ga Sa Ga
     ✓ $Ha \& Sa$ ✓ $Ha \& Sa$ ✓ $Ha \& Sa$ x
       Ha Ha Ha
       Sa Sa Sa
          x   x   x

15. $*(\forall x)(Hx \supset Gx)$
        $*(\forall x)(\sim Gx)$ | $(\forall x)(\sim Hx)$ ✓
                                   $\sim Ha$ ✓

                          Ha
                    ✓ $Ha \supset Ga$
                       ✓ $\sim Ga$
                          Ga
                  Ha | Ga
                  x   x

20.   *(∀x) [Hx ⊃ (Bx & Wx)]

      ✓ (∃y) (~By)   (∃z) (~Hz)*

          ✓ ~Bc

                              ~Hc ✓

                              Bc

                    Hc

        ✓ Hc ⊃ (Bc & Wc)

            Hc   ✓ Bc & Wc

            x   Bc

                              Wc

                                x

Exercise 9.5

Natural Deduction:

1.   1.   ~(∀x)(Sx)/(∃x)(Sx ⊃ Px)

      2.   (∃x)(~Sx)   QE 1

      3.   ~Sb   EI 2

      4.   ~Sb v Pb   Add 3

      5.   Sb ⊃ Pb   Imp 4

      6.   (∃x)(Sx ⊃ Px)   EG 5

5.   1.   ~(∃x)(Ax)/(∀x)(Ax ⊃ Bx)

      2.   (∀x)(~Ax)   QE 1

      3.   ~Aa   UI 2

      4.   ~Aa v Ba   Add 3

      5.   Aa ⊃ Ba   Imp 4

      6.   (∀x)(Ax ⊃ Bx)   UG 5

9.   1.   (∃x)(Px v Gx) ⊃ (∀x)(Hx)   6.   ~(Ps v Gs)   UI 5

      2.   (∃x)(~Hx)/(∀x)(~Px)   7.   ~Ps & ~Gs   DM 6

      3.   ~(∀x)(Hx)   QE 2   8.   ~Ps   Simp 7

      4.   ~(∃x)(Px v Gx)   MT 1,3   9.   (∀x)(~Px)   UG 8

      5.   (∀x)~(Px v Gx)   QE 4

Trees:

1. ✔ ~(∀x) (Sx) | (∃x) (Sx ⊃ Px) *
   | | (∀x) (Sx) ✔
   | | Sa
   | | Sa ⊃ Pa ✔
   | Sa | Pa
   | x |

5. ✔ ~(∃x) (Ax) | (∀x) (Ax ⊃ Bx) ✔
   | | (∃x) (Ax) *
   | | Aa ⊃ Ba ✔
   | | Aa
   | Aa | Ba
   | x |

Exercise 9.6

Natural Deduction:

1. 1. (∀x)(Ax ⊃ ~Bx)

    2. (∀x)[(Bx ⊃ (Hx & Ax)]/ (∃x)(~Bx)

    3. | ~(∃x)(~Bx)   AP

    4. | (∀x)(Bx)   QE 3

    5. | Bp   UI 4

    6. | Bp ⊃ (Hp & Ap)   UI 2

    7. | Hp & Ap   MP 5,6

    8. | Ap   Simp 7

    9. | Ap ⊃ ~Bp   UI 1

    10. | ~Bp   MP 8,9

    11. (∃x)(~Bx)   IP 3-10

5. 1. (∀x)(Hx ⊃ Qx)

    2. (∀x)(Hx ⊃ Rx)/ (∀x)[Hx⊃(Qx&Rx)]

    3. | Hs   AP

    4. | Hs ⊃ Qs   UI 1

    5. | Qs   MP 3,4

    6. | Hs ⊃ Rs   UI 2

    7. | Rs   MP 3,6

    8. | Qs & Rs   Conj 5,7

    9. Hs ⊃ (Qs & Rs)   CP 3,8

    10. (∀x)[Hx ⊃ (Qx & Rx)]   UG 9

10. 1.   (∃x)(Px) v (∃x)(Qx & Rx)
     2.   (∀x)(Px ⊃ Rx)/(∃x)(Rx)
     3. │ ~(∃x)(Rx)   AP
     4. │ (∀x)(~Rx)   QE 3
     5. │ ~Ra   UI 4
     6. │ ~Ra v ~Qa   Add 5
     7. │ ~Qa v ~Ra   Comm 6
     8. │ ~(Qa & Ra)   DM7
     9. │ (∀x)~(Qx & Rx)   UG 8
     10. │ ~(∃x)(Qx & Rx)   QE 9
     11. │ (∃x)(Px)   DS 1,10
     12. │ Pc   EI 11
     13. │ Pc ⊃ Rc   UI 2
     14. │ Rc   MP 12, 13
     15. │ ~Rc   UI 4
     16.   (∃x)(Rx)   IP 3-15

15. 1.   (∃x)(Px v Jx) ⊃ ~(∃x)(Px)/ (∀x)(~Px)
     2. │ ~(∀x)(~Px)   AP
     3. │ (∃x)(Px)   QE 2
     4. │ Ps   EI 3
     5. │ Ps v Js   Add 4
     6. │ (∃x)(Px v Jx)   EG 5
     7. │ ~(∃x)(Px)   MP 1,6
     8. │ (∀x)(~Px)   QE 7
     9. │ ~Ps   UI 8
     10.   (∀x)(~Px)   IP 2-9

Trees:

1.

$$* (\forall x) (Ax \supset \sim Bx)$$

$$* (\forall x) [Bx \supset (Hx \& Ax)] \quad | \quad (\exists x) (\sim Bx)*$$

$$\sim Bc \checkmark$$

$$Bc$$

$$\checkmark \ Ac \supset \sim Bc$$

$$\checkmark \ Bc \supset (Hc \& Ac)$$

Ac $\checkmark \ \sim Bc$

Bc   Hc & Ac   Bc

x   Hc   x

Ac

x

5. $* (\forall x) (Hx \supset Qx)$

$* (\forall x) (Hx \supset Rx) \quad | \quad (\forall x) [Hx \supset (Qx \& Rx)] \checkmark$

$Ha \supset (Qa \& Ra) \checkmark$

Ha $Qa \& Ra \checkmark$

$\checkmark \ Ha \supset Qa$

$\checkmark \ Ha \supset Ra$

Ha   Qa

x

Qa   Ra

x

Ha   Ra

x   x

Exercise 9.7

Natural Deduction:

1.   / $(\forall x) [(Hx \lor Px) \lor \sim Px]$

    1. $\sim (\forall x)[(Hx \lor Px) \lor \sim Px]$ AP

    2. $(\exists x) \sim [(Hx \lor Px) \lor \sim Px]$ QE 1

    3. $\sim [(Ha \lor Pa) \lor \sim Pa]$ EI 2

4. $\quad$ ~(Ha v Pa) & Pa $\quad$ DM 3

5. $\quad$ ~(Ha v Pa) $\quad$ Simp 4

6. $\quad$ ~Ha & ~Pa $\quad$ DM 5

7. $\quad$ ~Pa $\quad$ Simp 6

8. $\quad$ Pa $\quad$ Simp 4

9. $(\forall x)[(Hx v Px) v \sim Px]$ $\quad$ IP 1-8

5. _____ / $(\forall x)(Px) \supset \sim(\exists x)(\sim Px)$

$\quad$ $(\forall x)(Px)$ $\quad$ AP

$\quad$ ~$(\exists x)(\sim Px)$ $\quad$ QE 1

$(\forall x)(Px) \supset \sim(\exists x)(\sim Px)$ $\quad$ CP 1-2

10. _____ / $(\forall x)[(Sx \& Hx) \supset (Sx v Hx)]$

1. $\quad$ ~$(\forall x)[(Sx \& Hx) \supset (Sx v Hx)]$ $\quad$ AP

2. $\quad$ $(\exists x)\sim[(Sx \& Hx) \supset (Sx v Hx)]$ $\quad$ QE 1

3. $\quad$ ~$[(Sa \& Ha) \supset (Sa v Ha)]$ $\quad$ EI 2

4. $\quad$ ~$[\sim(Sa \& Ha) v (Sa v Ha)]$ $\quad$ Imp 3

5. $\quad$ $(Sa \& Ha) \& \sim(Sa v Ha)$ $\quad$ DM 4

6. $\quad$ Sa & Ha $\quad$ Simp 5

7. $\quad$ Sa $\quad$ Simp 6

8. $\quad$ ~(Sa v Ha) $\quad$ Simp 5

9. $\quad$ ~Sa & ~Ha $\quad$ DM 8

10. $\quad$ ~Sa $\quad$ Simp 9

11. $(\forall x)[(Sx \& Hx) \supset (Sx v Hx)]$ $\quad$ IP 1-10

Trees:
1. $\quad$ $(\forall x) [(Hx v Px) v \sim Px]$ $\quad$ ✔
$\quad$ (Ha v Pa) v ~Pa $\quad$ ✔
$\quad$ Ha v Pa $\quad$ ✔
$\quad$ ~Pa $\quad$ ✔

Pa

$\quad$ Ha
$\quad$ Pa
$\quad$ x

5.

	$(\forall x) (Px) \supset \sim(\exists x) (\sim Px)$ ✔
* $(\forall x) (Px)$	$\sim(\exists x) (\sim Px)$ ✔
✔ $(\exists x) (\sim Px)$	
✔ $\sim Pa$	
	Pa
Pa	
x	

10.

	$(\forall x) [(Sx \& Hx) \supset (Sx \lor Hx)]$ ✔
	$(Sa \& Ha) \supset (Sa \lor Ha)$ ✔
✔ Sa & Ha	Sa $\lor$ Ha ✔
Sa	
Ha	
	Sa
	Ha
x	

Exercise 9.8

Natural Deduction:

1. 1. $(\forall x)(\forall y)(Hxy \supset \sim Hyx)$

    2. $(\exists x)(\exists y)(Hxy)/ (\exists x)(\exists y)(\sim Hyx)$

    3. $(\exists y)(Hay)$ EI 2

    4. Hab   EI 3

    5. $(\forall y)(Hay \supset \sim Hya)$ UI 1

    6. Hab $\supset \sim$Hba   UI 5

    7. $\sim$Hba   MP 4,6

    8. $(\exists y)(\sim Hya)$ EG 7

    9. $(\exists x)(\exists y)(\sim Hyx)$ EG 8

5. 1. $(\forall x)(\exists y)(Sx \& Py)/ (\exists y)(\exists x)(Sx \& Py)$

    2. $(\exists y)(Sa \& Py)$ UI 1

    3. Sa & Pb   EI 2

    4. $(\exists x)(Sx \& Pb)$ EG 3

    5. $(\exists y)(\exists x)(Sx \& Py)$ EG 4

Trees:

1.   * (∀x) (∀y) (Hxy ⊃ ~Hyx)

            ✓ (∃x) (∃y) (Hxy) | (∃x) (∃y) (~Hyx) *

               ✓ (∃y) (Hay)

                    Hab

          * (∀y) (Hay ⊃ ~Hya)

             ✓ Hab ⊃ ~Hba

              Hab   ✓~Hba

               x   Hba

                        (∃y) (~Hya) *

                        ~Hba ✓

               Hba

                x

5.   ✔ (∀x) (∃y) (Sx & Py) | (∃y) (∃x) (Sx & Py) *

        * (∀x) (Sx & Pb)

                     (∃y) (Sa & Py) *

            Sa & Pb

                   Sa & Pb

             x

Exercise 9.9

Natural Deduction: *Trees:*

1.   1. Aa ⊃ Ha   1. ✓ Aa ⊃ Ha

     2. ~Ha   ~Ha

     3. a=b/~Ab   a = b | ~Ab ✓

     4. ~Aa MT 1,2   ✓ Ab ⊃ Hb

     5. ~Ab Id B 3,4   ✓ ~Hb

                                Hb

                            Ab

                     Ab   Hb

                     x   x

Exercise 9.10

Natural Deduction

1. 1. (∀x) (Cx ⊃ Mx)

    2. (∀x) (Fx ⊃ ~Mx)/ (∀x) (Fx ⊃ ~Cx)

    3. Ca ⊃ Ma   UI 1

    4. Fa ⊃ ~Ma   UI 2

    5. Ma ⊃ ~Fa   Trans 4

    6. Ca ⊃ ~Fa   HS 3 5

    7. Fa ⊃ ~Ca   Trans 6

    8. (∀x) (Fx ⊃ ~Cx) UG 7

5. 1. (∀x) (Mx ⊃ Ax)

    2. (∀x) (Ax ⊃ Dx)

    3. (∃x) (Hx & Mx)/ (∃x) (Hx & Dx)

    4. Ha & Ma   EI 3

    5. Ma ⊃ Aa   UI 1

    6. Aa ⊃ Da   UI 2

    7. Ma ⊃ Da   HS 5, 6

    8. Ma   Simp 4

    9. Ha   Simp 4

    10. Da   MP 7, 8

    11. Ha & Da   Conj 9, 10

    12. (∃x) (Hx & Dx)   EG 11

10. 1. (∀x) (Ex ⊃ Px)

    2. (∀x) (Bx ⊃ Ex)/ (∀x) (Bx ⊃ Px)

    3. Bc ⊃ Ec   UI 2

    4. Ec ⊃ Pc   UI 1

    5. Bc ⊃ Pc   HS 3, 4

    6. (∀x) (Bx ⊃ Px)   UG 5

    Note: Ex: x is an event in a person's life

        Px: x possesses eternal significance

        Bx: x is a person's birth

15.　1.　$(\forall x)(Dx \supset Lx)$　　　　　5.　$Lp \supset \sim Fp$　UI 2

　　2.　$(\forall x)(Lx \supset \sim Fx)$　　　　6.　$Dp \supset \sim Fp$　HS 4, 5

　　3.　$\underline{Dp}/ \sim Fp$　　　　　　　　7.　$\sim Fp$　　　MP 3, 6

　　4.　$Dp \supset Lp$　UI 1

Trees:

1.　　* $(\forall x)(Cx \supset Mx)$

　　　* $(\forall x)(Fx \supset \sim Mx)$　　$(\forall x)(Fx \supset \sim Cx)$ ✓

　　　　　　　　　　　　　　　　$Fa \supset \sim Ca$ ✓

　　　　　✓ $Ca \supset Ma$

　　　　　✓ $Fa \supset \sim Ma$

　　　　　　　　　　Fa　|　$\sim Ca$ ✓

　　　　　　　　　　Ca

　　　　　| Ca　　　　Ma

　　　　　　x　　　| Fa　　✓ $\sim Ma$

　　　　　　　　　　　　　　　　Ma

　　　　　　　　　　　x　　　　　x

5.　　* $(\forall x)(Mx \supset Ax)$

　　　* $(\forall x)(Ax \supset Dx)$

　　　✓ $(\exists x)(Hx \& Mx)$　$(\exists x)(Hx \& Dx)$ *

　　　　✓ $Hs \& Ms$

　　　　　　　　　　　　$Hs \& Ds$ ✓

　　　　　　　Hs

　　　　　　　Ms

　　　　　| Hs　　　　Ds

　　　　　　x　　✓ $Ms \supset As$

　　　　　　　　✓ $As \supset Ds$

　　　　　　　　| Ms　　　As

　　　　　　　　　x　　| As　　　Ds

　　　　　　　　　　　　　x　　　　　x

10.　　* (∀x) (Ex ⊃ Px)

　　　　* (∀x) (Bx ⊃ Ex)　　(∀x) (Bx ⊃ Px) ✓

　　　　　　　　　　　　　　Be ⊃ Pe ✓

　　　　　　　　　　　　Be　Pe

　　　　　　✓ Ee ⊃ Pe

　　　　　　✓ Be ⊃ Ee

　　　　　　　　　Ee　　　　　Pe
　　　　　　　　　　　　　　　　　　x
　　　　　Be　　　　Ee
　　　　x　　　　　x

15.　　* (∀x) (Dx ⊃ Lx)

　　　　* (∀x) (Lx ⊃ ~Fx)

　　　　　　　Dp　　　　~Fp ✓

　　　　　　　　　Fp

　　　　　✓ Dp ⊃ Lp

　　　　　✓ Lp ⊃ ~Fp

　　　　　　Dp　　　　Lp
　　　　　x
　　　　　　　　　　Lp　　✓ ~Fp
　　　　　　　　　x　　　　　　Fp
　　　　　　　　　　　　　　　x

INDEX

Universal Instantiation
From a universal quantification, you may infer any substitution instance.

Existential Generalization
From a singular sentence P containing one or more constants, you may infer any existential generalization of the sentence, provided that the variables you insert does not already appear in P.

Existential Instantiation
From an existential quantification, you may infer any instantiation, provided that the constant you instantiate with does not appear in the premises or on any earlier line of the proof and does not appear in the conclusion.

Universal Generalization
From a sentence P containing one or more constants, you may infer any universal generalization of P provided that: (a) the constant in P that is replaced by a variable does not appear in the premises and does not occur in any previous line that was obtained by E. I., and (b) the variable you insert does not already occur within P.

Quantifer Exchange Rule
(a) Switch one quantifier for the other
(b) Take the opposite of both sides of the quantifier